Evelyn Turner

Queensland Nurses
Boer War to Vietnam

by

RUPERT GOODMAN

Produced on behalf of the Returned Sisters Sub-Branch
of the Returned Services League, (Queensland)

BOOLARONG PUBLICATIONS

By the same author: A Hospital at War

First published in 1985
by Boolarong Publications,
24 Little Edward Street, Spring Hill, Brisbane. Qld.

National Library of Australia
Cataloguing-in-Publication data

Goodman, R.D. (Rupert Douglas), 1915-
Queensland nurses.

Includes index.
ISBN 0 908175 94 9.

1. Military nursing - Queensland - History.
I. Title.

355.3′45′09943

BOOLARONG PUBLICATIONS,
24 Little Edward Street, Spring Hill, Brisbane. Qld.
Design, reproduction and photo-typesetting by
Press Etching Pty. Ltd., Brisbane.
Printed by Apple Printing Co. Pty. Ltd., Brisbane.
Bound by Podlich Enterprises Pty. Ltd., Brisbane.

PROUDLY PRINTED IN QUEENSLAND

Cover design: Archie Robertson, Victoria Barracks.

Rupert Goodman

The Author

Dr Rupert Goodman is a graduate of the University of Melbourne and of The Australian National University in the field of education. He is the author of many publications including*Secondary Education in Queensland 1870–1970, Toowoomba Grammar School 1875–1975, A Centenary History* and *A Hospital At War, The 2/4 Australian General Hospital 1940–1945*. His career includes practical experience as a teacher with the Victorian Education Department, Headmastership of Malvern Grammar School and Readership in Education at the University of Queensland. He served with the 2/4 Australian General Hospital as a nursing orderly during World War II where he first became acquainted with the problems and conditions of nursing. Later he became the Education Officer for 2/4 AGH. After the war he continued his interest in the nursing profession, as a member of the RANF Committee of Inquiry into Queensland Nursing Education and as a member of the Board of Governors of St Andrew's Hospital, Brisbane. Since his retirement from the University he has specialized in military history. Among other appointments Dr Goodman was one of the Education Advisers to the Parliamentary Select Committee on Education in Queensland (The Ahern Committee). He was also the Queensland Commissioner for the Australian Broadcasting Commission (1977–1983). Currently, he is a member of the Senate of the University of Queensland and President of the Australian National Flag Association of Queensland.

Preface

When I was invited by the Returned Sisters Sub-Branch of the RSL (Queensland) to write a history of the contribution made by Queensland Sisters in all the wars from the Boer War to Vietnam, I did not realize the breadth of the canvas! However some limitations were imposed by restricting the story to those nurses who were born in Queensland, who were trained in Queensland hospitals, or who were associated in a large measure with Queensland. Even so, the story could not be restricted to those who served with the Australian Army Nursing Service (later the RAANC) as a significant contribution was made by those who served with the RAAF Nursing Service, as well as the RANNS. Again, the field spread to include those who enlisted in Queen Alexandra's Imperial Military Service and with Princess Mary's Royal Air Force Nursing Service.

While this is the story of Queensland's nurses, it must be acknowledged that they were but a part of the Australia-wide nursing service, about whom similar histories will be written.

This story includes particular reference to several hundred Queensland nurses, no doubt there are others who might have been mentioned. Every attempt has been made to locate them but with the passage of time and the paucity of records, this has proved difficult. Likewise there are problems with the spelling of names. In some cases, names vary on nominal rolls, official records and on war memorials. Again every attempt has been made to obtain the correct spelling. It must also be mentioned that a decision was made to restrict the names to the service name, so no reference is made to any married name. Inevitably there will be mistakes and omissions and for these an apology is made.

When a request was made through local and provincial presses for information I was literally deluged with information. Diaries, letters, newspaper clippings and photographs were made available from family records. I thank all those families, too numerous to mention, who have made this material available. It has proved a very rich source of human history. The story is not merely one about nurses, it is the story of Queensland families over almost a century, some of whom sent four, five and even six daughters off to the war. It is the story of generations of nurses, some born in the 1880s, some in the 1950s and 1960s, who have contributed in no small measure to the medical services of the nation at war.

I must also acknowledge that there were many other Queensland servicewomen involved in the medical services, working in hospitals alongside the nurses. I refer to the medical officers, physiotherapists, Red Cross workers, VAD's and AAMWS, occupational therapists and other professional groups. Their contribution is recognized but their story is another history, yet to be written.

More recent events, from Korea onwards, must inevitably lack the authenticity which comes from official records, not yet released. Sufficient evidence exists, nonetheless, to show that the high traditions of military nursing set in both World Wars have been maintained in all the bitter skirmishes and 'little wars' since.

I must make particular mention of the valuable assistance received from the Australian War Memorial, the Australian Archives (Melbourne), the Australian War Graves Commission and to members of the Returned Sisters Sub-Branch of the RSL (Queensland). Jean Dickson, Dorothea Harrison, Nan Hamilton and Margaret Hamilton have been tireless workers in searching out information about Queensland nurses. Without their assistance this story would not have been possible. Finally I record appreciation for the services of my wife who did all the typing and constant checking, without whose help this book could not have been produced in the allotted time.

RUPERT GOODMAN
Brisbane 1985

Foreword

GOVERNMENT HOUSE,
BRISBANE, Q. 4001

I have read with very great interest the proud record of Queensland nurses over almost a century of wars in which Australia has been involved.

It is a record that needed to be written so that future generations may know of the part the nurses played in many theatres of war. It is a story of courage in times of great stress as casualties flowed in from Gallipoli and France and in times of great personal danger when nurses were under fire in Greece, Crete and Malaya.

Nurses were also casualties of the wars, some paying the supreme sacrifice in Egypt, Malaya, in Australia and in the sinking of the "Centaur". Throughout it all nurses did not forget their dedication to the noble cause of caring for the sick and wounded.

I commend this book to all who are interested in our history and in the part our Nursing Sisters played, so unselfishly, in serving their country.

It is a proud record indeed.

Janet Ramsay.

Lady Ramsay

AANS CASUALTY LIST

First World War 1914–1918

Sister Charlotte **BERRIE** — QAIMNS. Died of illness in Jerusalem, 8 Jan 1919. Buried in the Jerusalem War Cemetery.

Sister Agnes **CORFIELD** — QAIMNS. Died of illness 2 Feb 1916 at Heliopolis. Buried Chatby Military Cemetery, Egypt.

Sister Letitia **MORETON** — AANS. 2 AGH. Died of illness in India, 11 November, 1916. Buried Quetta Government Cemetery, Baluchistan.

Sister Norma **MOWBRAY** — AANS. 1 AGH. Died of illness Heliopolis, 21 Jan 1916. Buried Cairo War Memorial Cemetery, Egypt.

Sister Rosa **O'KANE** — AANS. Died of illness 21 Dec 1918, while nursing patients at the Quarantine Station, Fremantle. Buried Woodman's Point Cemetery.

Second World War 1939–1945

Sister Ellenor **CALNAN** — AANS. 2/10 AGH. Drowned when *Vyner Brooke* bombed and sunk by the Japanese, 14 Feb 1942. Name appears on The Singapore Memorial.

Sister Pauline **HEMPSTED** — AANS. 2/13 AGH. Died of illness at Banka Is while POW, 19 Mar 1945. Buried Jakarta War Cemetery, Indonesia.

Sister Margaret **JACKSON** — AANS, 117 AGH. Died of illness 30 Sept 1942. Buried Toowoomba General Cemetery.

Sister Gladys **McDONALD** — AANS. 2/13 AGH. Drowned when *Vyner Brooke* bombed and sunk by the Japanese 14 Feb 1942. Name appears on The Singapore Memorial.

Sister Pearl **MITTELHEUSER** — AANS. 2/10 AGH. Died of illness at Loeboek Linggau, Sumatra while POW, 18 August 1945. Buried Jakarta War Cemetery, Indonesia. Name appears on The Bundaberg Memorial.

Sister Nita **PRIDEAUX** — AANS, 116 AGH. Died of illness 19 October 1943. Buried Charters Towers War Cemetery.

Sister Lilian **SMITH** — AANS, 2/7 AGH. Died of illness 17 June 1945. Buried Lutwyche Cemetery, Brisbane.

Sister Heather **STEVENSON** — AANS, 116 AGH. Died of illness 25 Nov 1945. Buried General Cemetery, Cairns.

Sister Cynthia **THIEDEKE** — AANS, 2/12 AGH. Died of illness 27 Sept 1942. Buried Colombo (Kanatte) General Cemetery Ceylon (Sri Lanka).

Sister Joyce **WYLLIE** — AANS, 2/3 Hospital Ship *Centaur*. Drowned when Hospital Ship *Centaur* sunk by the Japanese off Brisbane 14 May 1943. Name appears on The Sydney Memorial and on The Bundaberg Memorial.

PRO HUMANITATE

Evelyn Turner

Decorations and Awards to Queensland Sisters

World War I

Cheesman, B.	MID	King, E.	MM, RRC
Derrer, M.J.	MM	Langford, R.J.	MID
Dunne, T.J.	RRC	Macdonald, S.	MBE, RRC, FNM
Everett, V.	MBE	Paten, E.	MBE, ARRC
Fisher, M.E.	RRC	Payne, W.	MBE
Hart, J.M.	MID, RRC	Sorensen, C.	MBE, MID, RRC, FNM
Kennedy, J.	RRC	Tolmie, A.	OBE
Keys, C.	MID (Twice), RRC	Twine, M.	Dunkirk Medal
King, A.	RRC	Webb, D.	RRC

World War II and Later Wars

Abbot, J.S.	RRC, FNM	Guilfoyle, B.M.	ARRC
Allen, E.M.	MID	Hanrahan, E.F.	OBE, MID (Twice)
Beattie, T.J.	MID	Hooke, H.D.	MID
Blanch, T.J.	RRC	Marshall, L.	FNM
Burbidge, B.E.	OBE	O'Neill, J.	AM
Burnett, D.A.	MID	Oxley, C.S.M.	MID
Carmody, M.C.	ARRC	Potts, L.K.	ARRC
Dickson, J.C.	ARRC	Purtill, S.	OBE
Doig, E.N.	RRC, FNM	Rule, L.	ARRC
Duffield, E.M.	MID	Schultz, I.B.	MBE, MID, FNM
Edwards, A.B.	RRC, FNM	Smith, V.E.	MID
Elliott, F.S.	MID	Spearritt, A.	MID
Elms, J.	ARRC	Wallace, M.	MID
Glasgow, M.M.	RRC	Wilson, G.M.	CBE, RRC, FNM MID (5 times)

Contents

Chapter One

Introduction:

The Evolution of Military Nursing

Nurses guilty of great neglect of duty or of getting drunk and using their patients ill or of stealing or concealing or taking away the effects of men who die in the hospital are to be immediately sent to the guard and reported to the Commanding Officer, that they may be tried by Court Martial and be confined, whipped or otherwise punished as the military law directs.

Dr Monro, Physician to George III's Army, 1763.

The practice of enlisting women as nurses in the organized medical services of an army at war is of relatively recent origin. Military nursing, as it is sometimes called, has its origins in the development of nursing stretching back over many centuries, a history in which religion, war and science have been interwoven in the development of the nursing profession. Civilian nursing has come a long way over more than twenty centuries of caring for the sick, at times stimulated by advances in war, at times facing periods of decline. In part, its development had to wait on advances in medicine and science to give it credibility, respectability, certainty and status. Through it all, religious orders gave the profession qualities of compassion, dedication and caring which are the very heart of nursing, whether in civil life or in the army. This story is concerned with the evolution of military nursing in the context of religion, war and medical science. In a sense it grew out of civilian nursing but in later years the two branches of the profession grew side by side, each gaining from the other.

Nursing is one of the oldest activities of mankind, for the care of the sick, the nourishment of the young and the welfare of the older members of society have always been of concern to the family and to particular individuals and groups in society. In primitive societies disease was often explained in terms of evil spirits and the role of the medicine man and witch doctor was to drive out these spirits by the use of charms, dances, drugs, potent remedies and magic. Superstition gradually gave way to rational thinking and questioning. Sixteen centuries BC there is evidence of the Sacred Indian Books of Medicine which refer to procedures in operations. They were conscious of the need for cleanliness and they had recourse to burning sweet smelling substances to prevent devils or germs from entering the wound. While they were reaching for rational explanation of disease, they were conscious of religious influences, for all operations were preceded by long prayers! A major advance in medicine and subsequently nursing came in the 5th century BC when the Greek physician Hippocrates taught that disease was not the work of spirits but of natural causes. The chief therapeutic agents of the Hippocratic school were diet, fresh air, medicinal waters and exercise. Many of his ideas such as observation of the patient and sponging of the patient were to become basic principles in nursing.

While such thinkers were fumbling their way to an explanation of disease, great armies were fighting each other across Europe, Asia Minor and Asia. The Assyrians, the Chaldeans, the Babylonians, the Egyptians and other civilizations were engaged in constant warfare. Little is known about any provisions for caring for the wounded in battle, but presumably the soldiers themselves had to do what best they could, assisted by the occasional "doctor" and priest. Armies realized very early that greater casualties were caused by plague and pestilence, although it was some time before the importance of proper hygiene within army camps was recognized. The all-conquering forces of Sennacherib were destroyed, overnight, at the gates of Jerusalem, when, in the words of Lord Byron, "the angel of Death spread his wings on the blast".

However, it is to the legions of Rome that we look for the first attempts to integrate health care into the army. The Romans were conscious of the importance of public hygiene, witness the great engineering feats in building aqueducts, drains, roads and later on, hospitals. As their armies roamed over Europe and Asia Minor where health standards were primitive, soldiers had to be given instruction in hygiene and what might be called "first aid". In time, each camp had a "valetudinarium", staffed by "orderlies", slaves and doctors to treat the sick and wounded. Early traces of army nursing dating back to the Roman occupation of Britain reveal evidence of hospitals but there is no evidence that women were used in that capacity. A fresco from Pompeii dating from the first century AD shows a physician treating a wounded soldier with a sympathetic female looking on, presumably to give him comfort and solace. As the wounded returned from the wars, they were frequently cared for in the homes of higher class ladies, assisted by slaves. Women were gradually being brought into the task of caring for the sick and wounded "back home", but the task on the battlefield was for male "orderlies".

The coming of Christianity had its impact on both civilian and military nursing. It was seen to be a sacred Christian duty to love and care for the afflicted. The Old Testament was full of exhortations and laws covering responsibility for helping the sick and maintaining good standards of hygiene, while the teachings of Christ extolled the virtues of the good Christian accepting responsibility for "his brother", in acting like "the good Samaritan". Some congregations, especially in Eastern Europe, appointed "deacons" and "deaconesses" to visit the homes of the sick. Bishops found themselves establishing hospitals, one of the earliest being built by Basil, Bishop of Caesarea (AD 370) while Fabiola built the first public general hospital in Rome (AD 390) and actually worked in it as a "Nurse". Charity, in its original meaning of love for fellow men as sons of God, encompassed a wide range of facilities, so that hospital, hostel, hospice, almshouse and orphanage provided for many in need. Paula, patrician friend of Fabiola's, built hospices and hospitals for the sick in Palestine.

Perhaps the most significant development at this time was the establishment of religious communities and religious orders which were concerned with the care of the sick and later with the care of those wounded in war. In the early centuries monks and nuns were able to give spiritual comfort to the sick but it is doubtful if their nursing skills showed a high degree of competence. Christianity was brought to Britain by St Augustine (AD 596), a monk of the Order of St Benedict, heralding a new approach to the care of the sick and wounded, whether in peace or war.

The period of the Crusades provided a major impetus to military and civilian nursing through religious Orders. These holy wars, blessed by the church, inspired many to take up the challenge of freeing the sacred sites of the Holy Land, especially Jerusalem, from the infidel Turks. As pilgrims, warriors and ill-organized armies made their way towards Palestine, many Christians formed religious groups to offer them spiritual and material assistance in the form of shelter, comfort and when necessary nursing care for the sick and wounded. While their members took the standard vows of poverty, chastity and obedience they were in effect warriors who dedicated their lives and arms to the services of Christ. In the year 1023 a group of Italian merchants obtained permission from the Caliph of Egypt to establish two hostels at Amalfi "for the use of poor and sick Latin pilgrims". With the capture of Jerusalem and the establishment of a Christian

Kingdom in 1099, the Order of St John, as it became known, was soon established as a quasi military order of priests, knights and serving brothers. It became a powerful body, providing for the treatment of the wounded, one of the earliest organized army nursing services, albeit by males. The Patriarch of Jerusalem invested every approved candidate with a black robe bearing on the breast an eight pointed white cross. The Order of St John of Jerusalem continued its military and hospital work under its Master, Raymond du Puy, becoming rich and most sought after. Their hospitals were well equipped, staffed by the nursing brothers and run on military lines.

However subsequent military successes by the Turks forced them to flee westwards, first to Cyprus then to Rhodes where they remained from 1310 to 1522. When it too was captured, the Knights Hospitallers, as they were also known, were granted refuge in Malta by the Emperor Charles V. There they set up their Order again, providing hospital care for the sick and wounded from 1530 to 1798 when once again they were turned out, this time by Napoleon. Branches were established in many countries, including England, but it was suppressed by Queen Elizabeth I. While the monastic aspect has long since disappeared, the ideal of nursing and helping the sick and wounded has remained. In 1888 the Order of St John of Jerusalem was granted a charter by Queen Victoria while an associated body, the St John Ambulance Brigade, was founded to render first aid to those injured in accidents and in providing courses of training in elementary care. These organisations played an important role in nursing, both in peace and war.

The influence of the Knights Hospitallers continued across many countries of Europe, particularly in Germany. Like the Knights of St John of Jerusalem, the German Teutonic Knights were dedicated to defend the holy shrines and to serve the sick and the poor. After their hospital in Jerusalem was destroyed in 1187 they continued to nurse wounded Christian soldiers during the siege of Acre a few years later. After their retreat to Germany they were asked to administer many hospitals and to assist in nursing care. Another Order, the Knights of St Lazarus nursed the lepers of Jerusalem while still undertaking military duties. When the number of lepers decreased, Pope Innocent VIII closed the Order and handed over the possessions to the Order of St John.

While most of the nursing was being done by men, there were parallel developments in which women were becoming increasingly involved in nursing, both civilian and military. There was in fact a female branch of the Order of St John and Sisters were nursing the sick and wounded in Jerusalem, but after their forced evacuation they appear to have ceased as an organization. However, major advances were made by those women in religious orders. The oldest purely nursing order of nuns was probably the Augustinian Order of the Sisters of the Hotel Dieu in Paris, but their work was confined to visiting the sick poor in hospitals and in their homes. As the Sisters were untrained and the hospitals overcrowded, the standard of nursing was very low indeed, nevertheless the Sisters brought with them dedication and compassion in their task. In the thirteenth century the bishops declared that all nursing orders should take the vows of poverty, chastity and obedience and should wear a religious garb. This but made their role more unattractive and led to even greater inefficiency. On the other hand Isabella, the Catholic Queen of Spain (1451–1504) was so shocked by the horrors of the wars of her time that she introduced tented camp hospitals to care for the wounded, nursed by "reliable women".

In the 16th and 17th centuries there was another flowering of religious and secular nursing orders, the most important being the Sisters of Charity founded by St Vincent de Paul in 1617. St Vincent and his chief aide Mlle le Gras toured the countryside of France, enlisting girls for nursing and giving them some basic training. They were encouraged to watch the doctors and learn. The Order grew rapidly and the Sisters were in demand for nursing work of all kinds, including care of foundlings and asylums for the insane. In 1654 at Sedan they undertook military nursing for the first time, establishing a tradition for which they were to become famous.

Meanwhile the development of nursing was following a similar pattern in Britain and the idea was gradually being accepted that women had an important role to play in nursing the sick. However there is little evidence to suggest that women were involved in military nursing until the Civil War in 1642. Then the established hospitals — Savoy, The Hospital of St John the Baptist, St Thomas's and St Bartholomew's took in casualties to be treated by the medical and nursing staffs. This was extended when nuns tended the wounded at Dunkirk during the Flanders campaign of Oliver Cromwell.

In those days women sometimes followed husbands and sweethearts to war and occasionally they were enlisted. Perhaps the most renowned was the legendary Mrs Christian Welsh who fought in King William's wars and was dangerously wounded at Ramillies (1706). She joined the army to track down her husband who had been kidnapped and forcibly enlisted 12 years earlier. When she found him she liked the life so much, she continued not merely in giving nursing and companionship but also as a cook. She achieved fame as a swashbuckling rouster, silver heeled sword at her side and ale pot in her hand. She subsequently retired as an army pensioner. Then there was Elizabeth Alkin who became known as Parliament Joan. Born in the reign of James II, she volunteered her services as a nurse when her husband was enlisted in the army. She threw herself heart and soul into the work and made many petitions to Parliament before she was given a grant to continue her work. Her services were officially recognized by the Admiralty for whom she served in the Dutch War of 1652.

In some respects the navy was ahead of the army in accepting the presence of women in the service, probably because the ships were at sea for long periods. True, most of these were wives of officers and men but they frequently turned their hands to what loosely can be termed. nursing. The navy was also to the fore in designating certain ships as hospital ships, as early as in the Spanish Wars of the 17th century. While these were at first staffed by surgeons and their "mates" there is evidence that in 1684 the hospital ship '*Welcome*' working with the fleet off Tangier, was staffed with women as nurses. It is recorded that in 1703 medical staff on hospital ships included "nurses and laundresses", paid at the rate of ordinary seamen, but it is not clear what the duties of the nurses were. Sailors' wives also served on hospital ships at this time but their predilection for drink led to their replacement by male nurses or orderlies.

The navy made provision for the care of the sick and wounded not only at sea but on land by the establishment of hospitals and homes in their major home ports such as Plymouth. However it was not until after the Crimean War that the service became properly organized.

The idea that women might be regularly used for the care of the sick and wounded in war, not only at home but in overseas campaigns developed very slowly in the 17th and 18th centuries. In his review of the army nursing service, Colonel A.G. Butler points out examples of the use of female nurses in war time in the 18th century. They were attached to British hospitals in Portugal (1762) and they were used in the wars in Ireland in the reign of William and Mary (1689–1702). There are reports of hospitals in Flanders (1742), nurses in North American campaigns (1757) and in the Spanish wars (1762), but mostly medical and nursing services were haphazard and inefficient. On the one hand the status and reputation of women suffered by the large numbers of camp followers, while medicine and nursing had not yet received a break-through with scientific discoveries which were to give legitimacy and authenticity to their work. A far-seeing report came from Dr Monro, physician to George III's army following the Seven Years War (1763). He advocated the establishment of military hospitals, run on military lines. Every such hospital should be staffed by female nurses, supervised by a head nurse, who should inspect the hospital daily. She had the responsibility for seeing that the hospital was clean, that wards were sprinkled with vinegar, that the windows were opened twice a day and that the nurses did their job. Chamber pots had to be emptied and cleaned and the nurses had to be punctual, tidy and sober.

Dr Monro was one of the first to realize that nurses ought to be part of the military system and indeed be subject to military discipline.

> Nurses guilty of great neglect of duty or of getting drunk and using their patients ill or of stealing or concealing or taking away the effects of men who die in the hospital are to be immediately sent to the guard and reported to the Commanding Officer, that they may be tried by Court Martial and be confined, whipped or otherwise punished as the military law directs, all followers of armies on foreign service being equally subject to Military Laws as the soldiers themselves.

The army moved slowly to accept some of these propsals. By 1800 Matrons and Head Nurses had been appointed to regimental hospitals at home, with instructions that preference be given to the wife of an NCO or soldier of the regiment. The story was different in overseas campaigns. Wellington's army in the Peninsula Wars had no nurses as such. Women were not encouraged to join their husbands, because of the difficulty in keeping up with an army on the move.

It was the work of Pastor Fliedner in 19th century Germany which contributed so much to modern nursing, both civilian and military. He believed that young women should be attracted to nursing, that they should be properly trained and that they should work in hospitals which should be efficiently organized and administered. In 1836 Pastor Fliedner and his wife founded the Kaiserswerth Motherhouse for Deaconesses, an institution which had a powerful influence on nursing. Theoretical and clinical instruction was given by a doctor but nurses also had to undertake cleaning and cooking duties. Frau Fliedner also wrote a small book, the first to be written by a woman on the training of nurses. "Never sacrifice the soul of the work, for its technique," she advised. As the influence of trained Deaconesses spread to hospitals throughout Europe, humanitarians and philosophers flocked to Kaiserswerth to examine the system. Visitors included Elizabeth Fry and Florence Nightingale. Deaconesses continued their work into private homes, nursing patients through epidemics of typhus, cholera and smallpox. In time they extended their experience into the Prussian army, tending the wounded in the many battles of the period, although officially they were not part of the army.

But it was in nineteenth century Britain where those events and developments took place which were to determine the direction of civilian and military nursing not only in Britain but in the colonies, such as Australia. Hospital and nursing care had passed through a grim period in the previous century, with little attention paid to the state of institutions such as hospitals and prisons where people were crowded together in unhygienic conditions. Medical and nursing care was at its lowest ebb. With the turn of the 19th century came a new wave a compassion and humanitarianism for the poor and oppressed, for slaves, prisoners, the sick and those in institutions. Reform was in the air, in politics, in education, in local government, in public health, in industry, in social legislation and in medicine. This was a period which saw a revolution in scientific knowledge about the human body and disease, a revolution which was to change the whole pattern of training for doctors and nurses, a revolution which was to have a marked effect on the patient's chance of recovery, whether in peace or war. A greater understanding of physiology and anatomy, the use of chloroform in surgical operations by Simpson, the beginning of the science of bacteriology through the research of Pasteur and Lister, together with new approaches to conquer major diseases such as scurvy, smallpox, cholera and tuberculosis were all milestones in the history of medicine and nursing. It was realised that much disease could be avoided if proper standards of public health were maintained, that better training for nurses was essential. In 1840, Elizabeth Fry (1780–1845) already famous for her work in prison reform, visited Kaiserswerth and realized the need for improvement in nurse training. She was instrumental in founding an Institute of Nursing in connection with Guy's Hospital.

About this time the Church of England, with its long tradition of helping the poor, the sick and oppressed had expanded its social service role by increasing the number of visiting nurses. They were upper class ladies, refined gentlewomen, for whom helping the sick in hospitals and at home was an expression of Christian charity. In 1848 the Church established its first purely nursing order at St John's House, the probationers working with King's College Hospital for one year. These nurses were to be a valuable addition to Florence Nightingale's team in the Crimean War.

In the midst of this period of discovery and challenge came the Crimean War (1854–1856), a testing time for the British army to see whether it had kept up with these developments in medicine and nursing. Florence Nightingale was to provide the answer. Neither army nor civilian nursing was to be the same again.

Florence Nightingale (1820–1910) had grown up at a time when hospitals were still places of wretchedness, degradation and squalor. She knew the stench and smell of overcrowded wards, the lack of training and discipline among nurses. Deeply religious, she received a call in 1837 but it was not until 1844 that she realized her vocation lay in nursing. While devotion and dedication were essential, she knew that only knowledge and expert skill brought relief to the sick. However, her parents were horrified at the suggestion that this young lady from the higher social class might undertake such a vulgar occupation as nursing. Another eight years passed before her dreams came to reality. In this period visits to Kaiserswerth only strengthened her resolve and kindled her ambitions. She spent time in Europe, visiting hospitals, infirmaries, almshouses and institutions and collected a large number of reports on hospital administration and nursing arrangements in all major countries. Three times she attempted to join the Sisters of Charity to devote her life to nursing but each attempt was frustrated as though she was being saved for some grander purpose.

Back in London in 1853 she became Superintendent of the Institution for the Care of Sick Gentlewomen in Distressed Circumstances, an experience which at best strengthened her resolve to improve hospital administration, especially when dealing with men! The following year saw the outbreak of a cholera epidemic in London and Florence Nightingale went as a volunteer to Middlesex Hospital to assist in the nursing of patients. Again she saw much which needed improvement in nursing training and patient care.

The year 1854 closed an era in her life for in that year England and France declared war on Russia and within a few months allied armies landed in the Crimea. The disaster that followed is now part of history and the major cause of that disaster was the inefficiency of the medical and sanitary organization, reflecting forty years of neglect since Waterloo. In the attack on Sebastopol medical equipment and stores had been left behind and facilities were inadequate to handle an outbreak of cholera. When the British army fell back to its base at Scutari the old Turkish barracks were taken over as a hospital but this could not cope, without bandages, splints, chloroform, morphia, with no administrative plan to separate battle casualties from those suffering from cholera and other diseases. Food, sanitation, basic hospital equipment, these were completely inadequate. The British army had been slow to learn about military nursing, indeed preventive medicine, sanitation, hygiene and the need to keep troops healthy were almost unheard of in the army despite the many wars Britain had fought in the past century. During the winter of 1759 outside Quebec, during the retreat to the Ems in 1797 and during the disastrous Walcheren expedition of 1809 the British army suffered enormous losses through sickness and disease. From Waterloo through to the Crimea the British army medical services deteriorated through neglect to the point where they could not provide the medical logistics to service an army in a major campaign.

These tragedies were little known to the people at home but the tragedies of Scutari and the Crimea were publicised by William Campbell, war correspondent of *The Times*. He wrote:

> Not only are the men kept, in some cases for a week, without the hand of a medical man coming near their wounds, but now it is found that the commonest appliances of a workhouse sick ward are wanting and that the men must die through the medical staff of the British army having forgotten that old rags are necessary for the dressing of wounds.
>
> There are no dressers or nurses to carry out the surgeon's directions and to attend to the sick during the intervals between his visits. Here the French are greatly our superiors, their medical arrangements are extremely good, their surgeons more numerous and they also have the help of the Sisters of Charity . . . those devoted women are excellent nurses.

The cry went up, Why have we no Sisters of Charity? Sidney Herbert, Secretary at War, with uncanny insight, wrote to Florence Nightingale, inviting her to go to Scutari, in command of a party of nurses, with the Government's sanction and at the Government's expense to see what might be done. Sidney Herbert had for some time thought of introducing female nurses into the British army but each time the army authorities argued that male orderlies were necessary to cope with the mobility of the action. But here at last was a stationary military hospital at Scutari. However, in this case, he wanted her to go, not in the first instance as an angel of mercy but as a hospital administrator and organizer. With the acceptance of this challenge by Florence Nightingale military nursing took a major step forward.

Some of the problems to be resolved concerned army administration, outside her control or influence. Three government departments were responsible for maintaining the health of the British army — the Commissariat, the Purveyor's Department and the Medical Department. There was overlap and confusion as to responsilities. The administration of the army itself was divided between the Secretary for War (financial administration) and the Secretary at War (the sick and wounded).

One of the first challenges confronting Florence Nightingale and her nurses was to win the confidence of the doctors. She instructed her nurses not to enter the wards until the doctors ordered them and not to treat the patients in any way unless instructed to do so by the doctor. Once the doctors were convinced there was a role for female nurses in the traditional all male military hospital, Florence Nightingale and her band of nurses (soon to total 125) set about the daunting task of cleaning up the hospital. At every turn, new problems arose. There was the question of laundry arrangements for hospital linen and patients' clothes, the physical task of cleaning the wards, there was the need for adequate cooking facilities and the supply of food, the washing and bathing of patients, and of course the nursing of patients. There was seemingly no end to the tasks to be done. 'Orderlies were wanting, utensils were wanting, even water was wanting', she wrote in her report. The battle spread far beyond the wards. She had to expose the weaknesses of army administration, the petty regulations, the red tape, the inadequacies of politicians. Nursing was but the least of her tasks, but despite these day-long battles, she always found time to visit the wards, to supervise the nurses, to add a little comfort to the sorely wounded and distressed soldiers. What Florence Nightingale and her nurses did in the Crimea has been well recorded. Gradually, improvements transformed the whole pattern of military nursing. Funds were established to buy comforts for the sick. Recreation and reading rooms were established. At her own expense she had fitted out a house for doctors to carry out their own laboratory research. When the mortality rate dropped from 42 per 100 to 22 per 1,000, army and government authorities were at last convinced of the value of the nursing staff.

In 1856 Florence Nightingale was given official status, with the grand title of General Superintendent of the Female Nursing Establishment of the Military Hospitals of the Army.

But with the end of the Crimean War in 1856 her work and influence was only beginning, for ahead lay almost half a century of fighting to improve the quality of nursing, both civil and military. There was such widespread appreciation throughout Britain for the work of Florence Nightingale and her nurses in the Crimea that a Fund raised over 40,000 pounds sterling. The Nightingale Fund was set up "to establish and control an institute for the training, sustenance and protection of nurses, paid and unpaid." So began a new era in the training of nurses.

But it is in the development of military nursing after the Crimean War that the influence of Florence Nightingale is often underestimated. The Crimean War caused the British authorities to reassess its military organization and the role of its medical services within the army. In 1857 a Royal Commission was set up to inquire into "The Health of the Army", especially the conditions in military hospitals. Florence Nightingale submitted her report, *Notes on Matters affecting the Health, Efficiency and Hospital Administration of the British Army*, and her observations obviously shook army authorities out of their complacency. The value of females as military nurses was admitted, grudgingly, although there was as yet no clear statement as to their role and status. It was clear that a new type of nurse was required for military duties, not only a nurse who would see the army as a career, but a nurse with character, education and intelligence. Never again would there be an image of a nurse, either in civilian or military hospitals, as a tipsy, promiscuous harridan. British army authorities began to realize that battles would be won, not merely by force of arms but by the health of the army and the avoidance of sickness among the troops. If troops were sick or wounded facilities must be available either to restore them to fighting strength as soon as possible or to evacuate them for adequate medical treatment. The army medical service, including doctors and nurses, took on a new meaning in the order of battle. An Army Medical School for training medical personnel was established in 1860 and new military hospitals were built.

A major breakthrough came in 1866 when a Royal Warrant authorized the appointment of nursing Sisters to all military hospitals. Nevertheless, the tradition of male orderlies was not lightly cast aside. Some were still appointed to do ward duties, mostly untrained and assigned to general duties requiring physical labour. A few who showed promise were given the opportunity of further nursing training. Gradually, it was accepted that, subject to the medical officer, the nurse or Sister would be in charge of the ward. Her duties and responsibilities were becoming clarified. Behind those changes was Florence Nightingale. Against some opposition from 'the old guard' who thought soldiers were being pampered, she arranged that nurses should be appointed to the General Hospital, at Fort Pitt, Chatham, to look after soldiers recuperating from the effects of the Crimean War — a kind of modern Repatriation Hospital. Other hospitals were established at Netley and at Woolwich, in which the soldiers were tended by a Lady Superintendent and up to ten nurses. It was becoming accepted that nurses would be part of the military scene, so that when the

Army Medical School was transferred to Netley Hospital an integrated medical and nursing service came into being for the first time. Army nurses were trained at Netley not only in their duties in a military hospital but in the training and supervision of nursing and ward orderlies. One of the first army nursing Sisters to achieve distinction was Mrs Jane Deeble who with seven other nurses was appointed to Netley for training staff (1869). In 1879 Mrs Deeble, with 14 nurses was despatched to serve in the Zulu War. It came as no surprise when, in 1881, the Army Nursing Service was formally established through a Code of Regulations and Mrs Deeble was appointed the first Lady Superintendent. Applications were invited from nurses to join the service. They had to be of good health and character, preference being given to widows and daughters of army officers. They were given a year's training at Netley.

As there was now an established Army Nursing Service they were called up for overseas duty in time of war. In 1882 twenty-four nurses went with Mrs Deeble to the Egyptian Campaign, in 1883 to the Sudan War, in 1884 to Khartoum and to the Nile in 1889. Not only did they serve in military hospitals but they were called upon by the navy to serve in hospital ships.

In this latter part of the nineteenth century there came into being an organization which was to have a significant impact on the welfare of the sick, wounded and needy, in peace and in war — the International Red Cross. A young Swiss banker, Jean Henry Dunant, was so distressed by the suffering of the victims of France, Austria and Sardinia at Solferino that he wrote a book about the circumstances (1862). He believed much more should be done to help the innocent victims of war, so, inspired by the work of Florence Nightingale he took up the challenge.

> Would it not be possible, in time of peace and quiet, to form relief societies for the purpose of having care given to the wounded in wartime by zealous, devoted and thoroughly qualified volunteers . . . There is need for volunteer orderlies and volunteer nurses, zealous, trained and experienced, whose position would be recognized by the commanders of armies in the field, and their mission facilitated and supported.

The upshot of his proposals was an international convention in Geneva, attended by the major European countries. The Geneva Convention was drawn up, an historic document, giving neutrality and protection to the wounded in war and to those personnel attending them. The red cross on a white background (the reverse of the Swiss flag) was accepted as the emblem to identify staff, hospitals, ambulances, stretcher bearers and hospital ships. National Aid Societies (later known as Red Cross Societies) sprang up across the world. They spread their activities into a wide range of humanitarian functions, both civilian and military. In 1881 the War Office in Britain granted the National Aid Society (the British Red Cross Organization) permission to train military probationers, providing another pool of military nurses in time of war. However the Red Cross gradually relinquished this activity concentrating on male and female Red Cross representatives in wartime and encouraging the training of VAD's (Voluntary Aid Detachments). Their major contribution remained in the establishment of the Geneva Convention, encouraging belligerents to accept certain rules in warfare to protect the non-combatants, the wounded and the prisoners of war.

Related to this was the foundation by Queen Victoria of the Royal Red Cross as the military order for members of the nursing service, 'for special exertions in providing for the sick and wounded soldiers and sailors of Our Army and Navy'. (1883). Florence Nightingale and Mrs Deeble were the first recipients, but awards were also made to several nurses who attended the sick and wounded in the Egyptian and Zulu campaigns. Subsequently two classes of the Order were created, members and associates.

Some indication of the acceptance of women as nurses within the army structure is seen by the creation of the Indian Army Nursing Service (1888). India at that time was an important jewel in the British Empire, but it required a large standing army with military stations in most major cities. It was an exacting and dangerous climate for those born in England and many regular Indian army officers had doubts about sending nursing Sisters there although the need for them was acknowledged. The caste system prevailing in India and the status system of the army required that the nurses be gentlewomen to be socially acceptable. The move was to be eminently successful and in time Queen Alexandra's Military Nursing Service in India became recognized as one of the elite sections of the army nursing service, one with which Australian nurses were to be associated during the First World War.

The value of good nursing in the army soon spread to the navy, which had for a long time encouraged women to assist in naval hospitals and in hospital ships. Mostly these were ignorant and untrained and could hardly be classified as 'nurses' after Florence Nightingale's work in the Crimea. In 1884 the navy decided to establish trained sick berth staff and to employ only trained nurses in all naval medical establishments, at sea and on the land. Their record in treating patients at Plymouth was so outstanding that they were posted to naval establishments abroad, from Malta to Hong Kong.

The testing time for the relatively new Army Nursing Service came with the outbreak of the Boer War (1899), but reorganization of the army medical and nursing services had been going on for some time. In 1897 Princess Christian founded the Army Nursing Reserve, selecting candidates with wide nursing experience, but also with character, education and social position. They had to volunteer for service anywhere in time of war and as it happened there were but 100 in the Reserve when the Boer War broke out. As some 1,800 nurses served in the South African War the problems of administration were immense. Twenty-two general hospitals were established but the nurses officially were not appointed to field hospitals. Already a policy was being developed that the nurses should not be employed too far forward, where there was a probability of injury or capture. The lessons of the Crimea had been well learnt — there was greater emphasis on cleanliness, the nursing and medical staffs were better qualified, anaesthetics were in general use and the Army Service Corps was able to keep up a regular supply of food and drugs.

Even before peace was signed bringing the war to an end in 1902, it was decided to reorganize the army nursing service. Queen Alexandra's Imperial Military Nursing Service (QAIMNS) came into being by Royal Warrant on 27 March 1902, with a Matron-in-Chief on the staff of the Director-General of Medical Services in the War Office. Much was learnt from the South African Campaign, in which Britain lost 22,000 men, some two thirds of whom died from disease. A Royal Commission was set up to report on the 'Care and Treatment of the Sick and Wounded during the South African Campaign'. Among other recommendations, it was argued that the Matron, Sisters and Nurses in army hospitals must have more authority in running the wards, thereby making the task of the Commanding Officer of the hospital so much easier. Promotion was to be on merit — and soon army nurses were posted to such British outposts as Malta, Gibraltar, Ceylon, Hong Kong and Singapore.

These developments had their repercussions in Australia. In fact, the first Australian army nurses were to appear in the Boer War.

References

ABEL-SMITH, BRIAN. *A History of the Nursing Profession* Heinemann, London, 1960.
BEITH, J.H. (pseu Ian Hay). *One Hundred Years of Army Nursing*. London, 1953.
BETT, WALTER REGINALD. *A Short History of Nursing*. Faber & Faber, London, 1960.
BINGHAM, Stella. *Ministering Angels*. Osprey, 1979.
BLACK, C. *King's Nurse. Beggars' Nurse*. London, 1939.
BOLSTER, EVELYN. *The Sisters of Mercy in the Crimean War*. Mercier Press, Cork, 1964.
BOWDEN, JEAN. *Grey Touched With Scarlet*. Robert Hale, 1959.
BULLOUGH, VERN LEROY. *The Emergence of Modern Nursing*. Macmillan, N.Y. 1969.
BUTLER, A.G. *The Australian Army Medical Services in the War of 1914–1918*. 3 Vols. Aust. War Mem. Canberra, 1943.
CALDER, JEAN McKINLAY. *The Story of Nursing*. Methuen, London, 1958.
COOK, SIR EDWARD. *The Life of Florence Nightingale*. 2 Vols. Macmillan, 1914.
COPE, ZACHARY. *Florence Nightingale and the Doctors*. Museum Press, 1958.
COWLISHAW, L. 'Hospitals and Nurses in Past Ages'. *The Australasian Nurses Journal*. 15 Sept 1926. pp 421–428.
DELOUGHERY, GRACE L. *History and Trends of Professional Nursing*. The CU Mosby Co. 1977.
DIETZ, L.D. *History & Modern Nursing*. F.A. Davis, 1963.
DOCK, L. *A History of Nursing*. Putnam N.Y. 4 Vols, 3rd ed. 1931. (Vol IV includes Chap. on Hist of Nursing in Aust.)
DOLAN, JOSEPHINE A. *Nursing in Society; a historical perspective*. W.B. Saunders, London, 1978.
DOLAN, JOSEPHINE A. *A History of Nursing*. W.B. Saunders, London, 1968.
DOLAN, JOSEPHINE A. *Nursing in Society. A Historical Perspective*. W.B. Saunders & Co. London, 1973.
GOODNOW, MINNIE. *Nursing History*. Ramsays.

GRIFFIN, G.J. & J.K. *The History & Trends of Professional Nursing*. Saint Louis 1973.
HALDANE, E. *The British Nurse in Peace & War*. John Murray 1923.
HUXLEY, ELSPETH. *Florence Nightingale*. Weidenfeld & Nicolson. 1975.
KIRKCALDIE, R.A. *Nursing Sister. In Grey & Scarlet*. 1922.
LAMMOND, D. *Florence Nightingale*. Duckworth & Co. London, 1935.
LAURENCE, E.C. *A Nurse's Life in War & Peace*. Smith Elder 1912.
MATHESON, ANNIE. *Florence Nightingale. A Biography*. Thos Nelson & Sons. 1913.
MORRAH, DERMOT. *The British Red Cross*. Wm Collins, London, 1944.
MUNTHE, A. *Red Cross & Iron Cross*. London, 1916.
NUTTING, ADELAIDE & DOCK, LAVINIA. *A History of Nursing*. G.P. Putnam's Sons. 1907.
O'MALLEY, I.B. *Florence Nightingale, 1820–1856*. (The Story of her life down to the end of the Crimean War). Thornton Butterworth, London, 1932.
PAVEY, AGNES ELIZABETH. *The Story of the Growth of Nursing as an Art, a Vocation & a Profession*. Faber & Faber, London, 1953.
PIGGOTT, JULIET. *Queen Alexandra's Royal Army Nursing Corps*. Leo Cooper, 1975.
PLUMRIDGE, Lt Col JOHN H. *Hospital Ships & Ambulance Trains*. Seeley Service & Co., London, 1975.
ROBERTS, MARY R. *The Army Nurse Corps: Yesterday & Today*. Macmillan Co. 1959.
ROBINSON, VICTOR. *White Caps*: The Story of Nursing. (1946).
RUSSELL, SHEILA MACKAY. *A Lamp is Heavy*. Angus & Robertson, 1954.
SEYMER, LUCY RIDGELY. *A General History of Nursing*. Faber & Faber, 1956.
STEWART, ISOBEL & AUSTIN, ANNE L. *A History of Nursing from Ancient to Modern Times*. G.P. Putnam's Sons, 1962.
THOMPSON, ALICE M. *A Bibliography of Nursing Literature. 1859–1960*, with an historical introduction. Lib. Assn. Lond. 1968.
THOMPSON, ANNE. 'Military Nursing Through the Ages'. *Journal of the RAMC*. October, 1947. pp 194–202.
THRING, E.T. 'Nursing From the Earliest Times'. *The Australasian Nurses' Journal*. 15 Jan 1927. pp 4–7.
TILTON, MAY. *The Grey Battalion*. Angus & Robertson, Sydney, 1934.
TOOLEY, SARAH A. *A History of Nursing in the British Empire*. S.H. Bonsfield, 1906.
WINTLE W.J. & WITTS, FLORENCE. *Florence Nightingale & Frances E. Willard. The Story of Their Lives*. Sunday School Union, London.
WOODHAM-SMITH, CECIL. *Florence Nightingale 1820–1910*. Constable 1950.
WOODGHAM-SMITH, CECIL. 'A Short History of the Order of the Hospital of St John of Jerusalem'. *The Australasian Nurses' Journal*, 16 May 1938, pp 106–110.

Chapter Two

Foundations of Service:

The Boer War to World War I

Our reception here (Johannesburg) was curious. On handing in my papers to the PMO, he groaned, 'My God, Australian Sisters, what shall we do?'

Report on the Boer War by Matron Gould.

For most of the first hundred years of its existence following the arrival of the First Fleet in 1788, Australia remained a military establishment. The presence of the armed forces was necessary to guard the convicts and to maintain law and order. Medical services were required both for the convicts and the armed forces and increasingly for the free settlers. The story of the nineteenth century for the army was the gradual replacement of imperial forces by volunteer defence forces of the colonies leading to the establishment of a permanent army of which the Australian Army Medical Corps was to be an integral part. Parallel to this was the development of civilian medical facilities, with major city hospitals and smaller district hospitals in country areas, supported either by the government, the community or by religious orders or by a combination of all three. Until after the Crimean War there was little acceptance of the use of trained female nurses in either military or civilian hospitals. None came with the First Fleet and convict women were pressed into service as 'nurses'. These female convict 'nurses' were described as 'usually a dissolute class and often came on duty intoxicated in spite of frequent punishments'. The British practice of using male orderlies was also followed in 1816, but there was no organized attempt to select and train either males or females as 'nurses' during the first fifty years of settlement.(1)

After 1848 when transportation of convicts ceased, hospital authorites and medical men had to look more closely at those selected to act as nurses. At the Sydney Infirmary the Resident Surgeon was charged with the responsibility of supervising the work of the nursing staff. As the standard of medical care improved, doctors put pressure on nurses to upgrade their standards too. In 1857 there were complaints about inefficient nursing at the Sydney Infirmary and inquiries revealed that these were 'chiefly elderly women . . . vacancies were often filled by promotions from among the servants'.(2) Brodsky sums up the deplorable nursing situation which derived from the military origins of the settlement.

> The nursing had been rough and haphazard, performed by convicts and dissolute women in an atmosphere of chaos and cruelty, military disciplines and inadequacies and general ignorance. It may be supposed that these were inseparable from a penal settlement.(3)

Lucy Osburn's Influence

The knowledge that trained nurses had much to offer in peace and in war soon spread after the work of Florence Nightingale became known throughout the British colonies. In 1866, Henry Parkes, then Colonial Secretary in the New South Wales Parliament, wrote to Florence Nightingale seeking her help to establish a training school for nurses in Sydney. In 1868 a Lady Superintendent, Lucy Osburn and five nurses arrived, thereby initiating a new era for nursing in Australia.(4) Poorly trained male orderlies were replaced and a system of training probationary nurses was begun.

The influence of Florence Nightingale and her principles for nurse training as developed by Lucy Osburn soon spread to all Australian colonies. The Lying-In Hospital in Melbourne (later The Royal Women's Hospital) introduced midwifery training at the same time as Florence Nightingale opened a maternity ward at King's College Hospital, London to train midwifery nurses (1861–62). The Benevolent Society of NSW introduced nurse training in 1866 in what was to become the Royal Hospital for Women in Sydney. Then followed Lucy Osburn at the Sydney Infirmary 1868. In Victoria a training school for nurses began at the Children's Hospital (1878), Alfred Hospital (1880), Prince Henry's, then the Homeopathic Hospital (1885), and at Melbourne Hospital 1890. South Australia and Western Australia also followed in the 1890's.(5) By the end of the century there were in Australia large numbers of nurses trained on the Florence Nightingale system serving in all the major hospitals and available for military service should the occasion arise and should the army see fit to accept them.

The development of an army nursing service in the Moreton Bay District of New South Wales followed a similar pattern. It had to wait on such factors as the development of civilian hospitals staffed by trained nurses — and that required recognition of the need for training and the provision of appropriate training facilities. Within the military sector the creation of a medical service to care for the sick could only develop as fast as it did in New South Wales and indeed as fast as it did within the British army, as for many years the district and later the colony were guarded by British army units. Hospital accommodation was provided in Brisbane in 1825 and in 1828 a hospital for sick convicts was established. The following decade saw the cessation of the penal settlement and an expansion of free settlers followed in the 1840's. It was not however, until 1859 that Moreton Bay was separated from NSW to become the new colony of Queensland. The need for a civilian hospital became evident leading to the earlier establishment of the Brisbane Hospital in 1849, near the site of the old Supreme Court. In 1866 it moved to the Herston site, changing its name from time to time, but mostly it was known as 'The General'.(6) The Florence Nightingale system of training nurses was soon accepted in Queensland and 'The General' became the centre of nurse training. Religious and private hospitals as well as district Hospitals added to the availability of nurse training. As in other colonies of Australia there was, by the end of the 1890's a large pool of qualified nurses. By the end of the century the nursing profession had matured to the point where national associations of trained nurses were formed. The Australasian Trained Nurses Association was formed in 1899 while Queensland in 1912 became the first State to introduce compulsory registration.

Sister Mary Elizabeth (Hercy)

During this latter part of the nineteenth century there lived and worked in Brisbane a Sister who served with Florence Nightingale in the Crimea. Her story is little known but needs to be recorded in this context. While much prominence has been given, and rightly so, to the work of the Sisters of Charity in nursing the sick and the wounded of the French army in the Crimean War, there has been insufficient recognition of the contribution made by the Sisters of Mercy. Founded in 1831 by Catherine McAuley in Dublin, the members of this religious congregation, dedicated to Our Lady of Mercy the following year, vowed themselves 'to the service of the poor, the sick and the ignorant'. It was not a happy time for Ireland with economic distress, agrarian discontent, widespread poverty and appalling conditions for tending the sick. An epidemic of Asiatic cholera in 1832 gave these Baggot Street Sisters an opportunity to minister in various Dublin hospitals where nursing standards had been very poor indeed. During the 1840's the Sisters of Mercy built a fine reputation for nursing, especially during the Great Famine. Meanwhile, the Order had been extended, to establish the first religious house in England since the Reformation in Bermondsey in the diocese of Southwark.

When an appeal went out for nurses to serve in the Crimea, few expected a response from Irish Sisters and fewer still expected a favourable response from the British government, in view of the no-popery campaign still raging. However, when it was realized that more than one-third of the British soldiers fighting in the Crimea were Irish, the need for Irish nursing Sisters was recognized. Florence Nightingale also realized that the Nuns would have more desirable qualities than many of those offering, who lacked the true vocational spirit she was seeking.

> For what training is there compared with that of the Catholic Nun? Those ladies who are not Sisters have not the chastened temper, the Christian grace, the accomplished loveliness and energy of the regular nun.[7]

So it happened that on 14 October, 1854 Bishop Grant visited Bermondsey and informed the community that the government might require their services as nurses in the military hospitals in the Crimea. Three days later five Sisters set out for Turkey, but were held in Paris to await Florence Nightingale and her group of secular nurses, supplemented by volunteers from Anglican Sisterhoods. They arrived at Scutari at a time when medical services were in a state of collapse following the battle of Inkerman. At the same time there were protracted negotiations taking place between the British War Office and the Catholic hierarchy for a second group of Irish Sisters to go to the Crimea. The Rev Dr James Quinn, (later to become Bishop of Brisbane), was appointed to seek out volunteers and he was successful in obtaining eleven Sisters who were known as the Kinsale Sisters. Evelyn Bolster, in her book, *The Sisters of Mercy in the Crimean War*, assessed their contribution:

> Generally speaking, their work has been practically overlooked by a forgetful world, the only records remaining being confined to unpublished private letters and unpublished official correspondence. Yet the usefulness of these Sisters in the English military hospitals cannot be gainsaid; nor can they any longer be denied their meed of praise.

One of the Kinsale Sisters was Sister Mary Elizabeth Hercy of Berkshire. She had taken the veil in Dublin in 1845 and she gave devoted service in the military hospitals of the Crimean War. In 1865 she came to Brisbane to care for the poor and the sick as Sister Mary Elizabeth Joseph of All Hallows' Convent. On the occasion of Queen Victoria's jubilee in 1897, the Queen decided to honour surviving Sisters of the Crimean War. Queen Victoria was pleased to send to Sister Mary Elizabeth, through her brother, Major-General Hercy, the Royal Red Cross, in recognition of her services as a nurse in the Crimean War. This Royal Red Cross now has an honoured place in the Mater Hospital in Brisbane.[8]

This is a lasting tribute to one who was probably Queensland's first military nurse who had seen action in battle.

Army Medical Services

During these decades the question of the appropriate defence force for New South Wales and other colonies became a matter for much debate in the British Parliament. As transportation of convicts to Australia ceased and a new free society developed, the presence of British army units had less relevance. The defence of the Australian colonies was in the first instance the responsibility of the British navy but beyond that it was expected that the colonies would contribute in increasing measure to the facilities needed for home defence. After 1859 Queensland, as with other colonies, became responsible for its own defence, electing on the grounds of cost, to establish a Volunteer Defence Corps.[9] Act 18 Victoria No 8, enabled the government to accept on behalf of the Queen the services of 'such of Her Majesty's loyal subjects as may be willing to enrol themselves in a Volunteer Corps'. By 1870 the last of the imperial troops had left the colonies, each of which passed various Defence Acts and acted independently in such matters until Federation. Queensland's formal annexation of New Guinea (1883) was a classic example. A Report of a Committee of Inquiry into Queensland's defence force (1882) had nothing to say about the use of female nurses in army hospitals. Indeed Johnson's survey of Queensland's defence forces (1860–1901) makes no mention of nursing services.

The employment of female nurses in the Australian army had in turn to wait for the development of the army medical service. Medical care in New South Wales was at first provided by regimental surgeons, on

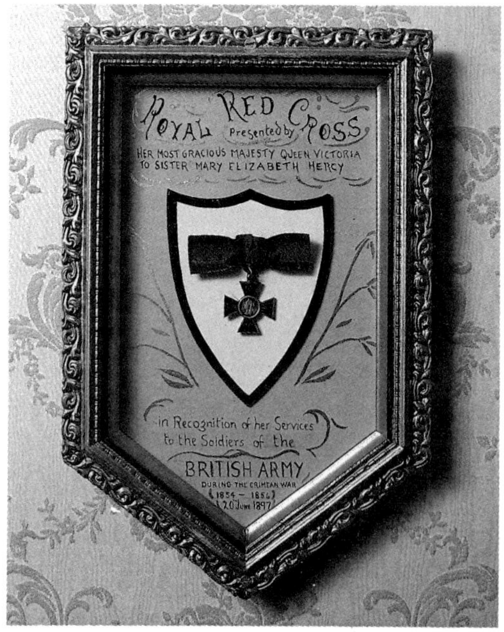

Sister Hercy's Royal Red Cross, Mater Hospital, Brisbane.

the British pattern and in the 1880's an Ambulance Corps was established, which was supplemented in 1880 by a volunteer Medical Corps. In 1891 the Medical Staff Corps was given official permanent status as the Permanent Medical Staff Corps. The first report in 1891 stated

> A permanent Corps of fifteen of all ranks was organized in 1891 for the performance of medical services of sanitation, working of the hospital, care of patients, medical and surgical field equipment and ambulances, attendance at shot practice and to act as a nucleus to the partially-paid branch of this service.[10]

At this time there was growing awareness of the vulnerability of the eastern coast of Australia to enemy attack and the French scare and the Russian scare heightened these fears. Stretcher bearers and ambulance waggons were introduced but no nurses, as yet. Principal Medical Officer William Williams (NSW) set about the difficult task of reorganizing the Medical Services, to give them more mobility, independent of the fighting units of the army. The shortcomings of the British army model were apparent in 1885 when Australia, for the first time sent troops to fight overseas, in this case to the Sudan.[11] Nevertheless the Ambulance Corps and the Medical Corps acquitted themselves with distinction. No Australian nurses were sent to the Sudan. By the 1890's the idea of Federation was taking firm ground and the question of defence was a vital issue. In 1898 the British army reorganized its medical services, formally establishing the Royal Army Medical Corps, (RAMC) as an integral part of the army. In the same year the Medical Staff Corps in NSW became the New South Wales Army Medical Corps (NSWAMC), with some 40 medical officers, but no nurses, although a recognition of their value was increasing. Undoubtedly Surgeon-General W.D.C. Williams was aware of the developments in the army medical services of the British army and of the valued service given by nurses in Egypt (1882), the Sudan War (1883), Khartoum (1884), and the Nile (1889). He would have been aware of the organization of Princess Christian's Army Nursing Reserve (1897) and of the Indian Nursing Service. It was inevitable that when the time came the English model would be the one for Australia.

The Army Nursing Service of New South Wales 1899

New South Wales, the oldest of the colonies, with a tradition of military service since its origins, was the first of the colonies to recognize the important role nurses could play in war-time. It was the vision of

Surgeon-General W.D.C. Williams that set in train a reorganization of the army medical corps, including the establishment of an army nursing service. He was in advance of this time in creating these facilities in the years before Federation. In May of 1899 an Army Nursing Service of NSW was established as part of the medical service, the Reserve consisting initially of 26 nurses under the control of E.J. Gould,[12] who was designated Lady Superintendent. The nurses were carefully selected, 'possessing the highest nursing qualifications and training'. They were required to maintain efficiency by attending a course of lectures on military organization. It was early recognized that while in civilian life they might be good nurses, they had much to learn about army life.

Unfortunately, there remained a great deal of confusion about the role and status of army nurses, a problem which was not resolved for many years in the Australian Army Nursing Service.

As the organization of the AANS in NSW, two years before Federation, became the model for the Commonwealth the plans of Colonel W.D.C. Williams should be noted. He kept the British classification system of Lady Superintendent, Matron and Nursing Sister and when called up for military duty they were to be paid £100, £60 and £40 per annum respectively. The first Lady Superintendent of NSW, Matron E.J. Gould recalled the circumstances of this foundation.

> In February 1899 Colonel Williams asked me to help form a Nursing Service in connection with his Army Medical Corps Service. In May of the same year, when the various branches were receiving the training necessary to make them militarily efficient, the little band of 26 nurses were "sworn in" — one Lady Superintendent, one Matron and twenty-four Sisters.[13]

The Boer War

When the Boer War broke out on 11 October 1899 the Australian colonies were individually responsible for their own defence policies. The Queensland Government had watched the drift towards war and in the event the Premier, J.R. Dickson, was first off the mark to offer a contingent for South Africa some three months before the commencement of hostilities. Subsequently, the British Government accepted the offer of two contingents from New South Wales and Victoria and one each from the remaining colonies. New South Wales with its longer experience had the most efficient force, including an army medical corps. Probably they were pleased not to be saddled with inefficient forces from the other colonies. During the nineties there had been various reports on the state of the defences, mostly showing up inadequacies and deficiencies. In 1899 the Commandant of the Queensland Defence Forces complained that his medical service was unsatisfactory. After setting out in detail the need for horses, bands, veterinary surgeons, cadet corps and the like he commented with respect to the medical corps,

> This Corps is not in a satisfactory condition and requires much more recognition than it has received hitherto.[14]

Nevertheless, he had to admit in his report that the provision for a permanent medical officer, an ambulance and other medical facilities were 'unavoidably excluded from the current year's Estimates'.[15]

The first group of nurses to be sent to South Africa sailed with the second contingent of troops in the *Moravian* on the 17th January 1900. With them went 52 horses, 5 ambulance waggons and 12 carts. The nursing group was composed of a Lady Superintendent, Ellen Julia Gould,[16] a Superintendent, Julia Bligh Johnstone, and 12 other Sisters

Anne Austin	Annie Matchett
Penelope Grater	Nancy Newton
Anna Garden	Elizabeth Nixon
Emily Hoadley	Mary Pocock
Elizabeth Lister	Mabel Steel
Marian Martin	Theresa Woodward[17]

As this was a NSW contingent and the two senior nurses were top nursing administrators and as it was the first time that Australian nurses were permitted to accompany Australian troops overseas, it is highly likely that the nurses were carefully chosen and probably well-known in NSW. It is extremely unlikely any were from Queensland.

The second group of nurses went with the third contingent of troops, for service under the Imperial Government, leaving on 17 March 1901. Along with 54 horses and 2 ambulance waggons went the following 10 Sisters.

Rawson,	Marianne (in charge)
Tiddy,	Diana
Smith,	Ethel Mary Bernhard
Langlands,	Eleanor Augusta Victoria
Smith,	Dorothy F.
Hines,	Frances Emma
Anderson,	Julia B.
Walter,	Ellen
Thomson,	Annie Eliza Helen
Ivey,	Isobel[18]

Again, it is not known if any of these were associated with Queensland. Murray's report concluded: 'These ladies were all single. They did excellent work in the hospitals, developing the best qualities of professional nurses'.

A third group of Australian nurses, anxious to serve in South Africa, enlisted with Princess Christian's Army Nursing Service Reserve. Records from Regimental Headquarters of the QARANC lists the following Australian nurses who served in this category.[19]

466	Amy Robinson Chutt, Alfred Hospital, Melbourne.
849	Dora Burgess, Melbourne Hospital.
194	Gertrude Fletcher, Prince Alfred Hospital, Sydney.
924	Rhoda Gwyer, Prince Alfred Hospital, Sydney.
151	Lucy H.M. O'Ryan, General Hospital, Hobart.
1003	M.M.A. Robertson, General Hospital, Hobart.
127	Mabel Gertrude Ashton Warner, General Hospital, Hobart.
33	Adelaide Teesdale, Government Hospital, Adelaide.
970	Kate Octavia White, Launceston Hospital, Tasmania.
844	Lily Dawson, Launceston Hospital, Tasmania.
1046	M.A. Grace, Adelaide Hospital, Tasmania.

While it is not known if any of these nurses had association with Queensland, it is probable that Gertrude Fletcher was Nurse G. Fletcher whose letters to her uncle, Mr C. Fletcher of Brisbane were published in the *Queenslander* during the Boer War period.[20]

These official groups were not the only Australian nurses to be involved in the South African war. In his account of the war, Wallace referred to several nurses not listed in the above contingents.

> These ladies were preceded in South Africa by at least one free-lance nurse from New South Wales. By paying her own passage when the initial applications were turned down, Miss Agnes Macready arrived in Durban to take up duties at the Fort Napier hospital in Pietermaritzburg.

Another was Mrs Betty Kennedy, a nursing Sister from Melbourne who was posted to a hospital at Estcourt treating the wounded from the final push to relieve the siege of Ladysmith.[21] He also refers to the extraordinary adventures of another Australian nurse, a Miss Rose Lina Shappere, who accompanied the commandos to the Natal border from Johannesburg at the start of the war. Eventually she arrived in Ladysmith just as the first bombardment began and was attached to the Military Hospital. Later she returned to England[22] on a troopship nursing the wounded and subsequently went back once more to Johannesburg.

Queensland Nurses in the Boer War

Although no Queensland nurses are listed in the official records of the Australian contingents to the war in South Africa, it is extremely likely that Queensland nurses found their way there in one way or another. It was a period of depression in Australia, while South Africa was booming with the discovery of gold. It is highly likely that Queensland nurses followed their families and friends to South Africa. There is certainly evidence that some did turn up and did become involved in nursing during the Boer War. The Official British History of the War in South Africa, 1899–1902 states:

> Nursing sisters from overseas colonies. They came from Qld, Vic, NSW, S Aust, W Aust, Tas, NZ and Canada. Others came with recommendations from their colonies and were engaged immediately on arrival in South Africa.(23)

Although there is a paucity of information about Queensland nurses in the Boer War in the official records, *The Queenslander* published a photograph of Nurse Huston with some of the boys from Clermont (31 March 1900) and another of Nurse Fletcher 'who has gone to South Africa'.(24) Beatrice Huston was known to have come from an old family at Clermont where she was a nurse at the local hospital. In a letter to Mrs J.M. Cross, published in *The Queenslander* she related how she sailed for Capetown in the *Salamis*, together with 7 nurses from New Zealand, sent and supported by the public of Maoriland. In Melbourne they were joined by a Melbourne nurse, paying her own way, as did Nurse Huston and in Fremantle a further group of 11 West Australian nurses joined the party. As an unofficial group it is not known what

Boer War — Nurse Huston and some of the Clermont boys in camp at Enoggera. The Queenslander 31.3.1900.

inspired them to offer their services at this time. The nurses were compelled to travel third class and the conditions were extremely primitive. The food was poor, the bread mouldy and the meat often bad, wrote Nurse Huston. She said she never expected to reach Durban alive. Her cheeks, well rounded when she left Queensland, were hollow and her eyes sunken after the voyage during which she lost 22 pounds in weight.

There was apparently a shortage of nurses in South Africa for they were engaged immediately on arrival and sent to the British Field Hospital at Rondebosch. For their services they were paid 49 pounds sterling per year plus extra allowances. One of her first impressions was the variety of uniforms worn by the nurses from different countries.

> Those from New Zealand wore khaki uniforms, white sailor hats, with a red cross on the arm and on the hat. The Canadians also wore khaki uniforms with gold buttons, white hats with a red thistle in front. The English nurses wore a bluey grey dress, with a little red cape and white hats. And I was in my blue dress, with a red cross on the arm and a white hat also with a red cross. We all wear Florence Nightingale caps for dinner and night duty.[25]

'Nurse Fletcher who has gone to South Africa'. The Queenslander. 21 April 1900.

Miss Miles and Miss Norman at the Jagersfontein Hospital. The Queenslander.
11 Nov 1899.

The *Queenslander* published several letters from Nurse Huston in which she described vividly the problems of nursing the Boer prisoners.

Arrived Cape Town on the 10th at 8 p.m. On 14th the Rondebosch Field Hospital, 500 patients, 300 tents, 9 doctors and 23 sisters. We sisters slept 4 in a marquee and all messed in a large marquee. I was put in the surgical division. The superintendent came and told 2 of us to be ready at 11.30 to go to Green Point to nurse the Boer prisoners. They were dying 2 or 3 a day and no one to nurse them. They were in tents in a filthy state and very ill. We had to stay at the new Somerset Hospital to sleep and meals. We had to go on night duty from tent to tent with a lantern and many a fall we got over the ropes in the little gullies. There were sentries all around us and the track where the

Military Hospital at Green Point. Nurse Huston (QLD) and some of her orderlies and patients. The Queenslander. 1 Sept 1900.

> Boer prisoners are enclosed, 1,800 of them with iron fences around them and barbed wire at the top and every 10 yards there are sentries on high stands. The Boers were very dirty, they did not like being washed or to take their clothes off, or boots, but got into bed with them all on. We nursed them for 3 weeks, but the rain was too often and they built tin huts. I was nursing 45 patients at night — Boers — in these tin huts for 3 weeks. One night I went to church and fainted and was brought home and put to bed for a week. In the same enclosure where Boers were patients there were 300 Tommies camped, nearly all sick. I was then sent to the new Somerset Hospital to be nursed and when I was better I was put on day duty. I had 12 wards, 45 patients and 4 orderlies to help me. Some had enteric fever, dysentery, pneumonia, diphtheria, jaundice, rheumatism, fractures of different limbs, bullet wounds, consumption and numerous other diseases . . .
>
> We have marquees to sleep in, boxes for washstands, nice jug and basin, waterbottle and a large bath in each tent which orderlies fill each night. We have good food as people around about us send us extras.[26]

The other nurse with Queensland connections who served in the Boer War was Sister Gertrude Fletcher, whose letters to her uncle, Mr C. Fletcher of Brisbane, were frequently quoted in *The Queenslander* of the time. She was selected to serve with Princess Christian's Army Nursing Service Reserve, on the basis of previous service and qualifications. Little is known about this, except that prior to appointment she was on the staff of Prince Alfred Hospital, Sydney. Describing the review by HRH The Princess of Wales, Sister Fletcher wrote,

> We all wore our blue serge dresses and cloaks with scarlet cashmere hoods, which form a square of scarlet in front, ugly little blue bonnets with white strings and our large silver badge of the ANSR on the right breast.[27]

Ambulance presented by the Women of Queensland for the South African War. The Queenslander. 10 Mar 1900.

After describing life on the troopship to South Africa, she related how she arrived at Capetown and endured a five hour train journey to the Karoo Desert where so little rain fell that a little African girl of three years cried and asked her mother if the stars were crying. Eventually she reached the Royal Yeomanry Hospital, some 4,000 feet above sea level, where she was placed in charge of the officers' ward. Most of her patients were typhoid victims. They received a great deal of help from local farmers.

In Perth, an independent group of nurses, sponsored by local citizens and supported by a general public appeal went off to South Africa in charge of a Miss Mary Nicolay. The following names are listed in a letter addressed by the Mayor of Perth in March 1900, to His Excellency, Sir Alfred Milner, KCMG, Governor of the Cape Colony, Capetown.

Tchan	Rogers
Plover	Emmins
Armstrong	Speers
Brooks	Milne
Naylor	Bole

Two of this group were Queenslanders, Sisters Susannah Armstrong and Miriam Plover, both trained at Brisbane Hospital. They sailed in the *Salamis* on 10 March 1900 on the same voyage as Nurse Huston. These Sisters were readily accepted for service in British military hospitals, the records showing Susannah Armstrong serving at Howick Military Hospital, Ladysmith.[28]

A vivid account of the conditions under which the nurses served in the Boer War and a first hand appreciation of the value of their service was given by Sir Frederick Treves. He was in charge of No 4 Stationary Field Hospital which followed for three months the Ladysmith Relief Column and subsequently

Boer War Nurses, 1900. Rear: Nurses Speers, Emmins, Armstrong, Bole, Rogers.
Front: Tchan, Naylor, Nicolay, Plover, Milne. University of Western Australia Press.

wrote an account of this episode.(29) The wounded came to the hospital by train, by ambulance or waggon and sometimes on stretchers. The hospital team included four sisters (nationaility not stated). At Chieveley they found only one house standing, the Station-master's and that had been looted. The waiting room at the station had been used as a stable by the Boers, nevertheless they had to make the best of it.

> The question of where to sleep was soon solved by the necessities of the position. These ill-housed women, as a matter of fact, were hard at work, all Friday, all Saturday and all Sunday night. They seemed oblivious to fatigue, to hunger or to any need for sleep. Considering that the heat was intense, that the thirst which attended it was distressing and incessant, that water was scarce and that the work in hand was heavy and trying, it was wonderful that they came out of it all so little the worse in the end.
>
> Their ministrations of the wounded were invaluable and beyond all praise. They did a service during those distressful days which none but nurses could have rendered, and they set to all at Chieveley an example of unselfishness, self-sacrifice and indefatigable devotion to duty. They brought to many of the wounded and the dying that comfort which men are little able to evolve or are uncouth in bestowing, and which belongs especially to the tender, undefined and undefinable ministrations of women.(30)

Wallace, in his definitive study of the contributions made by Australians in the South African War refers to the fine work done by Miss Gould and the nurses from NSW. Butler, in his summary, quotes a report from Miss Gould

> Our reception here (Johannesburg) was curious. On handing in my papers to the PMO, he groaned "my God, Australian Sisters, what shall we do?" On my asking the reason, he said, they did require help, but he understood we could not work with the RAMC Sisters.(31)

However, the difficulties were smoothed over and the Australian Sisters worked happily and efficiently within the British army.

When it was decided to send Australian nurses to the Boer War the matter of an appropriate uniform for the nursing service had to be settled quickly. According to a report by Matron Gould it was decided by the medical officers, Surgeon-General Williams and Lt Col Vandaleur Kelly on the lines of that used by the British services.

> A red cape was obtained from the War Office as a pattern, also the regulation cap (muslin hemstitched square). The dress was of dark grey serge (made in Australia) with in addition to the red cape, a three quarter length cape of same and a bonnet for outdoor wear. The plain trimming was of brown as worn by the RAMC. Working uniform was of grey zephyr. Washing capes (scarlet) and the muslin square caps. These capes were made for us at the army tailors. Before the matter was finally settled, I was called to go to see General French and he approved of the arrangements to be made for the uniform. However when the Boer War broke out we had to add serviceable rain coats. All buttons used were as worn by the military forces in NSW.

Matron Gould reported at length on her experiences in the Boer War and undoubtedly her comments greatly influenced the establishment of the Australian Army Nursing Service and its development between 1902 and 1914. She was indeed a powerful and determined figure whose influence spread far beyond her initial appointment as Lady Superintendent of the Army Nursing Service in the Colony of New South Wales. Referring to the Boer War she wrote,

> Our place in this magic organization was clearly defined, and twenty years later, the Australian Army Sister still reaps the benefit of those preliminary instructions for the nursing spirit of the New South Wales Section (of which alone I can speak with authority) has always kept in time with the rest of the Corps and worked harmoniously with the medical officers, orderlies and others.[32]

Matron Gould was speaking with the knowledge that the Boer War Sisters were 'sworn in' and therefore brought under Queen's Regulations.

Recognition of the valuable contribution made by the Australian army nurses in the Boer War came from many sources. Indeed some were decorated for outstanding service. Nurse Shappere was mentioned in despatches by Sir George White for devotion to duty; Sister Rawson was awarded the Royal Red Cross; Sister Ivey was mentioned in the Commander-in-Chief's despatches; Sister Pocock was mentioned in despatches; Sister Nixon was awarded the Royal Red Cross as was Sister Bidmead. Sisters Bidmead and Glenie were also presented with Devoted Service Crosses. This was a remarkable achievement for the first small contingent of Australian nurses serving overseas.

More importantly, the work of the nurses in the Boer War, whether from Princess Christian's Reserve or from Australia or from independent sources made a tremendous impression on medical and army authorities. There grew a realization that military hospitals could not function without them. They were well trained by the standard of the time, dedicated to the care and comfort of the sick and wounded, physically tough to stand up to the rigours of warfare in a difficult climate and mature enough to accept the discipline of an army at war. Undoubtedly, all of these factors hastened moves to establish a permanent army nursing service in Australia.

Foundation of the AANS

The lessons of the Boer War made it obvious that the Australian armed services, especially the medical services, needed to be re-organized. It was clear that the various colonies could not severally provide uniform defence policies to protect Australia, at a time when Britain was passing to the colonies the responsibility for their own defence. These considerations fortunately came to the fore at the very time when the Australian colonies agreed to federate into a Commonwealth and the new States agreed to pass to the Commonwealth Government the responsibility for defence.

The credit for the formal moves to establish the Australian Army Nursing Service must go to Surgeon-General of the army medical services, W.D.C. Williams, Director-General of Medical Services (DGMS) on the first Commonwealth HQ staff. He submitted the original scheme to Major-General Sir Edward Hutton, the responsible General Officer Commanding, entrusted with the conversion of the defence forces of the various Australian colonies into the Commonwealth Military Forces (CMF) after Federation.

The Australian Army Medical Corps

The change from colonial to Federal control of all defence forces of Australia was gazetted from 1 July 1902, with the control of all AAMC personnel vested in the Director-General who was responsible to the GOC for the drill, discipline, efficiency and statistics of the medical service.

General Order No 123 of the Military Forces of the Commonwealth 1902 formally brought into being the Australian Army Medical Corps for the Permanent Army, the Militia Army and the Volunteer Army and under this umbrella the Army Nursing Service Reserve. The following are the relevant sections of that Order.

(1) It is essential that, in order to meet the requirements indicated in the Minute on 'The Defence of Australia' dated 7th April 1902, the existing medical services of the various States should be reconstructed and organized so as to form one Corps. Under the present circumstances it is, moreover, impossible to exercise the necessary supervision and control or to ensure economy in the administration of the Medical Services generally. It is also necessary for the effective administration of the Medical Services of the Field Force and Garrison Force respectively in each State that one uniform system of organization should be adopted.

The following reconstruction and reorganization of the State Medical Services is provisionally approved as hereafter mentioned and will, subject to the existing State Defence Acts and so far as local circumstances will admit, take effect from 1 July.

(2) The whole of the existing Army Medical Services of each State will, subject to the conditions stated in para. 1, be dealt with as one Corps, which will be styled the 'Australian Army Medical Corps'.

(3) The Australian Army Medical Corps will be composed as follows:

A. Permanent Army Medical Corps
B. Militia Army Medical Corps
C. Volunteer Army Medical Corps
D. Army Nursing Service Reserve

(4) The Director-General, Army Medical Services, is responsible to the General Officer Commanding for the Administration, Command, Drill, Discipline, Efficiency and Statistics of the Army Medical Services.

He is also responsible that the supply of Medical and Surgical Field Hospital transport is adequate and in good order as far as the funds at his disposal will permit.

He is further responsible for the preparation of the yearly Estimates of his Department, taking into consideration the recommendations of the State Commandants.

(5) The Permanent Army Medical Corps will consist of such small cadres as may be locally required in the several States, consistent with the funds available. They will carry out the medical duties in connection with the Permanent Troops of their State, and they will further act as an Instructional Corps in Medical Duties generally, for all arms.

The Militia or Partially-Paid Army Medical Corps and Volunteer Army Medical Corps will be organized into Bearer Companies and Field Hospitals for service with the Field Force and for such medical services as may be required for the Garrison Force respectively.

(6) In accordance with the Minute on Defence above referred to, the troops composing the Field Force are required to be completely organized, equipped and trained. The portion of the Australian Army Medical Corps allotted to this duty will accordingly be formed into Field Units, namely Bearer Companies and Field Hospitals, for which 'Peace and War Establishments' will be laid down hereafter.

The Troops composing the Garrison Force consist for the most part of Volunteers, and will be required in connection with the defence of cities and towns, etc. The detachments of the Australian Army Medical Corps allotted to the Garrison Force will not, as a rule, be required for service in the Field or at distances from their own localities.

The foregoing constitute the portion of the Australian Army Medical Corps required for general medical services with the Field or Garrison Forces respectively.
(7) Certain officers of the Australian Army Medical Corps (Militia and Volunteers) will be attached for duty with certain specified Regiments and Corps. These officers, together with the Regimental Stretcher Bearers, etc. constitute the Regimental Medical Service.
(8) A Reserve of Officers will be created from Medical Officers who have previously held commissions in the Army Medical services of the various States, and also from duly qualified and registered members of the Medical Profession who are willing to become additional members upon a National Emergency.
(9) *An Army Nursing Service Reserve will be organized from those trained nurses who are qualified and willing to serve as such with stationary Field Hospitals and Base Hospitals when required upon a National Emergency.*
(10) A Principal Medical Officer will be appointed in each State. This officer may be detailed to perform the duties of Staff Officer for Medical Services in addition to his administrative duties.

In cases where an officer of the Army Medical Corps permanently employed exists, he will perform the duties of Staff Officer for Medical Services in addition to his medical and other duties. He will further be responsible for the command of the Permanent Medical Unit and for instruction generally of all branches of the service in medical duties.
(11) Regulations and instructions will be issued for the Department at an early date, which will define the duties of all branches of the Corps.
(12) The organization of the Army Medical Corps into *Bearer Companies and Field Hospitals* for service with the Field Force, in pursuance of certain fixed 'Peace and War Establishments', which will hereafter be laid down, will be carried out gradually in accordance with the principles in the Minute on 'The Defence of Australia'.

Certain units will similarly be allotted for duty with the Garrison Force.
(13) A Gradation List will be published hereafter, laying down the seniority of officers in accordance with the date of their respective appointments.
(14) All Warrant Officers and Non-Commissioned Officers will rank amongst themselves according to the date of their respective appointments.
(15) Departmental Orders for the conduct of the Australian Army Medical Services, will be issued from time to time by the Director-General, who, upon purely departmental detail and routine will, in accordance with the GO, para 6 (C), correspond direct with the responsible medical officers in each State.
(16) The Civil First Aid Societies of the various States will at an early date be invited to consider a scheme of organization whereby they may be associated with the Australian Army Medical Corps, for the purpose of supplementing this Army Medical Corps, and rendering medical assistance, in time of a National Emergency, to the Military Forces of the Commonwealth.

The AANS 1902–1914

From 1902 when the Australian Army Nursing Service was established, until 1914 when the First World War broke out, much administrative machinery to extend the nursing service to all States was necessary. Miss L.C. Marks was Lady Superintendent from 1904–1906, but was forced to resign when she found the travelling from Sandgate too difficult. She was followed by Mrs M.E. Trundle 1906, Miss M.J. Waldie 1908 and Miss A. Perry 1910.(33) In the same period in Queensland Matrons appointed were Miss C.W. Smith, Miss E.E. Bishop and Miss A.C. Isambert. A reserve was built up in each State, being regarded as 'a volunteer portion of the Medical Services of the Commonwealth'. Some attempts were made to encourage nurses to become 'efficient' especially in the field of military discipline and organization. Sister Agnes Isambert(34) attended a course of lectures in 1904 on the organization of military hospitals, hygiene and military surgery and qualified as being efficient. In 1910 and 1911 as a provisional Matron and later Matron she qualified in further courses. Those who were on the Reserve constituted a seniority list, a factor which was to cause a great deal of friction when they were sent overseas in 1914. While the 1 July 1902 may be taken as the birthday of the AANS, their official position remained unclear. Unlike other members of the

AAMC they were not enlisted into the armed forces, but remained an auxiliary force. In the words of the official statement,

> The Australian Army Nursing Service is a voluntary body and is formed for the purpose of supplying trained and efficient nurses under an organized system, available for duty at Base Hospitals and Stationary Field Hospitals in times of national emergency.[35]

It was to be many years before the position of the AANS was clarified and their role and status written into official army procedures.

The questions of rank and uniform caused some confusion and frustration, even though these were based on English models. While at first the titles of Lady Superintendent, Matron and Sister were commonly used, these reverted before World War I to Matron-in-Chief, Principal Matron, Matron and Sisters. The latter were graded on the grounds of efficiency into Staff Nurses and Sisters, although all were given the title of Sisters. In 1902 a Grey and Scarlet uniform was adopted, based on that of the NSW Nursing Service which in turn had accepted the British uniform when the Reserve had been established in 1899.

As the Australian Army Nursing Service inherited much from its British counterpart, Queen Alexandra's Imperial Military Nursing Reserve (QAIMNR) and as it was to be closely associated with them during World War I it is relevant to note the differences between them at that time. Miss Grace Wilson,[36] a Queenslander, Acting Matron-in-Chief of the First AIF, pointed out that the QA's were part of a permanent force, thoroughly trained in military administration and their role and status was thoroughly known and accepted in the army medical corps. The AANS, a voluntary, part-time reserve group, had no army training, were not paid and their status still to be determined and even their role in the medical corps was still uncertain. Moreover the QA's were used to working with male nursing orderlies whose training was little less than that given to nurses. In the Australian Army Medical Corps male nursing orderlies had generally very little training.

This first decade of the twentieth century was a time when women were breaking out into new roles. It was inevitable that in the case of army nurses there would be, as Beith claims, 'passion and prejudice in high places'.

Walking out uniform AANS World War I 1914. Sister M.A. O'Brien 2nd on right.

Group of Queensland Members AANS, c.1914.

The tradition that trained male orderlies for example, could do a more effective job in caring for the sick and wounded in battle than trained nurses persisted into the second decade of the twentieth century. Major-General Sir Neville Howse, DMS in the First AIF, is quoted as saying that the (female) Army Nursing Service as a substitute for trained male orderlies does little towards the actual saving of life in war, though it may promote more rapid and complete recovery.[(37)]

He held that its purpose was chiefly one of humane alleviation and support, physical and moral, to the sick and wounded.

In retrospect, it is clear that the period from 1902 to 1914 was a time of lost opportunity for the Australian Army Nursing Service Reserve. While the Commonwealth was divided up into Military Districts roughly corresponding to States, Queensland and Northern Territory were included in No 1 Military District. The duties of the Principal Matron in each Military District were minimal and no pressure was exerted on nurses to join the Reserve and indeed for those so accepted, there was no pressure on them to undergo any training.[(38)] It is significant that at the outbreak of World War I the total number of nurses on the Reserve was less than 200 and few of these understood the role of the medical services, including nursing, in war-time. It was inevitable that on the outbreak of war there would be large numbers of nurses with civilian training and experience eager to volunteer for military service, but without the training to turn them from nurses into 'army sisters'. Added to this was lack of action by the military authorities in making a clear statement on the status of nurses within the Commonwealth Military Forces. These problems had not been resolved when the First World War broke out in August, 1914.

MILITARY FORCES OF THE COMMONWEALTH.

Certificate of Efficiency.

Australian Army Nursing Service.

REGISTERED NO. 14

District Head-Quarters,

30th June 1904

This is to Certify that AGNES CATHERINE ISAMBERT

of the Australian Army Nursing Service in Queensland has attended a Course of Lectures on the Organization of Military Hospitals, Hygiene, and Military Surgery, and has qualified for a Certificate in accordance with the instructions laid down for "Efficiency."

Alfred Sutton Lieut-Colonel

District Principal Medical Officer.

Date, 10 July 1909

W. D. C. Williams. S.G.

Director General Medical Services.

Q 15817.

Certificate of Efficiency

Form No. 9—CONTINGENCIES.

DEPARTMENTAL REGISTRATION NUMBER.

THE COMMONWEALTH OF AUSTRALIA,

Dr. to Staff Nurse
6th Australian General Hospital
(Insert Address when necessary.) Kangaroo Point Brisbane.

Financial Year 191

Div. No.

Subdiv. No. Item No.

*191			£	s.	d.
	1	Grey coat & skirt with grey blouse (tailor-made)	6	15	-
		Grey mackintosh	2	-	-
	2	Red Capes @6/- each		12	-
	1	Pair grey gloves		2	-
	1	Grey felt hat with chocolate band		10	-
	2	Sets buttons @ 3/- each		6	-
	2	Badges @ 3/6 each		7	-
	4	Linen squares @ 2/6 each		10	-
	1	Commonwealth brooch		5	6
	6	Collars @10d each		5	-
	6	Pairs cuffs @10d each		5	-
	3	Zepher dresses grey @ 16/- each	2	8	-
	4	Grey aprons @3/3 each		13	-
	8	White aprons @4/6 each	1	16	-
	3	Grey belts @6d each		1	6
	3	White belts @6d Each		1	6
	1	Holdall	1	-	-
	1	Pair black boots	1	1	-
		Trimmings	[illegible]	10	-
		Rank badges		1	6

Total Nineteen pounds

Ten shillings pence.

Signature of Claimant. £ 19 10 -

* Insert in this column—Date of Supply or Period of Service.

Contract No. Approved Requisition No.

"Uniform costs, 1914"

References

1. AUSTIN, M. *The Army in Australia 1840 1850*. AGPS. Dec 1979.
2. BRODSKY, ISADORE. *Sydney's Nurse Crusaders. A Century of Trained Nursing in Sydney*. (Sydney. Old Sydney Press 1968).
3. supra, Introd.
4. OSBURN, LUCY (1835–1901). (i) There are numerous variations of her name, Osborn, Osborne, Osbourne, but her signature in the letter to Sir Henry Parkes suggests Osburn is correct.
   (ii) b. Leeds, Eng. d. William and Ann Osburn; well educ. and spoke several langs; worked 4 months at Kaiserswerth Hosp. and visited many hosps in Europe; 1866 attended Nightingale Training School, attached to St Thomas's Hosp; 1867 completed midwifery course at King's Coll Hosp; Henry Parkes appealed to Florence Nightingale for trained nurses for the Sydney Infirmary and Dispensary 1866; Lucy Osburn appointed and she and 5 Sisters arrived in Sydney, 1868; 1873 following criticism of her work a Royal Comm on Public Charities set up. Lucy Obsurn was completely vindicated; 1884 she resigned having successfully set up a training school based on Nightingale principles; returned to Eng and d. Harrogate 1901.
   (iii) For details of the Royal Comm which exonerated Lucy Osburn see V and P (NSW) 1873–4. Vol VI.
   (iv) The following are useful references to this section: ATNA 'Nursing in Early Australia' *Australasian Nurses Journal* 1950, v 48, 84, 120.
   Cope, Zachary, Sir. *Six Disciples of Florence Nightingale*. Pitman Med Pub Co Lond. 1961.
   King G.A. 'Sydney Link With Florence Nightingale'. *J & P* RAHS, 1953, v 39, p 150.
   MacDonnell, F. *Miss Nightingale's Young Ladies*. Sydney, 1970.
   Macmillan, D.S. 'Sydney Link with Famous Nurse'. *UNA*, Nurses Journal, 1955, v 53, p8.
   Savage, Ellen. 'Fifty Years of Nursing Progress in Australia'.
   UNA Nurses Journal 1952, v 50: 11(Vic), also Australasian Nurses Journal 1951, v 49: 202.
   Seymer, Lucy, *Florence Nightingale Nurses*. 1960.
   Skirving, Robert Scot, 'The Development of Modern Nursing' *Australasian Nurses Journal*. 15 Aug 1923, p 372.
   Susman, M.P. Lucy Osburn and her five Nightingale Sisters. *MJA* VI No 18, 1 May, 1965, pp 633–642.
5. ARMSTRONG, DOROTHY MARY etal. *First Fifty Years. A History of Nursing at the Royal Prince Alfred Hospital Sydney 1882–1932*. (Syd Royal Prince Alfred Hospital Graduate Nurses Association 1965).
   BUTLER, L. *One Hundred Years of Nurse Training at the Royal Women's, Melbourne*. 1963.
   Centenary of Nurse Training in Australia. 1862–1962. Melbourne. Melbourne Royal Women's Hospital 1963.
   HOBBS, VICTORIA. *But Westward Look. Nursing in Western Australia 1829–1979*. Univ of Western Australia, Press for the RANF (WA Branch), 1980.
   HUGHES, J. ESTCOURT. *A History of the Royal Adelaide Hospital*. Adelaide, 1967.
   INGLIS, K.S. *Hospital and Community*. Melbourne 1958.
   MILLER, D. *Earlier Days: A Story of St Vincent's Hospital, Sydney*. Sydney 1969.
   Nursing in South Australia. The First Hundred Years. 1837–1937 Adelaide. The South Australian Trained Nurses Centenary Committee. 1939.
   SAYERS, C.E. *The Women's: A Social History*. Melbourne 1956.
   TEMPLETON, J. *Prince Henry's: The Evolution of a Melbourne Hospital*. Melbourne 1969.
   WATSON, F. *History of the Sydney Hospital*. Sydney 1911.
6. JACKSON, E. SANDFORD. 'Historical Notes from the Records in the Brisbane Hospital'. 1825–1850. *MJA*. 1922 1 : 685. 1923 1 : 281.
   PEARN, JOHN & O'Carrigan C. (eds). *Australia's Quest of Colonial Health*. v 7, Univ of Qld Press, 1983.
   POWELL, O.W. 'Early Development of the Royal Brisbane Hospital. 1848–1867'. *MJA* 1. 685.
   SUMMERS, H.J. 'Brisbane and Some of Its History'. *MJA*. 1950 1: 90.
7. Florence Nightingale to Henry Manning, July, 1852, as quoted in Bolster, E. *The Sisters of Mercy in the Crimean War*. Mercier Press, Cork, 1964, p.
   BOLSTER, EVELYN. *The Sisters of Mercy in the Crimean War*. Mercier Press, Cork, 1964.
8. HERCY, Sister M. ELIZABETH (Sister Mary Elizabeth Joseph). For details regarding Sister Mary Elizabeth Hercy I am indebted to Sister Jean-Marie, of All Hallow's Convent, Brisbane who has faithfully searched the archives at the Convent for relevant information. For an obituary of Sister Mary Elizabeth Joseph see *The Queenslander*, Mar 9, 1901.
9. JOHNSON, D.H. *Volunteers at Heart*. The Queensland Defence Forces 1860–1901. Univ of Queensland Press, 1975.
10. As quoted by Gurner, J. *The Origins of the Royal Australian Army Medical Corps*. The Hawthorn Press, Melb. 1970. p 19.
    MCINTOSH, A.M. 'Army Medical Services in NSW prior to Federation'. *MJA* 1948, VI 485.
11. Ibid, p 21. Volunteers went to the Maori Wars in 1863.
12. GOULD, Ellen Julia (1860–1941) RRC, b Wales 29 March 1860. Educ Portugal & England & Germany. 1884 trd as nurse at Royal Prince Alfred Hosp NSW; 1891 Matron & Superintendent of Trg School at Sydney Hosp; 1898 NSW Pub Health Dept; Matron Rydalmere 1898–1900; 1900–1902 Boer War; Princ Matron 2 MD, AANS Reserve 1902–1910; 1914–1919 AANS World War I; RRC 1916; d 19 July 1941; *ADB*.
    WILLIAMS, Surgeon-General Sir WILLIAM CAMPBELL. KCMG 1916, CB, Director-General of Medical Services. b. Sydney 1856. Served in South African War and the First World War. d 10 May 1919.

HUTTON, Lt Gen Sir EDWARD THOMAS HENRY, b. England 1848. Served Boer War. First general officer commanding the Military Forces of the Commonwealth, responsible for integrating the forces of the different Australian colonies into the CMF. Returned to England 1907. Served in WWI, d. England 1923.

13. As quoted in Butler, A.G. *Official History of the Australian Army Medical Services, 914–1918.* Vol III, Special Problems & Services, p 534, AWM, Canberra, 1943.
14. Department of Defence, (Qld). Reports on State Military Forces, 1887–1901 Queensland: Queensland Military Forces: *Report of Committee for the year 1899–1900.* p 7.
15. Ibid.
16. GOULD, ELLEN JULIA. See fn 12. See also Gould, Matron Ellen J. Notes on Australian Nursing Sisters of the Boer War, Jan 1900 to Aug 1902 and on the History of the Australian Army Nursing Service with reference to nursing enlistments in the war of 1914–18. AWM, Canb.
17. MURRAY, P.L. Lt Col (ed). *Official Records of the Australian Military Contingents to the War in South Africa, 1899–1902.* Dept of Defence, Melb. 1911. p 14.
18. Ibid, p 241.
19. Details supplied in correspondence from Regimental Headquarters, QARANC, Aldershot, 27 April 1982.
20. *The Queenslander.* April 21, 1900 p 748
    April 28, 1900 p 800
    May 5, 1900 p 848
    May 12, 1900 p 895
    May 26, 1900 p 994
    June 2, 1900 p 1044.
21. WALLACE, R.L. *The Australians at the Boer War.* The AWM and the AGPS, Canb, 1976, pp 168, 232/
22. Ibid. pp 50, 57, 58, 294.
23. MAURICE, Maj Gen Sir FREDERICK. *History of the War in South Africa. 1899–1902.* V 4, Appendix 7.
    HMSO, Great Britain and Imperial Commissions. *Report on the care of the sick and wounded during the South African Campaign, 1901.*
24. HUSTON, BEATRICE. The Huston family was well-known in Clermont and links with South Africa continued after the war. Beatrice Huston was born in Clermont in 1868 but there is no evidence to show where she did her training. Most probably it was at Clermont. After the war she remained in South Africa as did her two brothers and one sister. Her son became a prominent businessman in Johannesburg. Her brother became Commissioner of Railways in South Africa. Another brother received the Military Cross in World War I. Beatrice Huston is also remembered by the 'Beatrice Battery', erected on the north Dam in 1897 to crush cement and wash from 'The Wild Cat', because she broke the traditional bottle of champagne over the flywheel in naming it. T.J.M. Higgins bought the Beatrice Battery in 1934, dismantled it and re-erected it at MICLERE, some 35 kms from Clermont.
25. For letters from Nurse Huston see *The Queenslander*, May, 1900 p 948. June 16, 900 p 1138.
26. Ibid. Sept 1 1900 p 470.
    FLETCHER, GERTRUDE. Trained Prince Alfred Hospital, Sydney 1891–94; Promoted Sister 1896. First in final year. Resigned 1897 'to go to Europe'. Believed to have taken up private nursing in Scotland. Corresponded regularly with relatives in Brisbane.
27. *The Queenslander.* April 28, 1900 p 800.
28. ARMSTRONG, SUSANNAH, (1860–1939); b 17 Sept 1860 Ireland; 1892–1894 trd Bris Hosp; 1898 pte nursing WA; 1900–1901 Service South African War; 1902 married St Luke's Anglican Church, Gin Gin Qld; pte nursing WA; 1913 midwifery cert St Helen's Perth; 1914–1918 service AANS WWI with overseas service; life member ATNA; 1937 Coronation Contingent; d 2 Aug 1939. (See Hobbs, V. *But Westward Look.* p 214 under Davern, Susannah).
    PLOVER, MIRIAM. Trd Bris Hosp 1893–1897; service in South Africa 1900–1901; Matron Topsham Ptd Hosp, WA.
    • HOBBS, VICTORIA, *But Westward Look.* Nursing in Western Australia 1829–1979. Univ of WA Press 1890. pp 15–16.
    • Information from material in the Battye Library, Perth.
34. ISAMBERT, AGNES KATHERINE, Sister (1874–1956). Trained at Brisbane Hosp 1897–99; Member of ATNA 1904; private nursing 'Walmer Nursing Home'; Colmslie Plague Hosp; Midwifery Cert. Bris Hosp 1913; Matron St Mary's Ipswich; Member AANS Reserve 1904–1914; Cert Efficiency. AANS World War I sailed on Kyarra, appd 1 AGH Heliopolis, 1916 Temp Matron.
35. For the founding of the AANS see; Brief History of the AANS. AWM 509/2/2; 492/4509/3. History of AANS, *Australian Army*, Vol 4 No 24, June 27, 1963, p 2. Reminiscences of AANS. *Mufti.* Dec 1939, pp 10–11; Jan, 1940 pp 10–11; Feb 1940, p 20; Mar 1940 p 10; May 1940, p 10. Our Nurses by Miss Conyers, *Reveille* July 1930, p 28; and Mar 1931, p 54. A Healing Hand. Brief History of the Nursing Service. *Army.* Apr 1943, pp 54–56. Pamphlet, Royal Australian Army Nursing Corps. *The Royal Australian Army Nursing Corps.* (n.d.).
    HART, DOROTHY. 'For Humanity' — Genesis of the Royal Australian Army Nursing Service. *Sabreteche.* Vol XXIII. No 2, April–June, 1982, pp 15–16.

KENNY, SUSAN. 'The Emergence of the Australian Army Nursing Service. 1914–1918. *Defence Force Journal.* No 6, Sept-Oct 1977, pp 19–25.

KENNY, SUSAN CORALIE. The Australian Army Nursing Service. During the Great War. A Study of its Growth and Development. Unpub thesis B.A. (Hons) Univ of Melb 1975.

Lest We Forget. Australian Army Nursing Service, Melb 1948.

Short History of the AANS. *Sabreteche.* No 2, 1958, pp 31–36.

Royal Red Cross. The first Australian nurse to be awarded the Royal Red Cross is believed to be Mary Jane West Lumley (nee Briscoe) on 27 May 1884 for her services in the Zulu War. She was also awarded a Zulu Medal. Mrs Lumley was a civilian working at Stafford House, a base hospital in Durban.

36. WILSON, GRACE MARGARET (1879–1957). Born Sth Brisbane, daughter of John P. and Fanny Campbell Wilson; Educated Brisbane Girls' Grammar School; Trained Brisbane General Hospital — Queen Charlotte Hospital, London; Matron Brisbane General Hospital; AANS Reserve prior to World War I. AANS 1914–1919. Principal Matron 3 AGH Egypt, Lemnos, France; Temporary Matron-in-Chief AIF; Mentioned in despatches 5 times; Florence Nightingale Medal; CBE (1919); RRC; Matron Rosemount Military Hospital; Matron Children's Hospital, Melbourne; Sister-in-Charge, Somerset House Private Hospital, Melbourne; Matron Alfred Hospital, Melbourne; Led AIF Military contingent to coronation 1937. AANS 1939–41; Matron-in-Chief AANS Service in Middle East and England 1940–41 when retired due to ill health; Executive Officer: Nursing control section, Manpower Directorate, Melbourne; Married Bruce CAMPBELL, England 1954; Died Melbourne, January 12th, 1957.

37. BUTLER, supra. VIII p 528, fn 3.

38. Standing Orders AANS 1913–1914, S 197.

Chapter Three

Establishing Traditions:

World War I (1914–1918)

Nor are there words in which to tell of the service of the splendid band of Australian nursing sisters who greeted the men from the front as they reached hospital and nursed them back to strength or softened the close of their soldier-life. No womanhood has ever presented a richer association of feminine tenderness and sheer capacity. They were true sisters of the fighting sons of Australian pioneers.

Sir Henry Gullett

With the declaration of war on 4 August 1914 there was an immediate loyal response by Australia in support of the Mother Country. Both political parties rallied to the cause and the Australian Prime Minister, Labor leader Andrew Fisher, promised to support Britain 'to the last man and the last shilling'. In the event a contingent of 20,000 men, to be known as the Australian Imperial Force, was raised immediately and after a brief period of training the volunteers were assembled for the voyage overseas, the first intention being to send them to England. At the request of the British War Office, Australia was also asked to raise certain Line of Communication medical units, which inevitably involved members of the nursing service. As stated earlier, there had been established since 1902 an Australian Army Nursing Service Reserve, with small voluntary groups in each State, liable to be called up in the case of war. Limited courses of training and examinations classified nurses on the reserve as 'efficient'.

The Principal Matron of the Australian Army Nursing Service, Miss Ellen Gould, warned members to be ready for home service and circularized them to ascertain how many would be available for overseas service, if necessary. There was also a prompt response from civilian nurses, many of them fully trained and qualified, with their certificates from the Australasian Trained Nurses Association. There was however, no publicity or recruitment campaign as it was anticipated that the number on the Reserve would be sufficient to meet the demands of a short war. Thus there was a long waiting list for acceptance into the AANS and many sought enlistment in the QAIMNS. The nurses who were selected to go with the first convoys all came from the Reserve. They had at least three years training and service in an approved hospital, they had to be between the ages of 21 and 40, unmarried or a widow. The AANS Reserve, for administrative purposes, required that applicants live in the metropolitan area.(1) They were given a capitation allowance of £1 ($2) per annum and had to provide their own uniforms. While four parades were held annually, it could not be said that many were fully trained in an army sense.

Many of these nurses came from country towns — Bundaberg, Clermont, Nanango, Mackay — where memorials now bear witness to their service and sacrifice. During World War I the provincial newspapers wrote glowing accounts of the careers of those nurses who had volunteered for active service. The Bundaberg Mail, for example, referred at length to Nurse Cheesman (Lady Chelmsford Hospital)[2] Sister W. Dods[3] (St Andrew's Hospital) and Sister M. Wilson[4] (Bundaberg General Hospital).

> The Three nurses are joining the noble band of self-sacrificing disciples of Florence Nightingale, who are now engaged in the work of caring for the sick and wounded on the field of the greatest war which the world has known. The best wishes of their friends, and they are numerous in Bundaberg and district go with these devoted nurses in their noble work, and may they be spared to return to their homes in good health is the wish of the citizens of Bundaberg.[5]

At the outbreak of war Sister Beatrice Bowes was nursing at Boonah General Hospital and was very keen to enlist, despite the fact that she had just become engaged to a well-known local cricketer, 'Tory' Cossart. Physically she was just over five feet in height and of very slight build. Enlistment conditions required that the applicant be not less than seven stone,[6] so it was with some trepidation she reported for her medical examination. The recruiting officer, well-known at the Boonah Hospital, kept his hand on her shoulder as she stepped on the scales and said, 'Yes, Sister Bowes, I think we can say you have passed this test satisfactorily'.

Enlistment procedures for members of the AANS were time consuming and expensive, particularly with regard to uniforms. At the beginning of the war the uniform was modelled on that of the British QA's but this was 'modernized' soon after enlistment. This included the red cape which was to cause some embarrassment later, as in the British Army it was only worn by the members of the Regular Army, while for the British Territorial Nursing Service the red cape was replaced by a grey cape with broad red border. The red cape was an important status symbol as Australian nurses were to find when they served in British hospitals. The initial problem for Australian nurses was the cost and secondly where to find a firm to make the uniform. Nurses on appointment were granted an outfit allowance of £21 ($42) plus £16 ($32) for maintenance. In Brisbane, nurses found Finneys to be most helpful in the tailoring of their uniforms.

Sister Ada Smith, who had trained at Warwick and whose family was well-known in that district, wrote in her diary about problems of enlistment and of uniform purchases.[7]

> I decided to enlist and join up with the Australian Army Nursing Service. This was early in 1915. To do so I had to resign my position in NSW and return to Queensland. At first my parents were rather against it, but when they gave their permission I had no idea whom to write to, so I addressed a letter simply to the Principal War Matron, Brisbane. Next morning a wire came telling me to go to Brisbane at once and see the Matron of Brisbane Hospital. I knocked at her door and on her asking me who I was, she just said, "Yes, we want you for Egypt". Needless to say how thrilled I was. Then followed a very busy time. First of all about fifty pounds was required to go, although the military handed me only £30. However, my brothers soon made up the deficiency. There followed visits to Finneys to have all uniforms and bonnets made and whole lists of things got — cabin trunks, kit bags, suitcases, all with our names and 2nd Australian General Hospital, A.I.F. and Red Crosses printed on them.

Sister Edith Avenell, who was born at Gympie but whose family later lived in Townsville, enlisted in April 1915 and also had to go to Finneys for her uniform. She described in her letters how she stood for two hours 'tired and dreary', then had to return twice more for fittings.

> I am getting a coat, grey and scarlet (out-of-door), it has a large cavalier cape and looks very nice. it is to travel in — cost 4 guineas, then our dresses are grey (pale) zephyr white aprons, scarlet capes, my own caps — all of this is allowed us at Finneys. I had to buy a cabin-trunk and carry all, leather, costing me £4 altogether.[8]

This was quite an expensive outlay, but the nurses were helped by donations from a fund organized by the Brisbane Courier.

EXTRACT FROM STANDING ORDERS

Syllabus of Qualifications necessary to become Members of the Australian Army Nursing Service

Qualifications of candidates.

198. A candidate for enrolment as a Sister must be between twenty-one and forty years of age, single, or a widow, and have not less than three years' training and service in Medical and Surgical nursing in a duly recognized civil General Hospital. She must be of British parentage or a naturalized British subject. The candidate will be required to fill in the declaration Form (C.M. Form 1). II.), which will be supplied to candidate by the District P.M.O., and to produce the following documents:—

(*a*) Certificate of registration of birth, or, if this be not obtainable, a declaration made before a magistrate giving the date of her birth.

(*b*) Certificate of training (in the original).

(*c*) A recommendation from the Matron of the civil hospital at which she was trained.

(*d*) A certificate from a duly qualified and registered medical practitioner that she is in good health and physically fit for duty in the Australian Army Nursing Service.

Age of retirement.

200. Members of the Australian Army Nursing Service will be retired on reaching the age of fifty-five years, but in special cases the age for retirement may be extended for a period not exceeding two years.

'Efficient' and 'non efficient'

201. Members will be classified as 'efficient' or 'non efficient,' to be reckoned from 1st July each year until 30th June following.

Requirements for efficiency.

202. In order to be classified as efficient, each member will be required to—

(1). Qualify in first aid.

(2.) Attend annually three out of the four lectures on Organization of Military Hospital, Hygiene, and Military Surgery.

Certificates.

203. Certificates of efficiency will be issued by the District P.M.O.

Discharge.

204. District Principal Medical Officers are required to bring forward for discharge such Sisters as they may consider to be medically or otherwise unfit for service.

Pay and allowance.

205. Rates of pay and allowance for members of the Australian Army Nursing Service, when called up for duty, will be as laid down in Financial and Allowance Regulations.

Uniforms.

206. Uniform, in accordance with Standing Orders for Dress and Clothing (Citizen Forces), will be worn upon all military duty, and for which a capitation allowance of £1 per annum may be granted to each 'efficient' during the financial year in which payment is made, subject to provision being made by Parliament.

Liabilities for service.

207. The Australian Army Nursing Service, or any portion thereof, may be called up for duty in case of emergency, and shall thereupon become subject to the like conditions as those prescribed for the Australian Military Forces, and remain subject to those conditions so long as they continue on active military service.

Supernumerary list.

208. Members of the Australian Army Nursing Service who are unable to comply with the requirements for 'efficiency' through residence in the country, but who have previously complied with the Standing Orrders, may, if they so wish, be placed on a supernumerary list as supernumerary to the Establishment. Vacancies so caused to be filled up by Nurses residing in the Metropolitan Area.

The following Orders are issued for the Australian Army Nursing Service, *First Military District*, in extension of Standing Orders for the Australian Army Medical Service, 1914:—

Admission.

(1.) Candidates for admission will be medically examined by the P.M.O. or a medical officer duly appointed by him. No standard of height or weight is required, but applicants must be of sound health and capable of doing hard work and undergoing fatigue.

Duly recognized hospital.

(2.) A duly recognized civil General Hospital will be taken to mean a hospital of eighty or more beds, in which lectures are systematically given to the Nurses.

Change of address and position.

(5.) Every Sister shall notify the Principal Matron of any permanent change of address or position.

Leave.

(6.) No Sister of the Active or Supernumerary List shall leave the District without obtaining leave.

A.M. McINTOSH, MAJOR,
Acting P.M.O., 1st M.D.

It was a massive undertaking embarking 20,000 troops into 28 transports, joining up with another 10 troopships from New Zealand and guarding them on the long voyage to Britain, as proposed. At this time German raiders appeared in the Indian Ocean, in particular the *Emden*, necessitating an impressive naval squadron to protect the convoy. So on 1st Nov 1914 the first contingent of the AIF sailed from Albany (WA), including the Australian light cruisers *Sydney* and *Melbourne*, the British armoured cruiser *Minotaur* and the Japanese warship *Ibuki*. The question whether nurses should accompany this first expeditionary force was hotly debated and in the end it was decided to send the Principal Matron (Miss Gould), three matrons and 20 nursing Sisters. Among the nursing Sisters who sailed with the *Omrah* from Brisbane on 24 Sept 1914, were four from Queensland, Miss J.M. Hart,[9] from Charters Towers, Miss E.M. Paten[10] (Brisbane), Miss C.M. Keys[11] (Brisbane) and Miss D.M. Williams[12] (Brisbane). Sister Williams had had ten years service in the AANS Reserve prior to the war, Sister Paten six years, Sister Keys eighteen months and Sister Hart sixteen months. Seniority and length of service were to become important issues later on. As Matron Isambert had been in the AANS since March 1904 and was one of the senior Sisters she was not allowed to go with the other four Queenslanders who left on the first troopship.

Sister Keys wrote home frequently, giving her impressions of life in army hospitals. From these and from her diaries comes a first-hand account of dramatic events in the war.

First excitement on the voyage over, came when the *Sydney* left the convoy to do battle with the *Emden* near the Cocos Islands. In her diary for Nov 9, she wrote:

> Got word this morning that the *Sydney* had destroyed the *Emden*. Great enthusiasm on board. Heard we were in danger last evening. *Emden* was only 40 miles away and only that we went the wrong side of Cocos we would surely have been destroyed. All lights out this evening . . . Hart and I powdered our noses by my little torch.

The following week she wrote home that 36 sailors and 14 officers from the *Emden* had been taken aboard as prisoners and some of them had been admitted to the hospital ward on the *Omrah*. They remained with the ship until its arrival in the Middle East, a change of plans having diverted the ship from its original destination, England. Sister Keys, in her diary noted:

> We lost our German prisoners a few days ago. I was sorry to see them go as I had grown to like them. I learned a few German words which I used to say on every occasion. They would laugh or try to talk English. One man we had in the hospital, Jacob Giebel, had a big lump of shell removed from his arm. He presented me with his cap band with *Emden* on it and I am very proud of it. Not many of the sailors parted with theirs.

The uniform problem, however, was not solved for some time. On board ship there were always formal occasions. Mess Dinner on the boat was always a bright scene. The Sisters wore red capes over grey uniforms, with white flowing captails! The officers wore full dress uniform and the effect was like some gay stage scene — contrasting with the drab effect of later service when gay uniforms gave place to khaki and grey.

Members of the AANS on these first troopships had been exceedingly busy attending to the usual illnesses which beset a body of troops on such a long voyage. The nurses also gave lectures and practical training to orderlies and ambulance men. Their presence justified the decision to send them. Butler, in his history, commented,

> The D.M.S., A.I.F., urged this step "in view of the length of the voyage to England". It may be noted here that, on being disembarked in Egypt, these nurses were for a time employed under a British matron, Miss Grierson, in the New Zealand Stationary Hospital. When they rejoined their units, this matron sent an almost enthusiastic report of their work to the Australian Director.[13]

These first members of the AANS overseas found themselves 'filling in' in various hospitals, as the enormous task of organizing the huge influx of allied troops and the ancillary medical establishments went on apace. Some went to Mena House Hotel, others to a former Egyptian Army Hospital at Abbassia and some to Heliopolis. These names were to become familiar to many Australian nurses in the months ahead. Sister Keys described events on 7 December:

The medical staff on H.M.T. Omrah. Reading from left to right: Sister Paten, Captain Conrich, Sister Hart, Captain Graham Butler, Sister Keys, and Sister Williams. AWM C2538

> Here we are in Heliopolis just outside wonderful Cairo with its Sphinx and Pyramids. We Queensland nurses together with three others from Melbourne are to be stationed at the Egyptian House Hospital, a fine big building, ten minutes by tram from the hotel. A good many New Zealanders are in hospital at present, they had a bad time coming over with no nurses on board. I think we will be very happy here, but there'll be plenty of hard work. But that's what we came out for.

One of their first problems was to adjust to the heat of the Egyptian sun. They soon found some parts of their Australian uniform were unsuitable for the Middle East. Sister Key's account reflects the bubbling happiness and friendliness of these Australian women in dealing with the local shopkeepers.

> We had to discard our military bonnets as they did not protect our eyes from the fierce glare when outside. Instead of them we bought panama hats lined with blue straw and found them very suitable. The French shopkeeper who served us seemed greatly interested in us, pointing to the blue lining and then to our red capes, saying — "Blue, white, red — French!" I laughed and said "Red, white and blue — English!" "Yes, yes," he exclaimed with a pleasant smile, "All the same — English and French, French and English".

On 4 Dec 1914, the second convoy, which included 161 nurses, left Australia for England, but it too finished up in the Middle East. One of the transports was the coastal steamer, *Kyarra*, (6,953 tons) well-known to Brisbane residents. It was painted white and lighted as a hospital ship. Enemy powers were notified in accordance with international conventions. The fact that it was used for transporting five medical

Hospital ship 'Kyarra', carrying first major contingent of Queensland members of AANS down the Brisbane River, November 1914.

units, with no hospital accommodation (this was provided in other ships of the convoy), drew some criticism later when the facts were made known.

As Queensland was in No 1 Military District those Queensland nurses who sailed in the second convoy were appointed to No 1 AGH. It was a brave farewell on 21 November 1914 as the *Kyarra* sailed down the Brisbane River carrying the Queensland nursing Sisters off to the war. The group included Srs Whipham, M.F., Croll, M.W., Nelson, N.M., Mowbray, N.V., Andrews, G.J., Ralston, E.A.V. and Scott, A.; as well, the following Staff Nurses are also known to have sailed on the *Kyarra*: Butler, E.E., Campbell, B.A., Campbell, S.F., Dalrymple, M., Dunne, T.J., Echlin, G., Farquhar, G.A.J., Forsyth, F.C., Geary, M.E., Gibbon, B.L., Grant, E.R., Heffernan, A.M., Hodgson, S., Langford, R.J., Lyons, Z.S., McPherson, A.I., Redmond, E.M., Relf, G., Scully, M., Snelling, L., Sorensen, C., Webb, D., and Wilson, M.[14] Sister Isambert described the emotional farewell from Brisbane:

> All my dear ones to see me off. Such a crowd on the wharf. Dear Mother kept up bravely.

Most of them appeared to suffer from sea-sickness on the very rough trip to Sydney, reaction from recent vaccinations and injections making things worse. At Melbourne and Fremantle, Sister Langford and her friends Sisters Echlin, Lyons, Sorensen and Isambert enjoyed the novelty of shopping in the big stores. Sister Isambert wrote in her diary of her day in Fremantle:

> *12 Dec 1914*
> Some fuss about going ashore . . . made first for the Post Office and sent wires, then made for the tearooms where we had a nice afternoon tea . . . went for a walk in East Fremantle. We Queenslanders thought the road with its gumtrees the loveliest we had seen — so homelike. Was very charmed with Perth . . . very pretty with imposing buildings, much larger than Brisbane.

So the convoy left Australia and pursued its way across the Indian Ocean. The nurses occupied their time with lectures, sports programmes, knitting and a mock trial by jury (a breach of promise case!) Some break in the monotony came with the sighting of Cocos Islands, but this time there was no *Emden* lurking there. Christmas Day was celebrated in an appropriate fashion, with the King's health being drunk, (King

Queensland Members of AANS on 'Kyarra' 21 Nov 1914. 1. Sister Scott 2. Sister M. Wilson 3. Sister Z. Lyons 4. Sister Langford 5. Sister Echlin 6. Sister McPherson 7. Sister Nelson 8. Sister Scully

George V). A day in Colombo was an exciting interlude, but the nurses and medical officers were beginning to feel the heat of the tropics a little trying.

New Year's Eve and the heralding in of a momentous year, 1915, was celebrated in traditional fashion but with unexpected results, as Sister Isambert wrote:

> The fancy dress masque ball was a great success and the costumes really wonderful out of the things to be had on board. Nightdresses, sheets and flags were the principal foundations. The officers provided supper which we had at 11.30 p.m. and at 12 saw the old year out and the new year in by singing and cheering. Lastly, the King's health was drunk by the O.C. and we sang God Save the King. Next morning a good many officers, some nurses and several crew were ill with ptomaine poisoning, but not seriously. My turn to inspect cabins — much tidier than usual. Also had to find out how many Queensland nurses had lifebelts.

So on to Aden and up the stifling Red Sea to Port Tewfik, then in convoy up the Suez Canal to Port Said. Here Sister Isambert noted the first real signs of war as the banks of the Canal were all trenched, with many fortifications and at night searchlights lit up the sky from the tower on top of de Lesseps statue.

There followed a week or so in the harbour at Alexandria, 'mucking about', as plans were changed and rumours abounded. Finally on 20 January 1915, all the staff for No 2 AGH disembarked for Cairo, more than eight weeks out of Brisbane. Sister Isambert was pleased to be paid, 'received £3 ($6), a godsend as I was stony broke!' A few days later No 1 AGH staff also disembarked for Cairo, their destination the Heliopolis Palace Hotel, taken over as a hospital. Sister Isambert's first impressions were indeed mixed.

> *24 Jan 1915.* Arrived at our palace about 1 p.m. but a very cold reception — nothing provided for our luncheon but

got some later on. Found to my surprise I had been made the Junior Matron — a great shock. I don't know that I am altogether pleased. Miss Bell transferred to Mena House, Miss Graham, Principal Matron, Miss Knowles, House Matron, then myself.

The first few weeks were indeed weeks of confusion, as Sister Isambert faithfully recorded. Two nurses, Forsyth and McKinnell went to Alexandria, 20 nurses went down to Suez, while another 8 (including Nelson, Langford and Wallace) were selected to return on a ship taking wounded back to Australia.

Rumours that all was not well with the administration of the AANS came with the announcement that Miss Bell was returning to take charge of No 1 AGH again. As Junior Matron, Sister Isambert appeared to spend much time in linen checks, control of Sisters and checking bed states. As night duty Sister, she found it extremely tiring, especially as sleep was difficult during the day. In the mornings she had to check that all night Sisters were in their rooms after breakfast. They were not to be up and about before 3 p.m. She was older and more experienced than many of the Sisters and it was her responsibility to enforce discipline. She recounts in her diary how she had to deal with those nurses who overstayed their late night passes and tried to sneak in unobserved. Punctuality was important and it took time for some nurses to accept the army routine. Five nurses were reported to her for being late for breakfast!

New Guinea — Rabaul 1914

Meanwhile one of the first overseas operations involving Queensland Sisters was in German New Guinea. Soon after the outbreak of war the Australian Government was advised by British authorities to seize German wireless stations in New Guinea, Yap in the Marshall Islands and Nauru. A naval and expeditionary force was raised immediately, including one infantry battalion, six companies of naval reservists, two machine gun sections, a section of signallers and a detachment of the Army Medical Corps. An Australian Naval Squadron made a preliminary visit to Rabaul to destroy installations but the major attack came on 11 September 1914. The full story of this first major Australian campaign against the Germans in the First World War is well documented in the official history, *The Australians at Rabaul*, one of the tragedies being the death of the medical officer, Captain B.C.A. Pockley.(15) On the afternoon of 13 September the British flag was hoisted at Rabaul and subsequently the former colony of German New Guinea was taken over by the Australian military authorities. One of the first tasks of the new administration was to get the hospital functioning. While casualties had been relatively light, malaria and dysentery began to take toll, not only among the Australian troops, but among the German troops and civilian population and among the New Guinea natives. Medical reinforcements including Nursing Sisters were rushed from Brisbane. In this first nursing team to serve overseas in the First World War were Matron Robertson, Sisters McLean, Nelson, Leatherbridge and Gibbon. Sister Gibbon's diary gives a first hand account of this unusual and dramatic experience for a group of AANS nurses.(16)

Four of us left Queensland in December, 1914, and after a fortnight on the boat, reached Rabaul, looking lovelier than we anticipated. Looking up the main street it was a blaze of colour. We were motored in a good old Ford up to the Hospital which we found full of our boys, mostly suffering from malaria, who were being cared for by our own AMC and two German Sisters.

Soon after our arrival the German Sisters went back to Germany via America and most of our patients had been evacuated. The hospital was about a mile out of town and known as the Namanula Hospital. It had been a hospital in the German time and all the equipment was marked K.G. — Kaiser Government — but the boys always said it meant King George!

The beds were not a bit like hospital beds, being large double ones, with a specially made spring mattress which was curved in the centre. So it did not matter where you might lie, there was always a hollow in the centre. There was also a horse hair mattress made in three pieces. We found it very hard to move sick patients in these large beds and in fact we had to climb on them whenever there was any moving to be done.

When we arrived the flies in the daytime and the mosquitoes at night almost drove us crazy. We got under out nets as soon as it was dark.

However the staff got to work cleaning up the rubbish and clearing out the water holes where mosquitoes were breeding. In a few months flies and mosquitoes ceased to bother us, although we always took precautions.

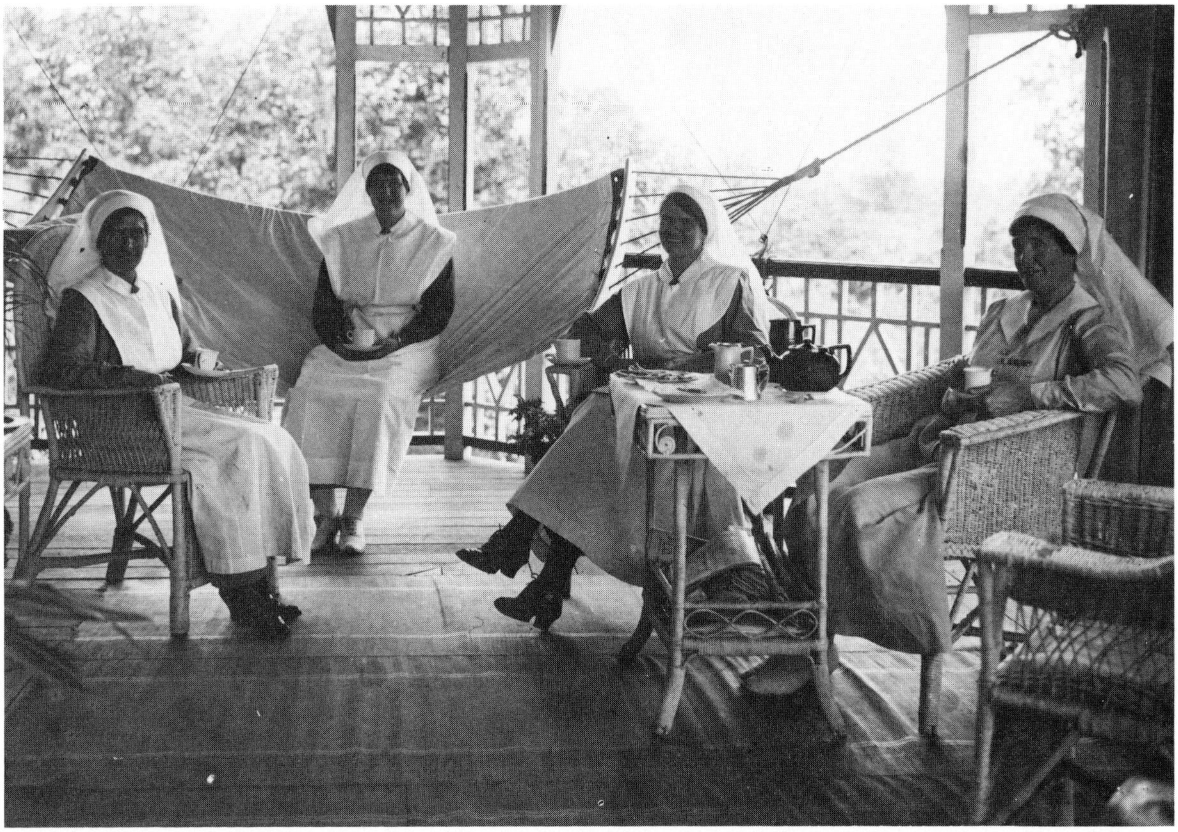

Sisters' quarters at the Military Hospital of Namanula, Rabaul. Matron Robertson, Sister McLean, Sister Nelson, Sister Leatherbridge. AWM J1866

During the three and a half years we were there our greatest worry was earthquakes. I remember one New Year's night sitting up from 11 pm to 3 am, during which time we had earthquake after earthquake. We had them three or four times a week but mostly they come and go before you have time to get frightened.

We sometimes motored down to where the boys who were killed in the landing are buried and put flowers on their graves.

The general health of the troops and the hospital staff was good, except for dysentery and malaria, but most of the malaria cases came from the outstations.

This brief episode was soon over and the army Sisters moved on to other campaigns. By a sad twist of fate the next generation of Sisters captured at Rabaul by the Japanese were not so fortunate.

Organization and administration of AANS

In his official history of the Australian medical services in the First World War, A.G. Butler stated very bluntly that

the AANS shared with the Australian Army Medical Corps the disastrous results of the failure of the Australian Government and Defence Department to make any effective provision for its internal administration, direction and discipline.[17]

Butler lays the blame for this administrative mess on the outlook of the first Director of Medical Services of the AIF, Surgeon-General Williams, who believed he would control the AANS as a member of the staff of General Keogh, Director-General of the Medical Services of the British Army. The AANS would then, in effect, have been controlled by the Matron-in-Chief at the War Office in London. But it didn't work

Recovering from Malaria fever — patients at the Military Hospital of Namanula, Rabaul. AWM J1870

out that way. On arrival in Egypt, members of the AANS found no administrative organization for their service. There was no Matron-in-Chief so inevitably Miss Bell and Miss Gould acted on their own initiative, assuming authority to make appointments and promotions, much to the annoyance of the DGMS in Melbourne. There was no seniority list and no regard was taken of the experience and qualifications of 'efficients' such as Sister Isambert. No one had set down clearly the duties and responsibilities of the various categories of nurses — Principal Matrons, Matrons, Sisters and Staff Nurses. These Sisters were senior administrators with years of experience behind them in large Australian hospitals. They had also been well trained in the Nightingale system. They were appalled by the way the RAMC ran the hospitals and they were shocked at the treatment meted out to the nurses.

General Fetherston, DGMS, reported on the problem of the status of Australian nurses. He pointed out that under Australian Regulations, The Army Act of Great Britain and other Imperial Regulations applied to Australian troops overseas in so far as they are not contrary to the local Defence Act or Australian Regulations. When the first nurses left Australia there were practically no Australian regulations governing the nursing service overseas. In practice they came under Imperial Regulations, as regards status and authority. They worked under these regulations for the first twelve months. While these Imperial Regulations were sufficient while nursing 'regular troops', they were soon found to be insufficient while working with 'citizen soldiers'. There were continual disputes as to the scope of nurses' powers and authority in the ward, in relation to MO's and to the NCO's and orderlies — and even Matrons found their authority over the Sisters often questioned.

In those hospitals with a firm OC in sympathy with the nurses all went well. With others not so firm there was trouble and the ward Sisters found considerable delay in finding out what they were required to do. In some cases conditions became unbearable and there were frequent appeals to higher authorites.

Col T.H. Fiaschi and his bride Sr Amy Curtis 19 Aug 1914 at Christ Church, Bundaberg.

Things came to a head very quickly at 1 AGH, in a clash between Miss Bell and the CO, Colonel Ramsay Smith. Miss Bell was an experienced administrator and a very determined woman, brought up on the Nightingale system that the nursing staff had a status and responsibility in a military hospital. As a result she resigned her position and returned to Australia. The Minister for Defence, Senator Pearce, intervened and ordered her re-instatement, but the problem went deeper than that. There was widespread concern among members of the AANS, including the Matrons, that their positions were fast becoming untenable due to lack of action by those in command. A position of 'Matron Inspectoress' was created for Miss Bell, with authority over all nursing staff except those in 2 AGH. She found her recommendations often disregarded and many actions taken with respect to nurses, without even consulting her. Even within her own hospital, 1 AGH, Matron Bell did not have complete authority over her own staff. She found, as one writer put it, 'passion and prejudice in high places'. Neither the DGMS, General Fetherston nor the Minister for Defence, Senator Pearce, supported her.

The news of this problem soon found its way to the War Office in London. General Keogh, Director-General of the British Army Medical Services, set up an independent inquiry into the medical and nursing arrangements at No 1 AGH. The subsequent inquiry strongly supported Miss Bell's stand and the report said that the CO Colonel Ramsay Smith 'had displayed a vexatious want of sympathy'.[18]

The report of the inquiry into the administration of 1 AGH, coming at a time when the public was becoming aware of the enormous losses in the Gallipoli campaign, sparked off an intense public and political debate on the efficiency of the army medical services and its ability to cope with the treatment of the wounded.[19] There were questions asked in the press and in Parliament about the role of the nursing service and generally disclosures awakened public sympathy that they were underpaid, under recognized and under used in the military service. The Director-General of Medical Services, General Fetherston undertook a special tour of inspection of the army medical corps,

> to inspect and report upon the working of the AAMC, AIF, generally and to pay particular attention to the treatment of Australian invalids in hospitals and on hospital ships and transports.

This was a difficult and delicate task as it involved some discussions with the War Office in London. Nevertheless he saw for himself the lack of organization in the disposition of medical units, particularly as between the Middle East and England. He also noted that despite this inefficiency the nursing service had performed well. General Fetherston who earlier had been lukewarm in regard to the role of the nursing service, was shrewd enough to realize that there was need for change. His Report to the government in January, 1916 recommended radical changes in the organization and administration of the AAMC and the AANS[20]. Susan Kenny, in her thesis on the AANS, saw this as a 'milestone' and a 'watershed' in the history of the AANS.[21] It is true that the Report gave greater autonomy to the AAMC in organizing medical care, free from War Office control and that it broadened the concept of treatment of the wounded to include their well-being and rehabilitation. Thus it opened up a range of ancillary services which, overall, helped in the patient's recovery.

A significant recommendation for the AANS was to double the number of nurses so that 'patients were nursed and not merely waited upon'. While this did not go the whole way, there was a limited improvement in organization and administration. The appointment of a Matron-in-Chief, AIF, in Egypt (Miss E.A. Conyers), on the staff of General Howse, DMS, AIF was the first step in a movement which was to give the AANS an identity and a definite role in the army medical services of Australia. In the beginning the Matron-in-Chief's responsibility had to be clearly determined and it is to the credit of Miss Conyers that through her administrative skill and ability that many of these early problems were resolved for the benefit of the AANS and for future appointees to the Matron-in-Chief's position. Much of her success was due to her personal knowledge of the nurses and intimate acquaintance with the difficulties, discomforts and occasional dangers in the day to day work of the nurses in army medical institutions. She was able to take up with 'higher authorities' many of these problems relating to status, rates of pay and conditions of service which were to lift the morale of the AANS. She worked well with her counterpart, the Matron-in-Chief of the British forces in coming to an understanding of the separate identity of the two nursing services, of the differences in structure and training between them and in the need for greater cooperation between them in the common task of treating the war wounded. The occasional appointment of Australian nurses to work alongside members of the QAIMNS in British hospitals was of tremendous value, both professionally and militarily, to members of the AANS and greatly appreciated by the British.

As the fighting ceased in the Middle East and the Australian troops moved to France, Miss Conyers moved her Headquarters to the Australian Headquarters in Horseferry Road, London. General Howse left more and more administrative detail to Miss Conyers, being content with a weekly report on such matters as posting, promotion, pay, discipline, leave, reinforcements and personnel matters concerning the AANS.

However, many major decisions concerning the AANS (and the AAMC) were made back in Melbourne by the DGMS, the Defence Department and the Minister of Defence. There they were subject to pressures from the press, the politicians and the public. There was still opposition from many quarters to any policy which exposed women to danger in battle. While these saw the role of nurses in military hospitals far behind the front line, the war in France was coming to the hospitals. While there was caution and conservatism on the home front, army medical administrators in battle areas were recognizing the value of nurses closer to the action, even in Casualty Clearing Stations.

At this time, nurses were given badges of rank. Under a Military Order promulgated on 18 April 1916, the Matron-in-Chief was recognized by a crown (rank of Major), Principal Matron or Matron three stars (Captain), Sister two stars, Second Lieutenant and Staff Nurse one star (First Lieutenant). There appeared to be mixed reasons for this move and certainly mixed reactions among members of the AANS to it. General Fetherston, probably influenced by the Canadian practice, thought they suffered through lack of recognition of their status as officers.

> The nurses had the privileges of officers, much as chaplains have but neither had any military rank. Nurses were not given the respect they were entitled to because their position is not understood either by officers or other ranks.[22]

Susan Kenny, in her thesis, points out that badges of rank were given to nurses to discourage familiarity with non-commissioned officers.[23] This argument was treated with derision and contempt by many nurses. Butler quotes a report from Sister B. Belstead:

> In 1916 an order was issued that nurses should wear stars. It was reluctantly obeyed. An order issued to maintain the dignity of a Sister and to protect her familiarity, surely an unnecessary precaution. Her profession, her uniform and her womanhood has already done both.[24]

Perhaps the best example of the indignation felt by some of the nurses is reflected in a letter home by Anne Donnell and recounted by her in *Letters of an Australian Army Sister.* [25]

> Matron had an order to give us, and when she had given it we felt as if a thunderbolt had burst. She said, "I give it to you this once, and once only; I shall never tell it again. Now that you Sisters have got the rank of Officers and wear stars, you are not to go out with N.C.O.'s or Privates, or speak to them, excepting on duty. And if you do so you will be sent away at once into a British Hospital". We were silent with surprise for a while, then one Sister asked if we could dress in mufti and go out with them? "No, certainly not". Another Sister got up and said, "Matron, I have a little brother fighting in France, that I haven't seen for six years; does that mean that if he gets his furlough and comes to England to see me, I musn't go out with him or speak to him?" The reply is, "That is the order, Sister."
>
> We never asked for stars — we have never received a commission from the King. We left Australia as Nursing Sisters and such we wish to be. Why afflict our freedom so, and with a threat that would, if carried out, cast a slur on us and the whole of the Australian Nursing Service abroad. Our boys have left home and country to give their lives, if need be, in a strange land. Could we slight them so? And for an unwritten order — no, we couldn't. In this instance our hearts rise above such unreasonableness. What will the consequences be? We dearly love our Unit, and pieces that appeared in the papers from time to time will tell you we have done good service. If a friend, or a relative, from the ranks came to see me, or any of us, and the threat were carried out, well I think we might just as well be recalled to Australia. We wonder if anything could be done; we don't like disobeying an order, but words fail to express our indignation.

Matron Grace Wilson had the final say on this matter of badges of rank. She did not favour the practice, for different reasons.

> As it is it means nothing. The Canadians wear it also, but they get the pay of the rank they carry on their shoulder equally with the officers of the same rank.[26]

Equality was a long way off in 1916!

Another issue which was a constant source of annoyance and friction was that of the nurses' uniform. Here they were the victims of the policy decision which placed them under British regulations. They were caught between traditional conservatism and their desire for individual Australian identity. May Tilton, writing in the *Grey Battalion,* described on Order which appeared on the Notice Board as they embarked on the Orontes.[27]

> The Imperial Government issues an order that all nurses must, in future, wear only the regulation uniform; grey cloth or cotton dress for outdoor, with a white stiff collar, to be worn *under* the collar of the dress, with a bonnet or helmet. No *veils* must be worn.

May Tilton went on to explain the effect this Order had on the Australian Sisters.

> An honest effort was made to obey the order, but two thick collars tying up one's neck in that climate were so drastic that most of the girls provided themselves with grey crepe-de-chine frocks with white peter-pan collars made of

> muslin and wore white panama hats with grey gossamers. This was too smart and didn't meet with approval. So the lives of the Sisters were made miserable endeavouring to dodge the matron.[28]

As members of the AANS tried to get round what they though were intolerable regulations their innovations drew forth new regulations in Routine Orders. Grace Wilson, as Temporary Matron-in-Chief had to point out that in the Uniform Regulations, 'furs are not included, but a plain grey woollen muffler may be worn if desired'. She pointed out white or grey silk blouses were not uniform and the regulation waterproof coat is grey and should not be of a 'fancy type'.

> It is requested of all members of the AANS that they will, at all times, try to uphold their reputation for looking well-dressed. Our uniform is one in which it is very easy to look well, but it is completely spoiled by touches of mufti.[29]

In November 1915 when a new group of Sisters arrived in the Orsova they were wearing a new regulation army hat — a grey felt with a stiff brim trimmed with brown and red. Grey bonnets which everyone disliked so much and which were quite unsuitable for the climate in Egypt were discarded. However the uniform issue was far from settled.

Dress for the AANS Sisters was a constant problem for nursing and army medical administrators, even after two years of overseas service. Matron R. Creal, Principal Matron, AANS, Egypt saw fit to clarify the regulations by writing to Miss Conyers, Matron-in-Chief, AIF, England, in November, 1916, in the following terms:[30]

> It was reported to me by the OC of this unit what the Commander-in-Chief had occasion to complain that Sisters in Egypt, (not necessarily AIF) were not always in authorized dress, when dining out at Hotels, etc. Instructions have since been issued to all AIF Sisters in Egypt that they must wear their red cape and white cap when dining out, at Hotels and private homes. I wish that definite instructions be forwarded to me re authorized dress for mess and outdoor wear, as some Sisters appear to think that they may wear stuff dresses and silk capes, or silk dresses, capes and caps or even print dresses and silk capes. It is very difficult to maintain discipline unless I know exactly what is to be worn. "Widows' collars and cuffs", silk hats, tan shoes and stockings are still being worn by some Sisters who have been on active service for over 12 months.

The matter of dress was but part of the general discipline required of the nursing staff at all times. Subsequently, the Matron of 14 AGH (Abbassia) posted instructions to Sisters and Staff Nurses. By modern standards, some of the following instructions might appear somewhat harsh![31]

- Uniform is to be worn on all occasions. Veils (except in summer), jewellery, fancy hat-pins and belt buckles are not to be worn. Canes are not to be carried. Only hemstitched caps may be worn. Cap pins must be invisible. Red capes (of Turkey twill) with rank badges are to be worn in the wards. Grey aprons are not to be worn in the wards. Cloth topped boots or shoes are not to be worn in the wards.
- No wine or beer of any sort is to be brought into the House.
- No Sister is allowed to dance while on active service.
- Every Sister in a Military Hospital is under the immediate supervision of the Matron and is directly responsible to her in all matters relating to conduct and discipline.

On the other hand she was responsible to the medical officer in charge of the ward for carrying out all his instructions regarding the treatment of the sick and wounded. Sisters and Staff Nurses had numerous other tasks and responsibilities with respect to linen, laundry, meals, diets, statistical returns, hours of duty, night duty rosters and the like. They worked long hours, with little help, in a framework of rules and regulations which would not be tolerated in the nursing profession today.

It was something of a cultural shock for Australian Army Nursing Sisters, coming from the informal atmosphere of country hospitals to find themselves in the midst of a conservative status-ridden military environment based on British traditions. They had to learn the hard way the niceties and intricacies of military etiquette but no matter how much they scoffed at it in the end they had to conform. One of these Sisters, recalling her reminiscences, remembered how they were lectured on military etiquette by one of the MO's, how they were told to stand at attention by the beds as the OC made his inspection and to say 'Good

Sisters of 1 AGH 1915. Sister Isambert centre front row.

Morning to superior officers. This Sister recalled some verses of the time which epitomised their feelings.[32]

Military Etiquette, by Jove,
Tis a red tape net well wove.
From the private to the Colonel,
Orders wait a time infernal.
"Good morning" to your superiors,
Cold stare at your inferiors.
That's correct. Stand erect!
Military Etiquette!
If poison be taken for a pill.
Do nothing for patient until
Your Orderly finds your ward MO
All over the camp he may go.
Orders are given, by superiors,
Obeyed by their inferiors.
That's correct. Stand erect!
Military Etiquette!

Gallipoli

When the main body of the AANS arrived in the Middle East early in 1915 there was still considerable uncertainty as to their role. Indeed there was some uncertainty as to where the Australian troops might be committed. While Germany was the major enemy, strategic plans from the War Office which were not always conveyed immediately to Australian leaders either in Government or in the field recognized the importance of the Suez Canal area as a springboard for attacking Germany's allies in southern and western

Europe. As it happened there were only a few short months of preparation before the first AIF involvement in what turned out to be the Gallipoli campaign. In conformity with accepted policy, nurses were to be attached to general hospitals and other medical units well behind the front lines. Hence the area around Cairo and Alexandria was first selected as a base in the line of evacuation, in close proximity to harbours for evacuation by hospital ships.

The task of establishing two AGH's was not easy. No 1 AGH (Matron Jane Bell) was allotted the Heliopolis Palace Hotel, No 2 AGH (Matron Ellen Gould) took over Mena House and later moved to Ghezirah Palace. As casualties began to flow in, more bed space was obtained by setting up Auxiliary Hospitals at Luna Park, The Atelier, the Sporting Club and other buildings. British and Australian authorities were forced to take over and adapt schools, hotels and similar places for hospital wards. This was not always easy nor convenient, the floor of the skating rink at Luna Park being converted to hold 700 beds! On the other hand the luxurious suite for the King of the Belgians in the Heliopolis Hotel made a splendid operating theatre. The Auxiliary Hospital at Atelier was previously a large furniture factory, a high roofed brick building with poor ventilation, most uncomfortable in the heat for 500 patients. Staff for these hospitals had to be in accord with the War Establishment Table (WET) but some amendments became necessary. Each division was allotted 2 AGH's of 520 beds, with 21 officers, 43 nurses and 143 OR's. As Australia had no large body of trained male nursing orderlies it was agreed that Australian hospitals would be staffed by a larger number of nurses.[33]

The plan for the evacuation of casualties from the Gallipoli campaign involved a coordinated plan by British and Australian army medical authorities. The proposition that Australian medical units would service only Australian troops had to be discarded as impractical. As events were to prove later, members of the AANS were called upon to serve in British medical units. As no members of the AANS were called

The Grand Palace Hotel, at Heliopolis, in Egypt, which was converted into the 1st Aust. General Hospital. AWM C2327

No 1 Australian General Hospital, Palace Hotel Heliopolis Medical ward 'F'. AWM H16957

upon to serve on the Gallipoli peninsula, the splendid work done by medical units there must be passed over. The first stage in the line of evacuation was to get the wounded on to ships lying offshore. In some cases 'black' transports were used, those normally returning empty after ferrying troops to the peninsula. Hospital ships were at a premium as they were needed transporting wounded from France to England. Butler was very critical of the lack of hospital ships which handicapped the whole process of evacuation.(34)

Many Australian Sisters worked on the hospital ships and transports in this campaign. In the *Sicilia* was an Australian consulting surgeon and four Australian nurses, with others from No 2 AGH, while to the *Gascon* went the senior surgeon of No 1 AGH and nurses from No 2 AGH.

Three Queensland members of the AANS, Sisters Amy King, Christense Sorensen and Zita Lyons were among a group of Sisters attached for duty on the hospital ship *Guildford Castle* during the Gallipoli campaign. The British Matron on the ship, Sister E.R. Collins, QAIMNS commended the Australian Sisters for their outstanding work in this episode.(35)

> Their devotion to duty was most marked, they are splendid medical and surgical nurses, and proved loyal and willing workers. Their discipline was good — they never once questioned an order given — and they were able to rise to emergencies and proved adaptable under varied conditions . . . All these ladies showed good common sense and judgment; they work well together and are not afraid of any amount of hard work. I cannot speak too highly of them, or of their care and devotion to the sick and wounded, and their loyal support to me at all times. I am more than sorry that I shall not have the pleasure of working with them again.

High praise indeed!

There were several options open to medical authorities once the wounded were taken aboard hospital transports. The nearest major base from Anzac Cove was Lemnos (60 miles, 96 kms), although many of

Group of nurses of No 1 Australian General Hospital. Palace Hotel, Heliopolis. AWM H16959

the wounded were cleared some 12 miles first of all to Imbros. According to Butler's official account, the evacuation arrangements centred on Lemnos were the most haphazard of the war. There were not sufficient hospital ships available and only nine of the twenty-five 'black' transports expected were available. There was administrative confusion as to the nature of the medical base at Lemnos, whether it should be merely a clearing house or a major treatment centre. It was expected that the August attack in the Gallipoli campaign would result in upwards of 20,000 casualties, so more beds had to be made available. By July the British had set to work to establish a major medical base at Lemnos, with Nos 15 and 16 British and 1 and 2 Australian Stationary Hospitals, each to expand to 1,000 beds, without additional staff.

Meanwhile No 3 AGH, whose Principal Matron was Queenslander Grace Wilson had arrived in Egypt on 25 July from England. They had left Sydney on the *Mooltan* on 26 May 1915 and included Queensland Sisters, among others, Avenell, Derrer, Leitch and McClelland. At first they were ordered to remain in Egypt. This order was countermanded and the hospital was ordered to Lemnos, but unfortunately its equipment was put aboard another ship! The CO Colonel T.H. Fiaschi, and the male staff of the hospital arrived there on the transport *Simla* on 29 July, but it was three weeks before the equipment arrived. On the 5th August the main body of 3 AGH landed on the bare and roadless hillside under circumstances without parallel during the war. Without tents or equipment, without water supply other than that from the ships, with no sanitary provision and little transport, the advance party could do little. Butler summed up the position[36]

> Of nursing equipment there was little, of eating utensils a bare sufficiency for patients. Having landed on the 5th, the male staff had by the 9th cleared the area of rocks, done some roadmaking and erected, somewhat promiscuously, a few bell tents and marquees and an operating tent. On this date the female nursing staff and the first batch of 150

No 1 Australian General Hospital Heliopolis. Supper room of Palace Hotel used for wounded from the Dardanelles, 160 beds. AWM H16956

> cases arrived together. Mattresses were available but not beds. Cooking and sterilizing were for a time done by spirit lamp, through lack of kerosene for the primus stoves.

The confusion was exacerbated by the arrival of more Canadian and British medical units and the movement of some British units to Gallipoli. Australian nurses replaced British nurses at one of the stationary hospitals as well as staffing the No 1 Australian Stationary Hospital. By the end of August there were over 5,000 beds available on Lemnos but by then casualties on the Peninsula had decreased, with fewer wounded and more sick patients arriving. To add to the problems, the site had been grossly fouled by the Egyptian labourers, resulting in an outbreak of muco-enteritis, while the myriads of flies helped spread the disease. These were appalling conditions for medical treatment and Butler paid a fine tribute to the work of the Australian nurses.

> It is clear, however, that the training in the nursing profession, severe beyond most in its standard of toil, self-discipline, and resource in compelling order out of chaos, enabled these trained women to adapt themselves (as they have often before) to circumstances, bend to clearly recognized ends such means as could be found, and in a short time obtain a comparative mastery of the situation. In the "wards" and operating theatre the medical officers soon found that, while some of the amenities which they had been accustomed to require for their cases were perforce lacking, the essentials of nursing had been carried out, to wit, cleanliness, care of the skin, attention to the calls of nature, careful feeding, dressings of wounds, and, withal, the ward discipline that makes effective ministration possible. Whatever arguments may be adduced to urge the impropriety of placing so complex a unit under such primitive conditions, inability of the nursing service to rise superior to the circumstances is not one.

The interior of No 2 Australian Auxiliary Hospital at 'Atellier', Heliopolis. June, 1915. AWM C693

Gaza, 1915. Nursing Sisters of the 3rd Australian General Hospital. AWM A5411

Lemnos, May 1915. Matron Margaret Grace Wilson 'doing a round'. AWM A5332

In a very critical letter home Matron Grace Wilson described the scene on Lemnos and the frustrations of the medical and nursing services faced with the inefficiency and disorganization of the army administration.

Things are just too awful for words. It is dreadful to think we have done nothing yet. Things seem so badly organized — some people just can't organize work . . . This is just about 40 hours from Alexandria — and about 40 miles from the firing line. We are the first women to come so far, except for the Sisters on the hospital ships. This waiting exhausts our spirits — and roughing it will be beyond description. If you can imagine at present a bare piece of ground, covered with stones, from the size of pebbles to boulders — the men in their clothes lying on it — waiting — and we Sisters imprisoned on a ship opposite, also waiting and doing nothing — and you have 3 AGH after touring for eleven or twelve weeks.

Sister Lillian Leitch, from Brisbane, graphically described the conditions at Lemnos:—

We as a hospital unit were chosen and given orders to proceed to a Greek island in the Aegean Sea called Lemnos, about 45 miles from the Dardanelles. The officers and orderlies of our unit travelled on a different ship and arrived at Lemnos on 5th August 1915. Some days later the Nursing Staff arrived and that same day 200 sick and wounded arrived from Gallipoli. Mismanagement in England put our equipment on another ship and this did not arrive until 3 weeks later. This was most unfortunate as we had to rely on the barest essentials. The Nursing Staff had included in their kit a few odds and ends such as small methylated spirit stoves, methylated spirits, cotton wool, bandages and a few instruments. Our Senior Surgeon had insisted on carrying his own surgical instruments — in that way we were able to do quite a lot of surgical work but under great stress and strain.

The male staff had cleared an area of rocks and marked out some paths. This was fortunate as our patients were nursed on the ground in the open and it was on this ground that the staff slept. How we all longed for a hole for our hips. Shortage of food and water was our main anxiety, particularly as by 13th August we were treating 900 patients. For some weeks we had water carried from a well in a Greek village 4 miles away. We had one water cart, which held four dixies each capable of holding 4 gallons of water. We commandeered a mule from a neighbouring Greek village and with this the cart travelled those 4 miles there and back, night and day. So often we were thirsty and the heat was intense. Our patients were thirsty too. How we watched for that cart coming down the hill and at times we were frustrated as the mule sat down and refused to move! However, within six weeks the engineers had completed a condenser which condensed the sea water and all was well.

A special block for enteric dysentery was opened and there followed a period of important work in this department of the hospital. By 15th Nov sixty per cent of the male staff of the hospital were reduced by sickness. This enteric dysentery was less with the nursing staff and a survey said we had taken greater care with our domestic cleanliness and hygiene. Conditions which the nursing staff worked under both then and some time later on Lemnos were more crude than any experienced during the war.

This relative freedom of nurses from intestinal disease as compared with male staff, including medical officers, was confirmed by studies made by a medical scientist, Dr C.J. Martin and reported by Dr A.G. Butler,[37] in his history of the medical services during the First World War. He believed this was due to the care and cleanliness in the preparation of the nurses' food while the messing arrangements for the male staff were far from satisfactory.

We were in the tender hands of Australian boys whose knowledge and practice was confined to what they may have picked up. The nurses' food stuff was protected from flies from time of delivery; ours displayed for the delectation and amusement of these insects. There was also the difference in the possibility of hand to mouth infection which would be less in the case of a kitchen looked after by a trained nurse.

Another Brisbane nurse, Sister Selwyn-Smith, who had also sailed on the *Mooltan* to England and subsequently to Lemnos also set down her vivid memories of this episode.

We arrived at Mudros Harbour on board the *Dunluce Castle* on the early morning of August 5th, 1915. On the next morning we were transferred to *HMT Simla*, at that time the medical depot . . . All day on the 7th August we waited ready to be taken off the *Simla* at any moment. It was not until the next night that 40 of the Sisters went ashore. I shall never forget that experience. Our C.O. was nothing if not a military man and we were met at the pier of West Mudros by a staff sergeant with bagpipes playing us into camp! I remember how desolate it was at that pier, the only person in sight was an English Tommy standing beside a pile of onions.

When we arrived at the camp we found some wounded had arrived at 6 a.m. and no equipment was there — nothing was ready for them. After a scrap lunch in a mess tent with just a board in the centre between poles for a table, with flies by the million over everything, we put on an apron and got to work. Several tents had to be erected while the patients lay on the ground and waited. To wash them (and they were all longing for a wash) we had to use salt water and any utensil we could find — basins were few and far between, if any. I washed my unfortunate patients with salt water out of a dixie and gave them tea from the same article. We used our own soap and tore up old clothes for towels. After we had washed and fed them we tried to make them comfortable.

Equipment in the hospital was very scanty for over a month but we managed to get a few beds for the very sick. We got a few crates for the cupboards and a little later boxes were useful. For about a month the work was very strenuous. It was difficult to care for patients lying on the ground. We did most of our cooking on a small methylated spirits stove and sometimes on a smokey Beatrice but they were very slow. The wind was very strong sometimes and the bread and jam were frequently blown off the plates and of course it was the usual thing for bell tents to be blown down.

We had been there about two months when the men began coming off the Peninsula for a rest. It was pathetic to see the first lot march, or rather straggle, past — most of them looked haggard or ill — numbers slipped out on the way and we got many of them in hospital. The cold weather came after a severe frost and we got many patients with frostbite — some so badly bitten that their feet had to be amputated.

> We left the island in January 1916, with mixed feelings. Most of us had adjusted to the conditions and wondered how we would face up to a normal hospital again. We looked a motley collection of females when we boarded the lighter for evacuation — our headgear was weird and wonderful, mostly balaclavas — as all our hats were worn out.

The irony of this episode was that as the Sisters of 3 AGH were leaving, construction began of a new hospital — the 27th British General, but it was not really needed. Another case of too little and too late. The Lemnos experience was a triumph for Matron Grace Wilson. By her example she inspired her staff to battle against the extreme difficulties. In turn, the Sisters loved, respected and admired her. Sister Selwyn Smith summed it up:

> At times I think we could not have carried on without her. She was not only a capable Matron, but what is more, a woman of understanding. She saw and understood many things without having to be told — and she was very human too.

The Director-General, AAMC, in a report on the work of Matron Grace Wilson and the nurses of Lemnos, commented:[(38)]

> She was the best of all our Matrons, was Miss G. Wilson, Matron 3 AGH and formerly of Brisbane Hospital. Had a desperately hard time at Lemnos with food, tent, mud and sickness, as well as great trouble with ____ who treated Nurses shamefully. I believe the hospital would have collapsed but for the Nurses. They all worked like Demons and were led and guided by Miss Wilson and one or two others. Everyone of these nurses deserves the greatest credit. They were the first women on Lemnos by a long way — lived in tents, lived on tinned meat and biscuits, no baths, no conveniences and when it rained, mud up to their knees. But they never grumbled.

Many important lessons were learnt by the army, the medical corps and the nurses from this Lemnos episode. It was the first time nurses were brought forward and they proved their value under extraordinarily difficult conditions. It was the first time a general hospital was set up fully under canvas under most primitive surroundings. Formality and military etiquette were of little moment under field conditions. The uniform problem was resolved in terms of necessity — so nurses wore sheep skin coats, riding breeches, sou'westers, gum boots, rain coats, men's woollen socks, whatever would keep out the cold and wet and enable them to carry out their duties. Matron Grace Wilson 'did her rounds' carrying an umbrella! They had to learn to improvise, but generally that came easily to Australian nurses. Without electric light, they had to fall back on hurricane lamps; without a regular supply of fresh food, they had to exist on bully beef and army biscuits. The army medical service was in dire need of reorganization, the line of evacuation of the wounded from Gallipoli to medical institutions needed re-thinking. Indeed, the whole philosophy of caring for and treatment of the wounded came in for critical re-appraisal.

But it was in and around Cairo where Australian nurses saw most of the action at this stage of the war. In the early stages of the Gallipoli campaign the trip from the front line to Egypt was taking almost a week, by which time many had died of their wounds. A review of the evacuation arrangements had brought in Lemnos, others were sent on to Malta, where occasionally Australian wounded were nursed by Queenslanders serving in the QAIMNS. They were also taken to two British General Hospitals in Alexandria, where too they caught up with members of the AANS reposted to serve with the British and again there were Australian and Queensland nurses serving with the QAIMNS.

Suddenly, members of the AANS in the Australian hospitals in Egypt came face to face with the grim realities of war. Sister Gertrude Andrews was in the operating theatre at Heliopolis when the first casualties from Gallipoli began to arrive on 28 April 1915. More than 2,000 cases were admitted between 5 pm and 11 pm on that day. Medical and nursing staff worked unceasingly. There were very few orderlies and no VAD's to help. Nurses had to undertake the physical tasks of removing the men's bloodstained clothing, then washing, feeding, treating and bandaging them. Every day men had to be moved out to make room for more urgent cases. Sister Ada Smith was also on duty in the operating theatre most of the time at Mena House and later at Ghezirah Palace. She was shocked by the large number of amputations.

> Some of the boys are terribly wounded and we have quite a lot of deaths. And there are such a lot of amputations. There'll be a great number of disabled people after this war.

Queensland Sister Ada Smith also painted a picture which was repeated many times over in the next few months.

> On arrival at GP we found we had come to the wrong place. Some 50 beds and supper were waiting for us at Heliopolis, some miles away. So there was a scramble to get supper for 50 hungry train travellers. As the place was full we spent the first couple of nights sleeping on mattresses in the lovely marble corridors of this magnificient place, with its marble bathrooms and huge hot and cold showers.
>
> The wounded were coming from Gallipoli and we were on duty ready to receive them. I have never forgotten the look of these men. Of course, the badly wounded were on stretchers, but the walking wounded were so tired out, they could have gone to sleep standing up. Many had not had a wash for weeks for constant shelling on the beaches stopped them from even using the sea. Once in the hospital they were quickly relieved of their grimy clothes, washed, their wounds dressed, then they were fed and put to bed. It must have been like heaven to them.

One of the most vivid descriptions of the hospital at Heliopolis and task confronting the nursing staff as the wounded came in from Gallipoli comes from a letter written home by Queensland Sister Constance Keys, who had come to the Middle East on the first convoy. She was very conscious of the impact of the casualties in so many Queensland homes, she felt very deeply about the sufferings of the soldiers and she was very proud of their bravery.

> There have been sad happenings since I last wrote. I don't know if the news is known in Queensland yet but the greater number of the men we came over with are either killed or wounded. The whole battalion was practically cut to pieces. We did not get any of the wounded in until Thursday 29 April. The hospital train came in right behind the Palace — nine long carriages painted white with the Egyptian star and crescent on the side.
>
> Then the unloading commenced. Those who could walk were shown the way in and those on stretchers were carried over in the waiting ambulance cars. I'll never forget it. To see these hundreds of men walking and being carried coming in one after the other in endless procession. There wasn't a sound except the noise of walking — and then the wards began to fill up. The work of getting these weary men washed and bathed and fed and wounds dressed was a big undertaking — but it was done.
>
> Remember the bad days I used to have in Brisbane at the Hospital, well the days here have been worse than that. It often seems hopeless trying to cope with the work, but when I think of those hundreds and hundreds of men — our own men — well I put on a spurt and try and work a bit harder. I don't think when all the news gets through there'll be many homes in Queensland not in mourning.

Sister Isambert, in her diary, gave a similar account of the dedication and sacrifice of the medical and nursing staff in tending the wounded.

> *1 May 1915*. Another train load of wounded arrived at 12 mn, about 100 very badly wounded, a great number of whom are going to die. Oh the horror of this war and the boys are so proud of themselves and they're so good about their wounds and dressings. So many are going to be cripples and lose their limbs. Doctors operated all night, not going to bed until 5 am — theatre nurses also, and most of the day staff stayed up until after midnight. It brings out the best in everybody.

These Sisters had to live with sorrow and tragedy every day; they could never wholly detach themselves from it. They formed brief friendships with some of the boys from back home, they went riding with them on camels around the Pyramids, occasionally they dined with them in Cairo or 'Alex'. A few weeks later they learned their friends had been killed in action. Again and again this happened. Deepdown they felt the hurt, the hopelessness of it all. No wonder Sister Avenell could write.

> We all get down to zero. I suppose it is living the whole time with war surrounding. Really, Mother, I don't know what will become of it all. I want to get back to Australia the moment it is all over. I'm fed up with the war and everything.

Even before the *Mooltan* had left Melbourne, Sister Avenell was homesick, her letter to her mother expressing it as only a Queenslander could:

> I don't think I'll stay a day longer after the war. I feel Queensland — sick right now.

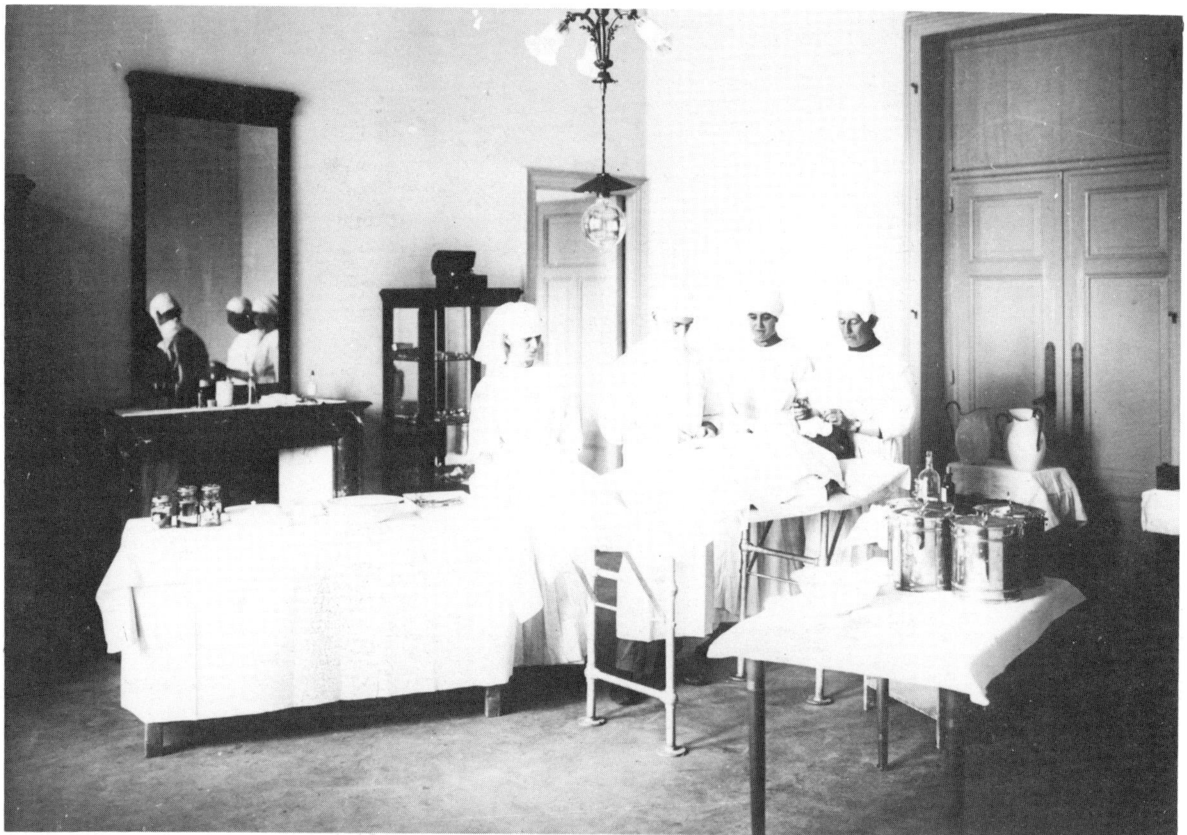

Scene in operating theatre, No 1 Australian General Hospital, Heliopolis. AWM H16961

Of another friend about whom she wrote feelingly to her mother, she put her innermost thoughts down in her letter.

> He was so good and nice to me. He went back to the front. I got three letters from him and this week his pal wrote to say he had been killed. I feel I can't meet any more of them now. I had just written him a letter. I feel very much down in the dumps. We go out and have such gay times before they go off to the front and then we never see them again. It's wicked and I hate the very mention of war. Although we never get away from it, it is with us day and night.

Yet Sister Avenell and other members of the AANS had to put those feelings behind them, as they faced up to day after day of serious nursing, always bright and cheerful for the sick and wounded who looked forward to that smile, especially for them.

Death of Sister N.V. Mowbray

Many of the Sisters found the conditions at the Hospital anything but pleasant. They worked long hours, the wards cold and draughty, their sleeping quarters bleak and cheerless. Little wonder that colds were prevalent and frequently developed into pneumonia. On the 21st January 1916 Sister M.V. Mowbray died of pneumonia. In her diary Sister Agnes Isambert recorded the sad event and the subsequent funeral arrangements.

> *21 January 1916*. Poor Mowbray died. Scott and I went into town and ordered wreaths. Her poor mother will be broken-hearted . . .
>
> *22 January 1916*. A miserable day — every one very depressed. Flowers began to come in about 10 am. At 2 pm the girls, officers and men going to the funeral went off in a special train arranged for them. Baker, Whipham, Scott,

Sister Norma Mowbray, AANS. Died of pneumonia 21 Jan 1916. Buried in Cairo War Memorial Cemetery, Egypt. AWGC.

> Gibbon and Echlin and myself remained behind to act as mourners and to walk beside the pallbearers. We followed the coffin covered with the Union Jack to the motor hearse. General Howse went in one car, Misses Conyers and Knowles in another, the pallbearers and mourners at the rear. Arriving at the cemetery we were met by a guard of honour. We walked beside the coffin, the Sisters and officers following, while the band played the Dead March. The two chaplains walked in front of the hearse to the graveside. After the brief service, the firing party fired three volleys, the Last Post was sounded and it was all over. It was a very trying experience.
>
> Back to duty at the hospital, feeling very tired. Bed state 723.

Sister Isambert continued to write up her diary, methodically recording the events of the day, the problems, the personalities, not sparing anyone in writing down her impressions. She noted down the deaths of known Queenslanders. As a devout Catholic she attended Mass regularly and always made sure that the R.C. Chaplain was called for dying R.C. patients. On 3 February 1916 she recorded briefly in her diary:

> Sister A.B. Corfield, a Queensland nurse with the Imperials (QAIMNS) died yesterday from influenza.[39]

The extraordinary nature of the fighting on Gallipoli was a factor in an increasing number of cases of nervous disorders, ranging from shell-shock to neurasthenia. Medical officers were ever on the alert for the 'malingerer' who might try to 'put it over' by feigning vague symptoms in order to be evacuated from the danger zone. Sometimes they might have erred in not assessing correctly the genuine causes of illness.

Sister Corfield's grave, Chatby Military Cemetery, Egypt. AWGC

Sister Isambert was most critical of the doctors' handling of such cases.

> Some men are sent in for nothing at all, while others are allowed to peg out before they are sent, especially those with dysentery. One man in particular reported ill and paraded 4 mornings in succession. He was ordered to do heavy duty, pleaded inability to do it and was sent to the RMO who gave him 3 pills, some house mixture and a big dose of castor oil. Some hours later he crawled to the officer's tent and was sent in here passing nothing but pure blood. Pigs they are!

While the nurses had every sympathy for 'the boys' who came back from Gallipoli, they were also upset by their disregard for rules and authority in the hospital. They commented adversely on those who went AWL and came back drunk. Sister Isambert had a simple solution:

> One boy of about 18 was brought in drunk. I told him he should be put across his mother's knee and smacked!

The large number of pneumonia cases entailed much hard nursing for the AANS in the hospitals in Egypt, following the Gallipoli campaign. Sister Isambert commenting on these cases regretted there was so little they could do for them. 'I am afraid they will all die', she wrote in despair, and most of them did. Another generation was to pass before the coming of the wonder drug penicillin.

Writing from Heliopolis during the Gallipoli campaign Sister Avenell constantly referred to the long and arduous work involved in changing the 'dressings' of the wounded.

You know, I have 36 men to look after and there are dozens of dressings to be done. Really our feet are simply dreadful by the end of the day — we all rub methylated spirits on them.

I am now nursing typhoid, pneumonia and dysentery cases — 200 of them, all back from the Dardenelles.

One of the most troublesome problems in Egypt was the ever present swarm of flies. Sister Avenell commented:

The damn flies are so bad, they creep and crawl all over one — they are quite tame. They swarm around the natives' eyes. Sometimes we have to have a man to swish them when we are doing dressings.

May Tilton in her reminiscences recalled the ditty they made up about the fly menace.

The new flies, the old flies, the flies in all our jam,
The blue flies, the blowflies, the flies like potted ham.
The dirty, filthy Egyptian flies
That made us all say "Dam"!

The nurses laughed about it, but they were conscioous of the dangers inherent in fly borne diseases. Unfortunately not all the troops were so careful or conscious of the problem and sickness from this source remained a constant factor in the casualty lists throughout the war. Indeed, the next generation of troops and nurses twenty five years later were to face the same problem.

One of the outstanding features in the letters and diaries of Queensland nurses serving overseas in the First World War was their identification with Queensland. Of course, in the first decade of the century

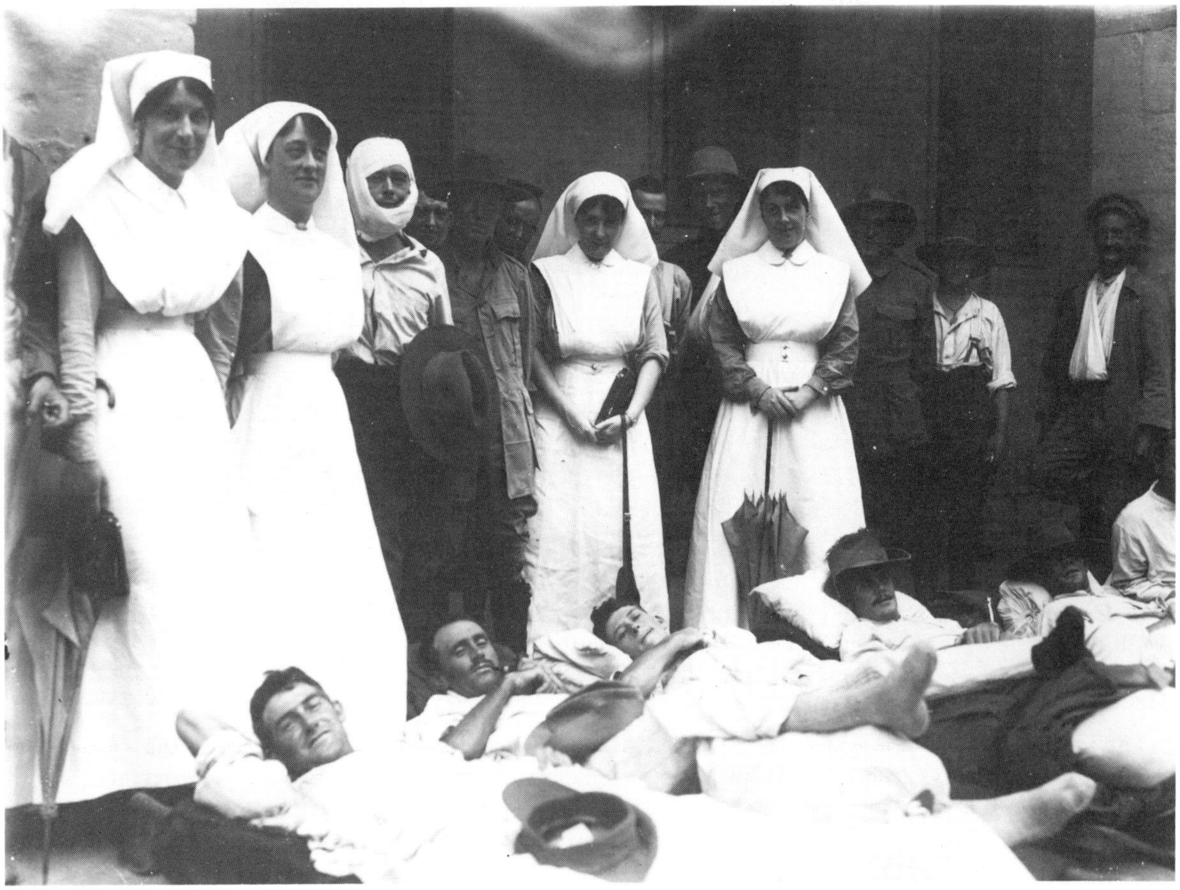

Evacuation of Patients Abbassia. AWM J1726

Sisters Milne and Dobson, A.A.N.S., who were attached to one of the hospital trains running from Heliopolis to Ismailia. AWM C357

Queensland was still a relatively small settlement — population 1910 only 600,000 — with smaller provincial cities where families and friends lived in a tight community.

Many of the nurses trained together or worked together in country hospitals. In the early stages of the war they were drafted to the same medical units. Naturally, they exchanged information about what was happening 'back home'. They heard about the droughts and the floods, they passed on the gossip they received about friends and acquaintances in Townsville, Rockhampton and Clermont. They heard about the tragic deaths of 'boys' they knew, while they scoured the patient admission list to pick up the names of any who might be known to them. The diaries and letters of the time refer to Sisters Jess Kennedy (Rockhampton), Florence Herbertson (Clermont), Georgina Farquhar (Bundaberg), Margaret Goggins (Warwick), Beatrice Bowes (Boonah), Mary Chataway (Mackay), Elsie Rose Grant (Clermont) and Maria McClelland (Rockhampton). The hospitals at Heliopolis, Mena House, Luna Park, Atelier, Abbassia, the Sporting Club, Choubra — these became household words among the nurses, as they sometimes moved from one to the other. The Queensland nurses constituted a tight club as they met and exchanged information, gossip and rumour. Throughout her diary Sister Avenell made frequent reference to her Queensland friends, Sisters (or Nurses) Echlin, Elsie Grant, Gibbon and Gifford. From No 1 AGH at Heliopolis she commented,

> Sister Webb from Brisbane is in my room. She has had German measles and is here for a month. She is such a brick and has cheered me up immensely already.
>
> Did I tell you I saw Nurse Derrer? She is over at No 2 and Chidgey arrived last week, so Mackay nurses are well represented.

Occasionally an element of dissatisfaction crept in, as some long serving members saw recent arrivals promoted over them. One nurse wrote to her mother to see if she could do anything in Australia.

Luna Park, Heliopolis, converted for use as an auxiliary hospital soon after the operations at Gallipoli commenced in 1915. AWM C4052

I was wondering, Mother, if you could do anything for my other star. I am only one of a hundred staff nurses, of course, but still why shouldn't we get promotion as well as others. We have tried over here but it can only be done from Australia. On May 15 I shall be two years in the army and you see I am still only a staff nurse. There have not been any promotions since the first *Kyarra* went over to Egypt.

But these nurses had to wait for the reorganization of the AANS.

By the end of 1915 the casualties from Gallipoli had ceased and there remained the long and difficult task of preparing some for the long voyage home, of caring for those slightly wounded or temporarily laid aside through sickness who might be helped to recover and take up the fight again. As Christmas approached the nursing staff joined in the festivities, hoping to bring some cheer to those in hospital. Sister Ada Smith wrote home a vivid account of the event, complete with a touch to gladden the hearts of all Queenslanders.

We decorated the wards beautifully with ferns, palms, potplants and flags. Ward 6 looked lovely. At nine o'clock the bugle called Officers, Nurses and Orderlies to attention in the square. The Major then addressed us nicely and gave us each a billycan. My billycan contained a Christmas cake (a bonzer), lollies, cough mixtures, leadpencils, postcards, one pair socks, 1 tin tobacco, soap and 1 toothbrush, toothpaste, etc. All the nurses here also got a present from the Lady Bridges Fund for nurses.

I got one tin of shortbread, powder, eau-de-cologne, toothpaste, toothbrush, 1 pair stockings and a pretty red cross brooch with 1915 on it. The sick got their billycans about 9 o'clock and soon their beds were covered with cigarettes, cigars, tobacco, socks, cough mixture, cakes, puddings, soap, razors, bootlaces and insectibane. Derrer and I were disappointed that nothing had come from Queensland. So imagine our delight when a Christmas Cake

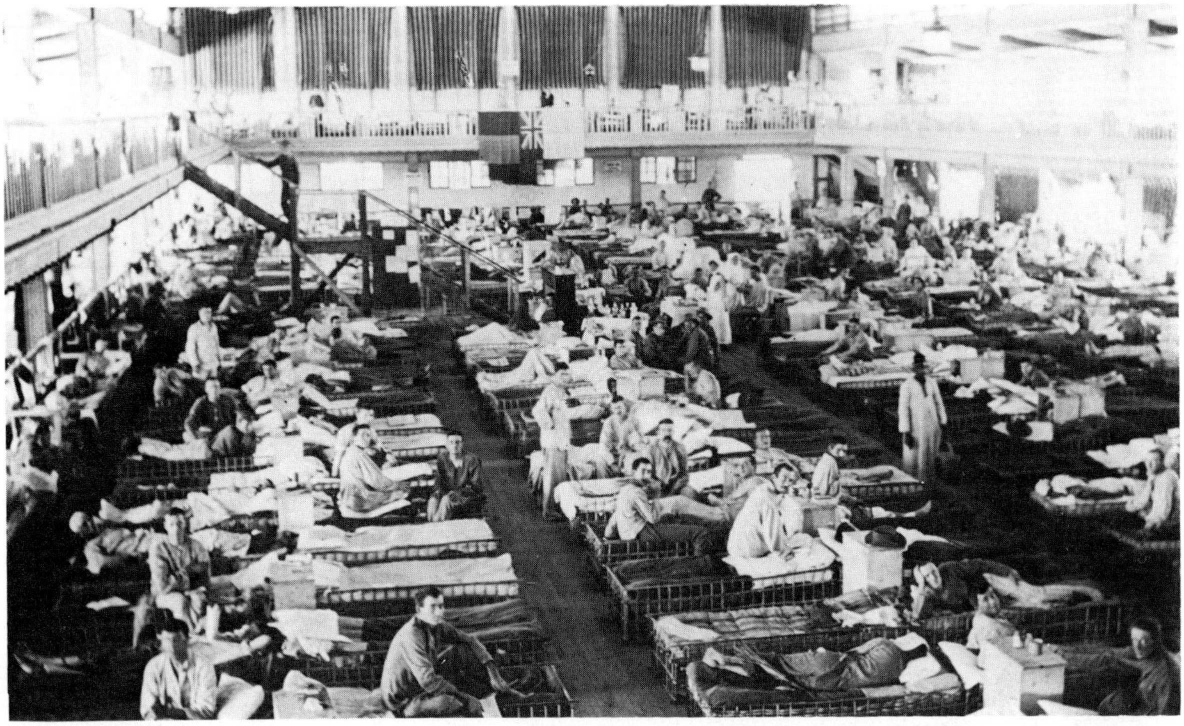

One of the wards at Luna Park, which was used as an Australian auxiliary hospital in 1915, and early, 1916. AWM C4052

arrived with "Queensland" written across it. It also had sprays of holly on it which were worn by the Queensland nurses.

The AANS after Gallipoli

When the Australian troops withdrew from Gallipoli to Egypt, it was anticipated that there would be time for re-fitting, re-grouping and reorganizing before the next battle assignment. However, in February 1916 the Germans launched a major attack on Verdun and Australian divisions were rushed to France. With them went medical teams including 1 AGH, 2 AGH and two CCS's in which AANS members were to serve for the first time. The important contribution made by the nurses in the Gallipoli campaign and the widespread public demand that greater use should be made of their services resulted in a rapid diversification of nursing activities by the AANS on all fronts. Some auxiliary hospitals and convalescent depots went to England, together with AANS staff. Nurses found themselves on hospital ships and transports or 'black ships', plying between the Mediterranean and the Channel to British ports or in some cases nursing the wounded on the long voyage home to Australia. Some were allocated to British hospitals, serving in Malta, Italy, and India. Others remained in Egypt for the Turks remained a constant threat. In each of these areas Queensland members of the AANS served with distinction.

Grace Wilson on the AANS

This was a time of stock taking and the most valuable summary available is that made by Matron Grace Wilson. Her comments deserve to be recorded as they were relevant to all aspects of medical service. There was too, a great deal of wisdom in it for the AANS.[40]

Matrons. Without any malice — as fortunately in my case I had no cause for complaint — I feel compelled to say, from what other Matrons have told me in confidence, that I do not think that on the whole the Matron of an

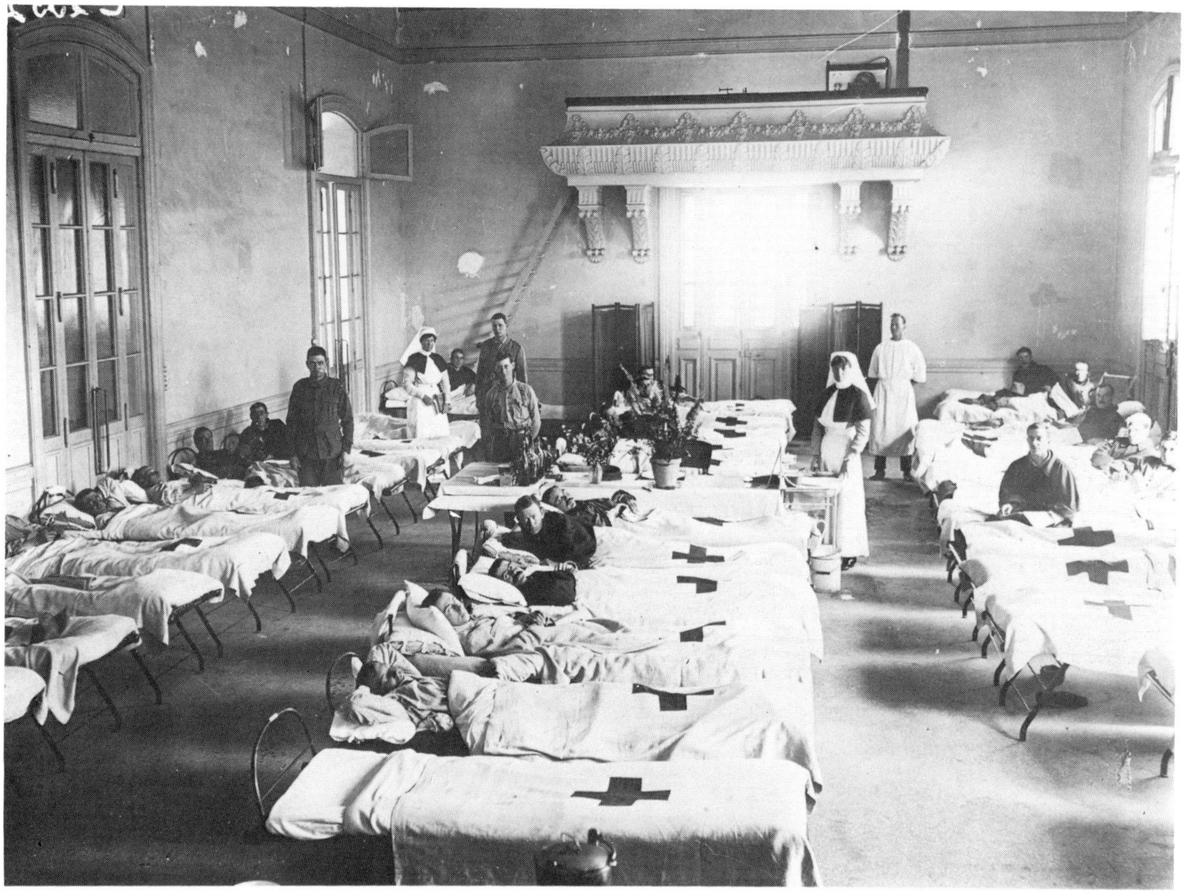

A ward at No 1 Australian Stationary Hospital, at Ismailia. AWM C2332

Australian Hospital was as much considered — or had as much respect shown her position — as the Imperial Matrons had.

Promotions. I do not think that Staff Nurses should automatically have been given promotion at the end of two years. Some nurses, who may do satisfactory work as Staff Nurses, are a thorn in the flesh to any administrator when promoted and put in charge of a ward. They just cannot do it.

Sisters. Speaking generally the members of the A.A.N.S. were very careless with hospital equipment. They improved very much in this respect towards the end of the war. Would suggest that Matrons and Sisters be clearly shown their duty in this respect — that the duties be laid down in black and white, not in a vague or general way.

I would suggest certain standing orders for the A.A.N.S. on all necessary points. At times it was very difficult in the A.I.F. to know exactly what applied to Sisters and what did not. When 'officers' were spoken of sometimes Sisters were included — sometimes not.

Baggage. Baggage should be limited, and this rule enforced. This question has often caused a great deal of worry to Matrons. Q.M.'s are often unable to supply transport for hugh quantities of baggage. I have seen Sisters in France travelling with ten large packages. One regulation cabin trunk, one large holdall, a hat bag and one large suitcase, and in addition a small suit case that can be carried in the hand are quite enough under active service conditions — unless in very unusual circumstances. This I have proved in 4 years of moving about in France and in the East.

Training of personnel. I think a list of trained Nurses should be kept showing those suited for administrative work, for surgical work, including special theatre experience, and for medical work including some with special experience in infectious work. All should be made acquainted with the various Army Forms in use — the routine of

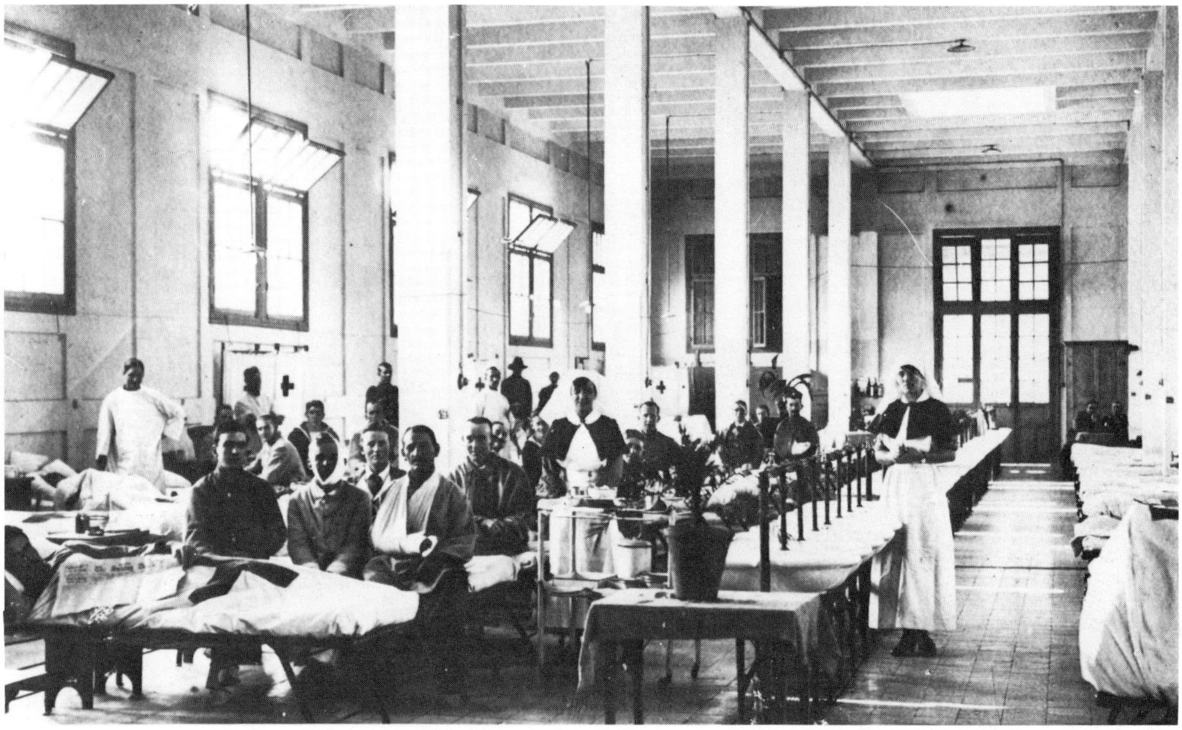

The interior of ward No 7, 14th Australian General Hospital, Port Said. December, 1918. AWM B655

ordering different things from the right store, etc. — so that no confusion occurs when they are plunged into a Military Hospital.

To a large extent untrained nurses might be utilised instead of orderlies — but a certain number of men would be needed for heavy lifting and carrying both of patients and stores. The women should be better because I believe they would take more interest. Ward work is not congenial to the majority of men.

Adequacy of staff. The establishment that we had — of 90 Trained Nurses to a General Hospital reckoning say 1,500 beds with possible expansion up to 2,000 — was good and workable, and allowed a margin for sickness. This is speaking of a tent and hut hospital; but when, as often happened, we were called upon to send Sisters to the Army Areas, and to work with a depleted staff of say 60, it was not possible to do the best for the patients.

Where large institutions were taken over for use as hospitals more staff was necessary — they were not nearly so workable.

Ward orderlies. I do not consider that (our) Matrons were given sufficient control over the Nursing Orderlies. Matrons have told me that they never knew from one day to the next what orderlies were available for nursing duties — or when they would go to a ward and find the orderly gone, or another in his place. The nursing orderlis should be as much under the Matron's control as far as their ward work is concerned as the Sisters — if the Matron is to make the best use of her staff. Several instances could be quoted from my personal knowledge to show this.

Bombing. I think that Nurses have proved themselves for danger zones. They have shown a high courage, and good nerve. If a nurse is at all nervous she should be encouraged to acknowledge this frankly without being made to feel ashamed, and moved elsewhere. Also regular leave should be insisted upon, the Nurse not allowed to refuse leave; and except in exceptional circumstances I think six months is long enough to leave a Nurse in the danger zone.

In September, 1916 the 14 AGH was set up at Port Said in somewhat unusual circumstances. The major AIF medical units had been transferred to France and those remaining found only a trickle of new patients coming from the Sinai. There were of course still a large number waiting transport back to Australia, under the six month's regulation, that their return as fit for further service would take longer than that. There

The hospital ship 'Kanowna' with Australian nurses on board, leaving Alexandria, Egypt, for England, Aug. 1916. AWM C1054

was considerable pressure from the DMS, AIF, to wind up all the Australian medical units in the Middle East and have them transferred to the major threatre of war in Europe and Britain. Although there were still large numbers of Australian troops in the area, including the Australian Light Horse, it was felt that any casualties could be handled by British medical services. This was quite a reversal of the long held policy that Australian patients should be treated in Australian hospitals by Australian medical and nursing staffs. The DMS's counterpart in the British forces in Egypt was not prepared to accept the loss of Australian medical units, partly because he appreciated their value and partly because he knew the difficulty of replacing them. In addition the Australian hospitals had a larger number of nurses to replace the British system of male nursing orderlies. As the casualties mounted on the Western front and Australian medical units there were in need of reinforcements so the pressures increased to move Australian medical units where they were most needed. This applied particularly to 3 AGH, one of the most experienced Australian hospitals after its work on Lemnos. It was, in Butler's view, merely cleaning up 'the dregs' following the Gallipoli campaign.[41] So a compromise was worked out. No 3 AGH (1,040 beds) was transferred to England, being replaced by a new hospital No 14 AGH, commanded by Col Walter Summons, with Principal Matron Creal in charge of the nursing staff. No 1 Australian Stationary Hospital was also transferred to England, leaving No 2 ASH in Egypt. These two hospitals plus the Infectious Diseases Hospital at Choubra constituted the major Australian medical force in the Middle East.

The war against the Turks in Sinai, Palestine and Syria was a long war of attrition which continued from 1916 through almost to the Armistice in 1918. It was a war in which the Australian Light Horse made its reputation, a war highlighted by General Allenby's entry into Jerusalem in December, 1917. While casualites were heavy in some engagements, the toll through sickness was constant — gastro-enteritis, and as the armies pushed northwards malaria entered the lists. Most of the casualties eventually came back to 14

AGH, with the overflow going to No 31 British General Hospital, which itself was partly staffed by Australian nurses. There were periods of intense activity, as for example, after the heavy fighting around Gaza when the bed state of 14 AGH shot up to 1,140, but for the most part the work of the AANS although constant lacked the dramatic quality of that being experienced by other members of the AANS in France and England or of their colleagues serving with British hospitals in other battle fields.

In general, Australian Sisters welcomed the opportunity of nursing in British hospitals. One Sister described her impressions:

> On my arrival in Egypt, I, with nine other Australian Sisters, was detailed for duty in the 27th General Hospital, Abbassia and we were attached there for eight months. I have nothing but the most pleasant recollections of my stay there, both Matron and Sisters being very good to us and I was glad of the opportunity of studying slightly different methods in nursing for the RAMC officers. Of course working in base hospitals one had practically every convenience, the chief difficulty being the trouble to obtain boiling water for sterilization purposes and the part the primus stove has played in the hospitals in Egypt will not be forgotten by Sisters and orderlies.[(42)]

Numerous Queensland nurses were stationed at these hospitals in Egypt, from time to time during the period 1916–18, mostly awaiting posting to other units. Some remained longer, but all were anxious to 'do the job' for which they had enlisted. Sister Annie Norton (Mt Morgan) recalled her work at the Choubra Infectious Diseases Hospital, while at 14 AGH the following names appeared, reflecting the wide range of Queensland nurses — Sisters Catherine Black (Rockhampton), Jane Parker (Charleville), Eva Coote (Brisbane), Catherine Monckton (Rockhampton), Grace Homewood (Mt Morgan) and Mary Scully (Bundamba). Elsie McLaughlin (Rockhampton) was one of many AANS Sisters attached to the 31 British General Hospital, while she and Mary O'Brien (Longreach) met at the British Citadel Hospital. There was tragedy, too, even after the formal ending of the war. On 8 January 1919 there died in Jerusalem Sister Charlotte Berrie, believed to have come from Mareeba. She was serving with the QA's when she contracted 'Spanish influenza'. She was buried with full military honours in the Light Horse Cemetery on the Mount of Olives, the only woman among the thousands of graves of allied servicemen.[(43)] A friend wrote of her:

> She was a marvellous nurse. She truly gave her life on active service.

The work of the AANS in these campaigns in Sinai and Palestine has been well praised by war historian, Sir Henry Gullett.

> Nor are there words in which to tell of the service of the splendid band of Australian nursing sisters who, under the inspiration of the late Miss Rose Creal, matron at the No 14 General Hospital, greeted the battered men from the front as they reached hospital and nursed them back to strength, or softened the close of their soldier-life. No womanhood has ever presented a richer association of feminine tenderness and sheer capacity. They were true sisters to the fighting sons of Australian pioneers.

The transport of sick and wounded back to Australia posed many problems. There were only two small Hospital Ships available, the *Kanowna* and the *Karoola* but these took some four weeks from Port Tewfik back to Australia. In a year each ship made about four round voyages, moving in all some 3,500 patients per year. As the Germans stepped up the submarine warfare these voyages, even in hospital ships were dangerous. Some 16 Hospital Ships were sunk by the Germans during the war, but fortunately none of these were Australian. As there were some 93,000 waiting to be transported home, some other means had to be found. Several large liners were commandeered, as 'black ships', sailing round the Cape and relying on their speed to avoid the submarines. Again, fortunately, none of these invalids was lost by enemy action, a remarkable tribute to the seamanship of the British navy. Members of the AANS accompanied the wounded on all these trips.

Among those Queensland nurses known to have been involved in sea transport, either on Hospital Ships or on 'black ships, were Sister Ida Axelsen (Maryborough) who served on the *Kanowna* and made nine trips between the Middle East and Australia, Sister Ethel Limpus (Bundaberg) who made three trips on 'black ships', Sister Mary Fisher (Gladstone) who was Sister-in-Charge on the H.T. *Ypiranga*, on which she met Sister Nea Low, Staff Nurse Mary Matheson and Staff Nurse Florence May Auld, all from

Sister Charlotte Berrie, QAIMNS. Died of illness, Jerusalem 8 Jan. 1919. Buried in Jerusalem War Cemetery. AWGC

Queensland.[44] Sister Cocking, who had previously had experience on the Hospital Ship *Kanowna* was selected for further service on the troopship *Milhades*, taking reinforcements for the 11th, 31st and 46th Battalions from Australia to England in 1916. The trip was very rough at first, many were sick and the heavy seas played havoc with the dispensary and the sick bay. As Sister Cocking described it in her letters:

> The ship's doctor and myself attended the sick parade. The doctor invariably prescribed bovril for the mal-de-mer and a pill (no 9). Our dispenser was ill, so were the orderlies, the dispensary was in an untidy mess, bottles were not stowed away safely and a lot of their contents, especially from a jar of castor oil were spilled all over the floor.

One of the AANS Sisters involved in sea transport was Elizabeth Kenny, who after the war became the centre of controversy over her treatment of infantile paralysis, as it was then known. Elizabeth Kenny was born at Warialda (NSW) and her earliest interest in nursing came from her friendship with Dr Aeneas John McDonnell, of Toowoomba, for whom she worked for a number of years. She claimed to have entered a hospital and completed three years of training for which she received a certificate, but there is no evidence to support this. For several years prior to the outbreak of the First World War she did bush nursing and very early became acquainted with infantile paralysis. It is not known whether she applied to join the AANS in Australia, but unless she had a certificate to verify her training she would not have been eligible. Armed with a recommendation from Dr McDonnell she sailed for England in the *Medina* in 1915 and reported to

Surgeon-General Williams. She was apparently accepted into the Imperial Army but in what capacity is not known, as these records were destroyed by bombing in the Second World War.

According to Australian Army records Sister Kenny enlisted in the Australian Army Nursing Service on 30 May 1915, appointed Staff Nurse and allotted for duty with No 1 Sea Transport Section. She claimed to have made 12 round trips to the Middle East. On one occasion she had to report for temporary duty at Enoggera Military Hospital. She was promoted Sister in 1918 and subsequently boarded out in 1919. From all the evidence Sister Kenny was a good nurse, a dedicated army nurse who served her country well during four years' war.[45]

Salonica

In the official Australian history of the First World War it is recorded that General Howse wrote to General Birdwood in 1917 seeking more Australian nurses to staff four British General Hospitals at Salonica. It was thought to be safer to bring them from Australia instead of risking British lives in bringing them through the dangerous waters of the Mediterranean. Too many ships had been torpedoed in this area in the past few months. Besides, there appeared to be a larger number of nurses available in Australia, waiting to be called up for service. As a result three units of 91 nurses each, embarked from Australia in June 1917. No 1 unit comprised Victorians, No 2 members came from NSW, while No 3 was a mixed lot comprising Queenslanders, Tasmanians, South and West Australians. No 1 was in charge of the Principal Matron, (Mrs McHardie White), No 2 by Miss Beryl Campbell (a Queenslander) and No 3 by Miss Uren, a South

Sister Elizabeth Kenny Enoggera 1919. Showing Sister's Badges of Rank and War Medal.

Australian. Subsequently a fourth unit was sent in charge of Miss J.R. Gemmell. The first three units took up duty in Salonica in August 1917, but the 4th unit was detained in Egypt and did not arrive in Salonica until the April–June period 1918.

On landing, nurses were taken in ambulances to the 52nd British General Hospital which was staffed by Canadians whom the Australians were to relieve. The Australian nurses were given biscuits and lemon drink and packed off to bed in an old hut until the Canadians had left. After settling down the nurses found they were being attacked by bugs in their thousands. They quickly transferred to their own stretchers elsewhere! A few nights later the city was bombed and set on fire. Most people thought it was a good thing, reminiscent of the Great Fire of London, as it cleaned up the breeding grounds of the germs and disposed of a lot of filth.

Among those Queensland nurses who went to Salonica were Linda Andrews (Southport) Marianne Dowling (Rockhampton), Helen Lawson (Rockhampton), Frances Grace Walpole (Brisbane), Jane McLennan (Bundaberg) and Christense Sorensen (Brisbane).[(46)] Sister Jane McLennan, from Bundaberg in a letter home gave a dramatic description of conditions at this hospital.

> Just before getting in to Salonica we saw the snow capped peak of Mount Olympus. To most of us this was our first glimpse of snow and we stood there and gazed at it for sometime. At 9.30 a.m. next day (Sunday, August 11, 1917) transports took us to our different hospitals. No 3 unit was taken to the 60th British General Hospital at Hortiach, about 20 miles from the town of Salonica. It was summer time and the trip up was hot and dusty and by the time we reached the hospital we were a dusty looking lot. The hospital was situated in a valley and was a very large canvas one, accommodating two thousand patients. We did not have to wait for any patients to come in as 1,000 were already in hospital. These had been attended to by the medical officers and orderlies. Most of the cases were malaria, blackwater fever, and dysentery, with a few surgical cases. The work was hard with few conveniences. Each day brought large convoys of sick and wounded. The malaria and blackwater fever patients often arrived in a comatose condition with temperatures up to 106° and 107°. They were given large doses of quinine intravenously and intra-muscular straight away and in a few days showed signs of improving. The dysentery cases were long ones and required very careful nursing but in spite of this many died.

A different view of Salonica came from the pen of Sister G. Walter.

> One was surrounded by a confusion of strange tongues, weird customs and diverse stinks. Children, numberless as the sands of the desert, swarmed all over the place, dressed, boys and girls alike, in a motley of rags and dirt — more dirt than rags perhaps. There is something sordid and mean pervading the whole scene, but it is all so appealingly human — those dirty-faced little urchins, the fruit-sellers, the money-changers, the boot-blacks, the famous Salonica porters, the dirty looking priests, the narrow dirty streets, the white minarets peeping up above the cypress trees. This is Salonica.

The personal conditions for nurses at Salonica were primitive indeed. Sister McLennan continued,

> The night staff had to take every precaution against mosquito bites and wore gloves, puttees, hat and net veil on duty but many of the sisters went down with the disease in spite of these precautions. The nurses quarters were of canvas and four sisters occupied one tent. Each sister was issued with a canvas bucket, enamel dish, camp stretcher with a small mattress and pillow, mosquito net and four grey blankets. Sheets were scarcely ever used. We used a suit case as a dressing table and we pinned our clothes to the side of the tent. A hand mirror tied to one of the tent posts was the only one we had and for a light at night a lantern hung in the centre of the tent. Our laundry we did in our spare time. There was a boiler and a good supply of water, except in winter time when the taps were frozen. There was a large tent used for a dining room but food was scarce and table appointments rough. Each sister was issued with a knife, fork and spoon, one enamel plate and one blue enamel mug as a drinking vessel.

Sister Kathleen Cowen (Brisbane) had numerous Bulgarian prisoners to nurse and many of them appreciated the kindnesses of the Australian nurses. Subsequently one wrote to Sister Cowen in such touching terms that the letter (in translation) deserves to be recorded.

> Dear Sister,
>
> The evil fate had brought me to be a prisoner during the best part of my life but I was so lucky to meet you — a kind lady from the farthest west that relieving me of all my troubles, for my grief and my joy was your joy because I forget such a care that you have taken of all of us. NO!

Matron Christense Sorensen, AANS W.W.1. Her reputation was made at Salonica.

When the peace is proclaimed and you go far away to your home, let it be known to you that there are some people far in the middle of pretty Bulgaria where the golden sun rises and sets so magnificiently and where the storms are so sharp and troubling there are some people there that will never forget you for your mother's care to them. Often the silent moon will turn its sharp tops pointing to the north, then you remember us for we and our mothers are sending to you warm and hearty thanks and blessings.

Gaincheff Neno

While several Queenslanders distinguished themselves serving with the British medical forces in the Balkan campaign, great credit was bestowed on Christense Sorensen who became Matron of the 60th British General

The sisters' quarters at the 60th General Hospital, Hortiach, near Salonica, where a number of members of the Army nursing Service were employed. AWM C4336

Hospital at Salonica. She had been one of the group of nurses who had sailed from Brisbane in the hospital ship in November, 1914. Her predecessor, Matron McHardie White said of her,

> I cannot speak too highly of Miss Sorensen's efficiency in nursing and hospital management. She maintained perfect discipline throughout without friction and was beloved by both patients and staff, due, I considered, to her gentle tactful manner and wonderful administrative ability.

Matron Sorensen was mentioned in despatches for 'Gallant Conduct and Distinguished Service', and in 1919 she received from His Majesty King George V, the Red Cross Medal, First Class, in recognition of her valuable services with the British forces in the Balkans. She also received from the President of France the Medaille des Epidemies en Vermeil for conspicuous services. Queenslanders everywhere were indeed proud of the nurse from Sandgate.

France

The escalation of the war in France with the opening up of the Somme offensive in July 1916 brought about marked changes in the role of the medical services, including the AANS. The overall direction of the war from the War Office emphasized the value of integration of allied resources. In such desperate times there was no place for parochial interests. Even within army groups there was a recognition of the need to integrate the medical units within the order of battle, so that the increased number of casualties could be handled with speed and efficiency. In April 1916 No 1 AGH under the command of Lt Col C.T. Champion de Crespigny was transferred by the Hospital Ship *Salta* from Cairo to France, with an establishment of 520 beds, with possible expansion to twice that size. It took over the huts and tents previously occupied by the No 12 British Stationary Hospital on the racecouorse at *Rouen* and within a few days was ready to receive casualties from the Western Front.

While ostensibly for Australian sick and wounded, in practice casualties from all allied forces were admitted. As part of the Lines of Communication system all Australian medical units were to function as

One of the dugouts at the sisters' quarters at the 60th General hospital, Salonica. AWM C4949

'imperial units'. The practical realities of dealing with casualties necessitated the adoption of this policy, even if at the political level there was constant clamour for a separate identity for the AIF.

No 2 AGH which was transferred to France and established at Moussot, near Marseilles, at this time had a unique role. In order to prevent the introduction of infectious diseases by troops moving from the Middle East to the Western Front, 2 AGH acted as a 'filter'. At one stage as many as 13 different diseases were under treatment, including smallpox, typhus, diphtheria, malaria, dysentery, measles and mumps. A section of the hospital was transferred to *Wimeraux*, near Boulogne, for a different role, as a critical base in the line of evacuation from the salient on the Western Front to the hospital ships ferrying the sick and wounded to Britain.

Early in 1917 as casualties continued to mount, No 3 AGH was brought from England to *Abbeville*, where Col B.J. Newmarch set up his 1,500 bed hospital. In time the centre which included the South African General Hospital and the No 2 British Stationary Hospital became an enormous clearing house for patients evacuating through Rouen or Boulogne.

As casualties increased on the Western Front, all able-bodied men, including nursing orderlies in hospitals, were rushed to man the trenches. This threw an extra heavy burden on the nurses whose limited numbers were already spread thinly through the allied hospitals. Towards the end of 1916 the Australian authorities, for the first time, conducted a recruitment campaign for nurses, netting in all some 659 recruits. In part this was a response to a request from the War Office for the services of 200 nurses for duty in France, India and Malta. The critical issue was whether they would be absorbed into the British nursing service on the same pay and conditions as British nurses, or whether they would retain an Australian identity, with Australian conditions of service. Ultimately the latter prevailed.[47]

Against this backdrop, members of the AANS carved out for themselves a new role for the army nurse. At first however, they were the victims of top level administrative disputes between the War Office and the

Salonica. Col Crawford OC 60 Brit. Gen. Hosp. SR Mears & Matron Sorensen (Rt).

Commonwealth Government. As the demand for nurses increased, in the light of enormous casualties, Australian nurses were temporarily reposted to serve in British hospitals. Sister Lillian Quinn, (Bundaberg) for example, served with No 8 British General Hospital in France and in her letters home reported that 'she had been bombed a lot'. Many were transferred to No 25 British General Hospital at *Hardelot* near Boulogne, including Sisters Paten, E.M., Selwyn-Smith, N.F., Pines, S.E. Norton, A.E., Norton, A., Pennefather, M.E., Homewood, M.M., Birt, M.T., Skyring, G.M., Bowe, M.A., Edwards, E.M., Rowe, A.M., Walker, E., Toft, E.M., Macdonald, S., Chataway, M.E., Webb, A.A. and Wilson, M.(48) Sister Norton, A.E. was temporarily reposted to No 7 Stationary Hospital at Boulogne where for the first time she observed the Carrell Dakin method of treating war wounds.(49)

Others who were 'lent' to the British for service in British hospitals included Sisters Maria McClelland (Mt Morgan) who was sent to 38 Stationary Hospital at Calais and Le Havre, Ruby Nicholls (Toowoomba) at 13 Stationary Hospital, Eva Clerke (Hampton) and M.A. Loosemore (Gympie) at 7 Stationary Hospital.(50)

Sister Georgina Farquhar who was attached to 2 AGH at *Moussot* near Marseilles, wrote feelingly of her impressions of the war in France. She was involved in the nursing of cases of infectious diseases brought in by the troops from the Middle East. While that was demanding it was a picnic compared with the nursing involved when that section of 2 AGH moved to *Wimeraux*. Soon the casualties came rolling in and the wounded lay crowded together, stretcher by stretcher in the hospital tents, the beds having long since been occupied. Then came the first winter, 1916–1917. Sister Farquhar wrote home:

> Snow abounding and we hadn't got to the shack stage, but were still in tents. Hot bottles were of no avail as the water in them was frozen by morning. We even took our boots to bed with us so they would be reasonably pliable

Some of the tent Wards at the 1st Australian General Hospital at Rouen, in France, photographed on Sept. 23rd, 1918. AWM E3447

> by morning. Luckily we had no time to dwell on anything but the convoys arriving constantly. The wounded were bodily drafted into the theatre for immediate operation and then shipped as soon as possible across the channel to England.

Bombing and shelling went on incessantly, much to the discomfort of shell shock victims, who always reminded her to 'Douse the glim'. 'Carrying a torch to watch for Haemorrhage was out of the question, so I relied on hand feeling'. Sister Farquhar tried on numerous occasions to join the Casualty Clearing Station nearer the Front but Matron refused to endorse the application, saying, 'You are too good to give away'.[51]

Sisters' letters and diaries refer to the bitter French winter of 1916–1917, the coldest in living memory. Raw eggs had to be sawn through with a knife, sheets were frozen hard when a hot water bottle burst. Nursing was very heavy, as many of the patients had gas gangrene, the operating theatre worked around the clock, with many patients lying on stretchers until another patient was evacuated to England or Australia.

The first ACC's in France did not contain any female nurses, as it was considered they would be too close to the front line. Eventually medical staff realized that a CCS should provide initial treatment for the wounded, instead of merely passing him down the line. The change in function was possible only if surgical teams of doctors and nurses were available closer to the front line. Old prejudices had to be forgotten. The CCS became an important part of strategic planning.[52]

Three Australian Casualty Clearing Stations were sent to the Western Front. No 1 ACCS took over the École du Sacre Coeur at *Estaires* in April 1916 and during the following thirteen months it was to be part of several major allied offensives. Butler, referring to Colonel Dick's diary on the contribution of this

The Matron and Sisters of the 3rd Australian General Hospital, near Abbeville, in France, photographed on June 23rd, 1918. AWM E2578

ACCS stated,

> This CCS, like many others, is virtually also a stationary hospital — owing to the long continued stage of stationary warfare such as has been necessary from trench warfare which resembles for evacuation purposes all kinds of siege warfare. No unnecessary moves of similar CCS's are carried out — the damage to equipment of a highly technical kind is always considerable in any move, and CCS's now possess such.(53)

No 1 ACCS also played a significant medical role in the Flanders offensive of 1917. At *Bailleul* some 50 marquees were erected and more than 300 yards of planking laid down to traverse the mud. During the Battle of Messines the CCS was heavily bombed and, among others, Sister Rachel Pratt was wounded. Consequently the ACCS was withdrawn to a relatively safer place near *Oultersteene*, where a tented and hutted hospital was erected to accommodate 750 patients.

No 2 ACCS established its reputation at *Trois Arbres*. This ACCS had to take over an undeveloped site at Trois Arbres near Steenwerck and in a matter of weeks had to be ready for the wounded coming in from the Battle of Fromelles.

> The laying out of the station; the erection of marquees on very rough ground; the making of an in and out road consisting of a plank track; the building of bridges and the sinking of wells to ensure a good water supply — involved very hard work. Two large tanks were installed on raised platforms giving a sufficient fall to enable water to flow freely where needed. A large Nissen was erected to serve for the operating theatre and x-ray department. Two long rows of double marquees were erected, and duckboards placed through the camp.(54)

Sister Ada Smith was nine months with the 2nd CCS at *Trois Arbres*, which according to her letters was but a collection of shacks, Nissen huts and tents, all connected by 'duckboards' so there was no mud or

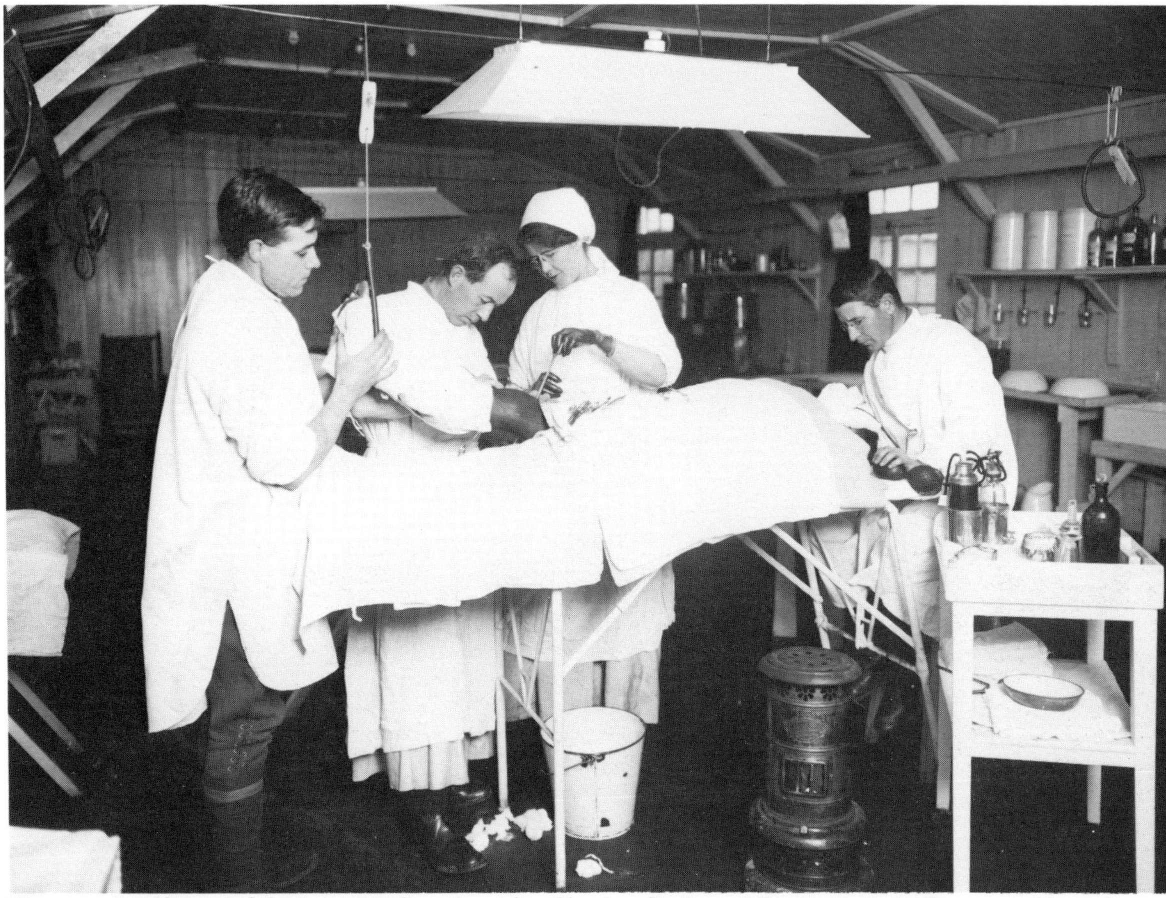

The operating theatre of the 1st Australian Casualty Clearing Station, at Outtersteene, in France, on November 23rd, 1917. AWM E1304

slush while you remained on the boards. In the hospital wards she heard the latest accounts from the battlefield — and the latest rumours. The patients always spoke about what would happen 'when we take that ridge'. They talked about it so often it became almost a laughing matter, but in all seriousness there was always another ridge to be taken before the enemy could be driven back. This was the period prior to the Battle of Messines, a ridge which was to mark a turning point in the war. In anticipation of large casualties, the facilities of the Casualty Clearing Station were enlarged. Sister Smith described the scene.

> Our dressing room where patients were admitted to have their dressings seen was enlarged to such an extent they were able to put 8 tables in it. Our theatre was enlarged to take 6 tables. Tents sprang up everywhere and duckboards were laid to them. All necessary equipment was installed in keeping with a busy ward. Primus stoves were issued and wards were given an extra primus in case one would not go. Then splints, bandages, pyjamas, blankets and the hundred and one things needed in a ward began to arrive, until we guessed that that last ridge was about to be taken. All bed patients were sent to base. Our nursing staff was increased from 7 to 20. More doctors, orderlies and stretcher bearers arrived . . .

No 2 ACCS was heavily committed during the Battle of Messines, when all members of the medical and nursing staff worked round the clock tending the wounded. On the first day of the battle 2,067 wounded were admitted and 2,000 evacuated in 24 hours! This CCS was also hit by enemy bombs suffering considerable damage and numerous casualties. Blackouts were rigidily enforced and sandbagging of tents was increased to give greater protection. Sister Margaret Goggins (Warwick) after long service with 2 AGH

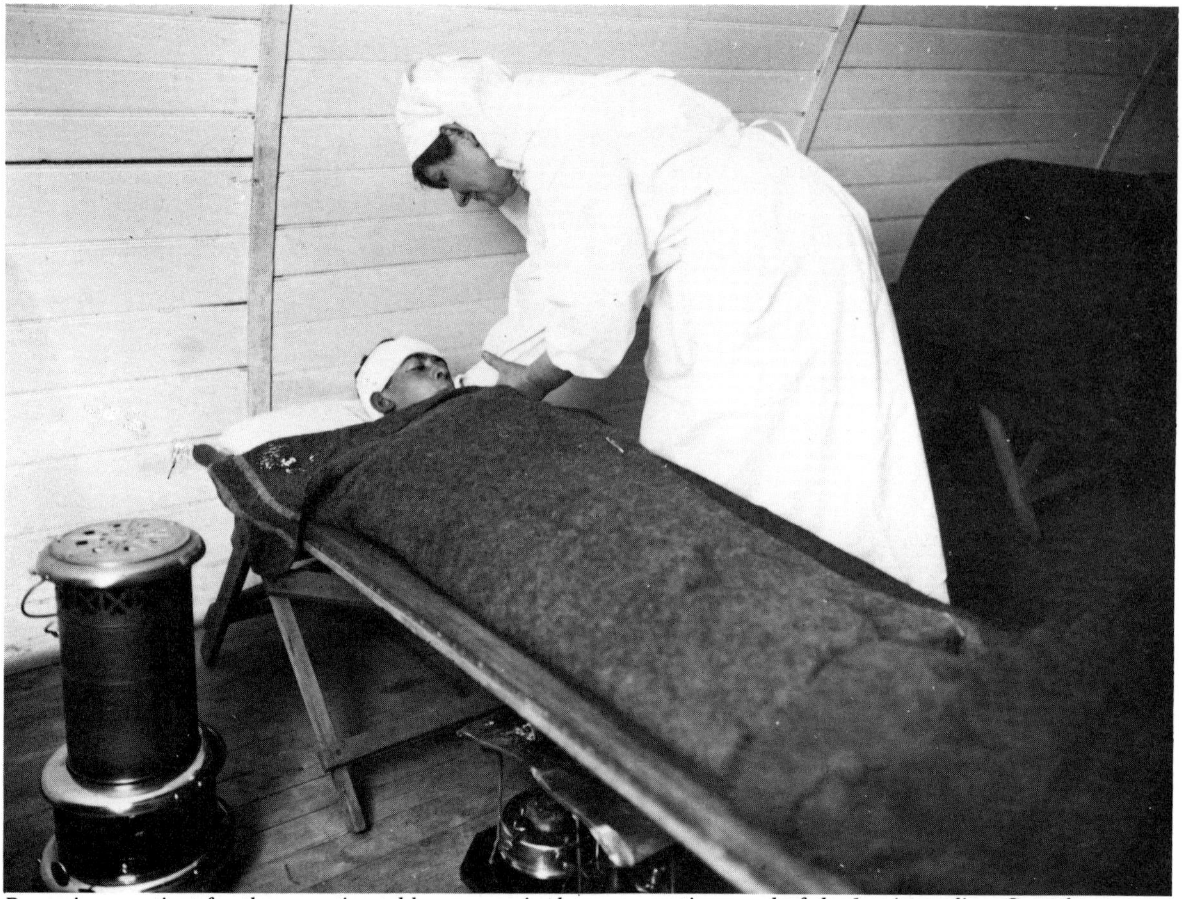

Preparing a patient for the operating table: a scene in the pre-operation ward of the 1st Australian Casualty Clearing Station, Outtersteene, in France, on November 24th, 1917. AWM E1305

in Egypt and France was one of those selected to go forward to the 2 ACCS. Unfortunately she was one of those wounded by shrapnel when the casualty clearing station was hit by enemy shells.(55)

Life was uncertain for everyone working in a CCS less than five miles from the front line. Shelling was a constant danger, particularly as the Germans were trying to locate a 12 inch gun kept on the railway line close to the CCS and camouflaged as a truck. Staff and patients had many an anxious moment as shrapnel burst close by. Nurses who had had a strenuous day in the wards found little rest at night as the bombardment continued. The Germans had also begun to drop bombs from aircraft (referred to as 'Taube' by the troops). On July 22nd 2 ACCS was hit by bombs, killing 4 patients and wounding 15. It was in this episode that Queensland nurse, Sister M.J. Derrer was awarded the Military Medal.

Another Queensland Sister who was decorated for bravery while serving in a British CCS at this time was Sister Eileen King, attached to the QAIMNS. According to newspaper cuttings in the possession of the family the circumstances for which she received the Military Medal are as follows:—

> Eileen King was serving in a tented field hospital in France when it was struck by a stick of bombs. She had part of her left thigh broken away and received other serious wounds but broke no major blood vessel. She remained on duty and managed to get her wounded out of the burning tent. Soldiers who knew her described her as one of the bravest women they had ever met.(56)

Sister Ada Smith's account of the events of those hectic days and nights at the CCS is worthy of record.

> At 7 am we heard the Ridge had been taken with "light casualties" and shortly afterwards the first wounded began

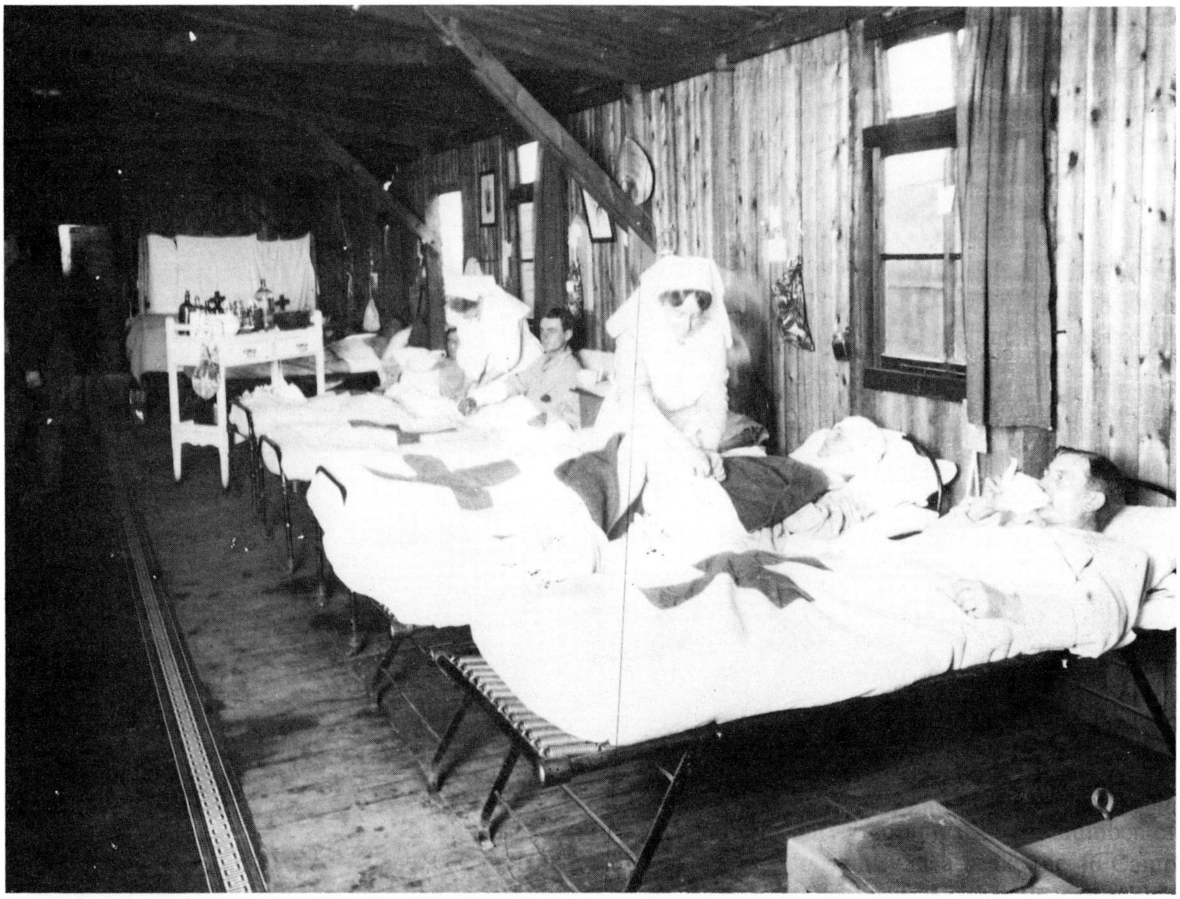

A ward in the 2nd Australian Casualty Clearing Station near Steenvorde in Northern France on November 30th, 1917. Most of the patients were wounded in the fighting of the third Battle of Ypres. AWM E4623

to arrive. I couldn't see where the light casualties came in, as all those strong, healthy men came in dead, dying, unconscious or moaning. We had a large marquee erected at the entrance to the CCS, much like a circus tent, where the wounded were examined by the MO. Their blood-stained khaki was cut away to allow inspection and dressing. They were then carried away on stretchers for further treatment, if necessary. In some cases, such as abdominal wounds they went straight to the operating theatre. Then we had a post-operative ward, a pre-operative ward, resuscitation, chest, abdominal, jaw, multiple wound wards and lastly a moribund or dying ward.

We had then no time to think of the hundreds of casualties, we only knew that work was waiting to be done everywhere and that men were suffering. Everybody worked many hours at a stretch. How could anyone go off duty with dying and wounded men all waiting for their turn for treatment. Everyone tried to snatch a little sleep whenever possible, just enough to keep going.

Often we were successful, sometimes our best efforts failed, sometimes it was hopeless from the beginning.

It was certainly a far cry from the quiet, relatively unexciting period of nursing at Warwick, 'back home'. Yet nurses at Messines faced the same heartbreaking problems that Florence Nightingale and her team faced at *Scutari*, the seemingly hopelessness of the task of saving human lives. Hour after hour, day after day, nursing the badly wounded and dying was a duty that called for the highest qualities in nurses — physical endurance, devotion to duty, dedication to the cause for which they had enlisted. There was no shrinking, no turning back, no room for the faint-hearted. Australian nurses at Messines exhibited the finest qualities of nursing. No one could have put it more graphically than Sister Ada Smith.

A group of Australian nursing sisters at the 3rd Australian Casualty Clearing Station, Grevillers, March 1917. AWM A2283

For some time I had charge of the Moribund Ward. It was a heart-breaking, hopeless place, the only time since leaving Australia that I felt things were almost beyond me. Rows of dying men, mostly Australians and New Zealanders, nearly all head cases and unconscious or semi-conscious or else raving in delirium and tearing their bandages off, none likely to live more than a few hours at the most, each on a mattress on a stretcher, mostly in their khaki. Every hour or so someone dying and being taken away, only to be followed by someone else, again and again. So many of them had discs with their name and rank on one side and their next of kin on the other, seemingly asking for some one to write to their people and let them know how and where they died.

The Sister-in-Charge of No 2 ACCS, (sometimes called Head Sister and occasionally Matron) was Sister 'Connie' Keys whose reputation in the AANS had already been well established. Her report on the actions of the unit in 1918–1919 are a model of detail and personal modesty.

Sister Keys wrote:

On 9 Feb 1918 I was sent as Sister in Charge to No 2 ACCS at Trois Arbres, Steenwerck, about 3 miles from Bailleul.

On Mar 10 the Germans began their attack and things began to look serious. Bailleul was badly shelled and some shells fell on the Steenwerck Railway Station. Late that night we received orders to pack and move out. Next morning, a freezing one — the patients were speedily evacuated — and at midday the 15 Sisters left by motor ambulances for No 10 Stationary Hospital at St Omer. A few days later we set up again at Ana Jana, outside Hayebrouck. Work commenced with a rush on the 9th April. The weather was very wet and cold, making nursing very difficult. The theatre and the dressing room worked at the highest tension while the evacuation of the wounded never ceased.

On the 12th April the area was again shelled and orders were again given to move back to a safer area. Things seemed hopeless, with wounded still pouring in. Very reluctantly we moved off. This time the conveyance was an old London double-decker bus. The road was bad and boggy, refugees were hurrying, carrying all they could of their possessions. The scene was heartbreaking. Here and there was an ambulance not knowing where to take the wounded. We arrived at No 10 Stationary Hospital as night fell to find 2 Australian Sisters and others from several CCS's there, all refugees like ourselves. That night we were again badly bombed, many people in St Omer were killed or wounded and we did not expect to come through safely.[57]

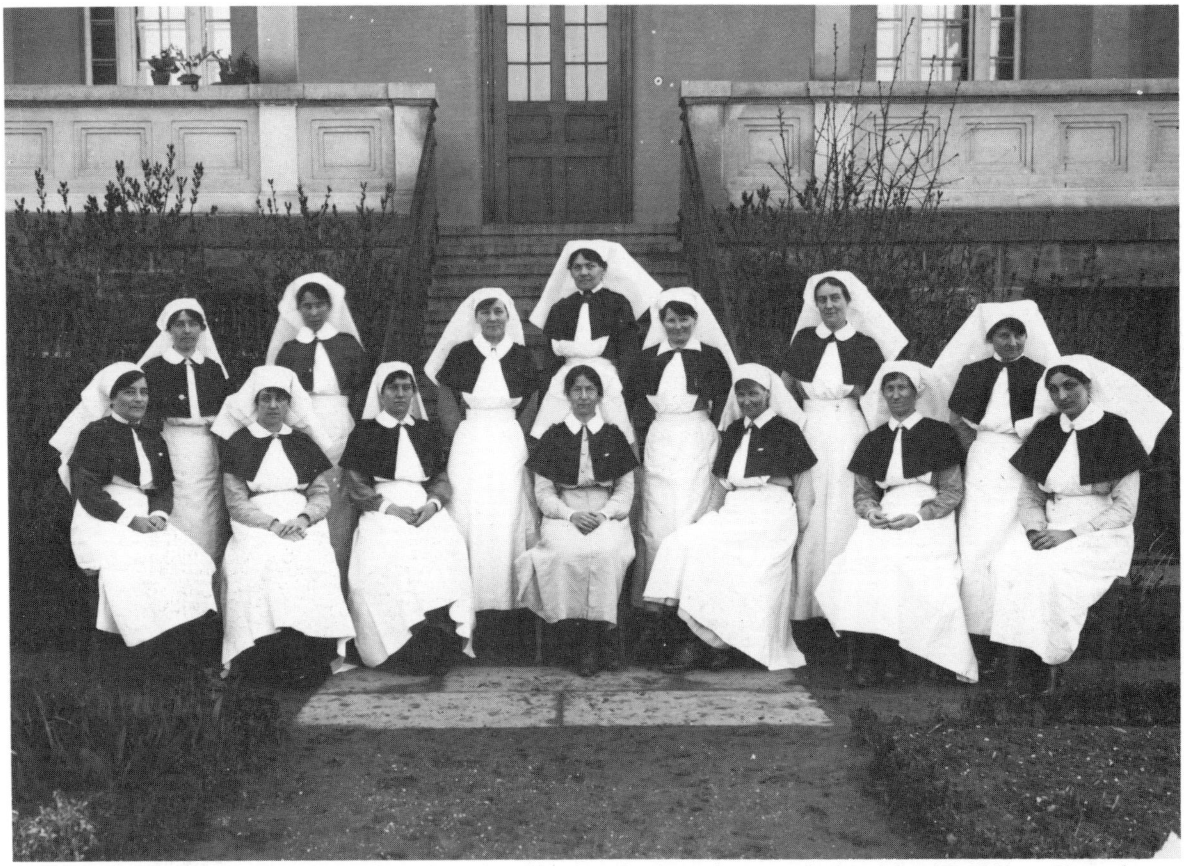

A group of sisters of the 3rd Australian Casualty Clearing Station, photographed on May 20, 1919, at Eustachen, on the Rhine, in Germany. AWM E5117

So 2 ACCS continued to follow the action, as the Germans were pushed back — *Klenderques. Ana Jana* again, *St Kenant, Estaires, Tournai, Ath* and on to the Armistice.

In her reports, Sister Keys commented on several aspects of nursing, the problems, the innovations, the achievements, the disasters.

Fowler Beds:

These are frames placed under the mattress. For patients suffering from penetrating chest and abdominal wounds and from pneumonia these beds are invaluable. With them there is none of the incessant propping up on back rests and pillows as in the old way. Once the head and knee pieces are raised, the patient lies comfortably and nursing is much easier. But for these we would have lost many cases.

Gassed patients:

We received many gassed patients. Inhalations of creosote were their chief treatment. The inhalers were made of perforated zinc with a piece of wool inside over which the creosote was poured. Many were so badly burnt by the gas as to be blistered from head to foot. In spite of our efforts several died.

Burns cases:

On 27 December there was a big ammunition explosion at the railway station and as a result of it many were admitted to the hospital, terribly burnt. Not much of their bodies appeared to have escaped. They were piteous sights and suffered terrible agonies. During the next few days 4 died and but for the ceaseless nursing care many others would have died too.

Australian sisters enjoy a few minutes rest before returning to duty. AWM C4083.

Matron Keys could not speak too highly of the work of the Sisters at the CCS, always under threat of attack, always under pressure to nurse the wounded, always under tension, not knowing when to pack up and move on.

> Of the work done by the Sisters I cannot speak too highly. In good times and in bad, they were always the same — cheery and full of energy.

At the end of this period Sister Keys was awarded the Médaille des Épidemies (en vermeil) in recognition of the valuable work done by this unit not only in caring for the wounded but for the civilians in the villages.

But there were other Queensland Sisters whose names appear regularly in the movement orders, as they were hastily posted to where the action was, to fill the gaps, to help out here, to meet an emergency there. They built a reputation that they could be relied upon to do a first class nursing job, under fire. These included Sisters Frances Walpole (Brisbane), Elizabeth Parker (Rockhampton), Mary Fisher (Rockhampton), Elizabeth Jessup (Brisbane), Louisa Parnell (Rockhampton), Jessie Kennedy (Rockhampton), the Macdonald Sisters and the Toft Nurses.[58]

No 3 ACCS achieved a measure of distinction by being part of a plan by Surgeon-General Skinner to take the surgeons and nursing staff even closer to the front line, to give more immediate treatment to the wounded. Unfortunately the site chosen at *Brandhoek* near the main *Ypres — Poperinghe* road, only 5 miles from the front, was too close to legitimate military targets. After much bombing and shelling it was moved back to 'Nine Elms', five miles behind *Poperinghe*.[59]

Elsie Grant (Clermont) was with 3 ACCS when it came under fire. Her sad and bitter letter home described the events.

> Those brutal Germans deliberately shell our hospital with all our poor helpless boys in it, but really God was good to us. There were only four killed. Our hospital now is a total wreck. Our tents were dug down three feet into the ground and sandbagged and even then they are riddled.

Sister Grant then goes on to give some clues as to where they were. Presumably this letter eluded the censors!

The sisters mess at the No 3 Australian Casualty Clearing Station at Eustachen, on the Rhine, in Germany, photographed on May 20, 1919. AWM E5111

Our hospital was in Belgium near a town that begins with the second last letter of the alphabet and it has five letters in it. You'll have to get your map out to find out where we were.[60]

No 3 ACCS first of all went to England with the 3rd Division AIF in July, 1916. A few months later it was sent to *Gezaincourt* in France to replace the No 11 British CCS. Its first patients were walking wounded from the Battle of Beaumont Hamel. In the first 48 hours 1,516 men were admitted and 1,400 evacuated, while from 14th to 30th October 1916, 4,130 were admitted with 413 operations.

Trains ran into the little siding loaded with wounded from the Somme battlefields. The weather was terrible and the mud near the line so bad that many came in for treatment literally covered from head to foot . . . Sometimes 3 and 4 trains per day with loads of 200–400 cases . . . In December, the sick began to come in numbers — trench foot, rheumatism, pneumonia. In spite of the terrible weather, the winter passed very pleasantly and with plenty of blankets and stores we managed to keep the cold out.[61]

During this bitter winter nurses had to endure very primitive conditions, particularly in sleeping and messing arrangements. Those who had come straight from the heat of Egypt were not prepared for the sudden change. At first stoves were at a premium, even in the wards. Grace Wilson, who at the time was acting Matron-in-Chief, AIF, investigated the nurses' complaints and subsequently found some improvements had been made.

Sister Mary Derrer, MM.

> In the sleeping huts it would be desirable to have more stoves. None of the hospitals have sleeping huts entirely without stoves although in some cases there may be only 1 or 2 in a hut containing 12 bedrooms.[62]

One of the most exciting experiences of nursing in France during this period befell well-known Brisbane nurse, Sister Florence James-Wallace who was temporarily attached to 61 British CCS during the German counter offensive of March 1918. Her story is well documented.[63]

> I was on Night Duty with 2 other Sisters for the Hospital.
>
> At about 4.15 am a terrific bombardment commenced, like continuous thunder rolling, then I could distinguish shells screeching through the air and guns going off with a deafening crash the whole place shook and trembled. I had never heard anything like it before, felt quite excited, the Orderlies said 'it's something big, Sister', I went out and found another Sister; she was quite mystified too as to what it all meant, we could not see the usual flash on the horizon from the guns the fog was so dense. The noise continued. The patients were unconcerned remarked that 'those were heavies' when a shell whistled and crashed. Seemed quite content as they were out of it all and in Hospital. We got some wounded in, but they talked of a British raid and couldn't tell us anything.

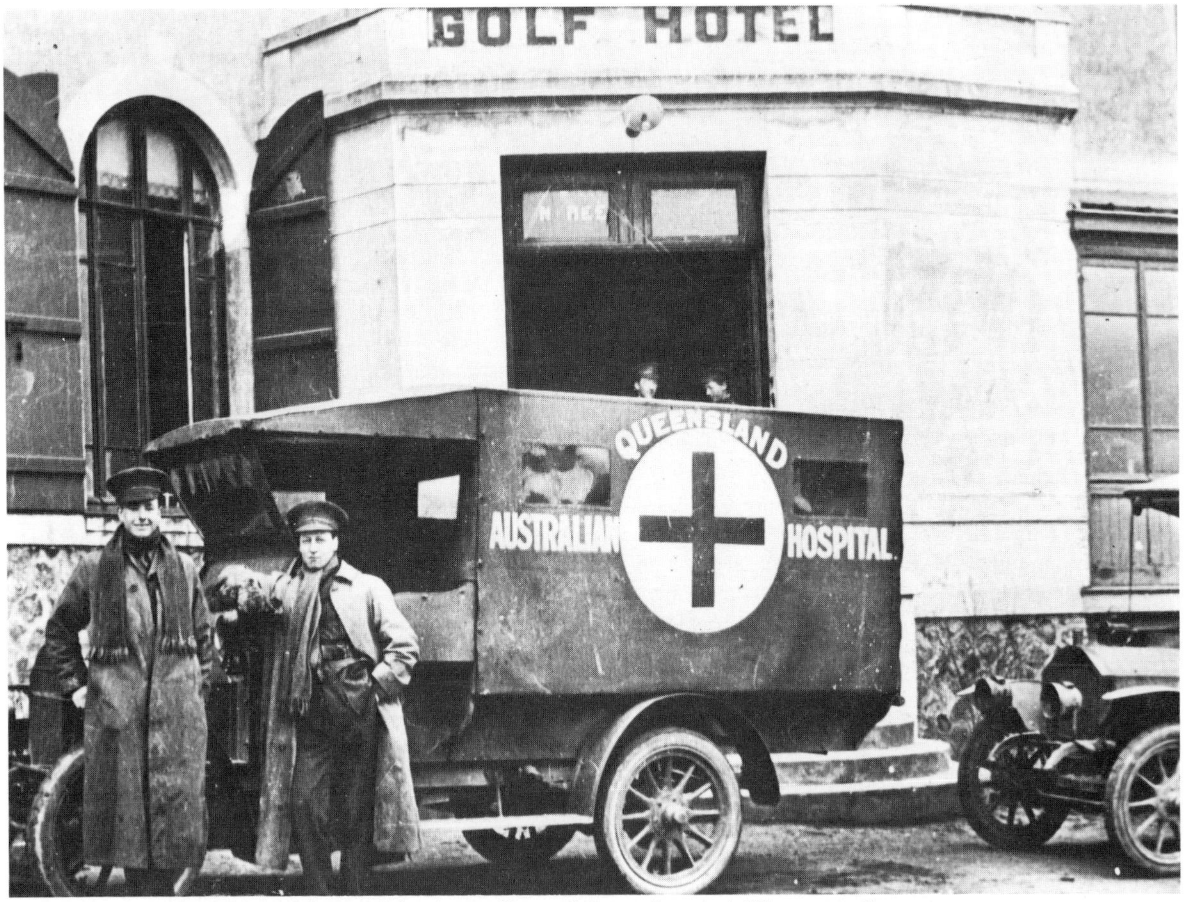

'Queensland Ambulances in front of the Australian voluntary hospital Wimereux, France.

I realized it was something unusual when I saw the M.O.'s appearing at 6.30 am instead of 9 am with tin hats on, and tearing round very excited. They began marking up the field cards, and getting the bedded Wards empty by filling the hangars with walking patients. Went to bed about 10 am with everything up side down in the Hospital. Up about 4 pm to find the nosie even worse, and 3 Anaesthetists for our C.C.S., and 18 refugee Sisters from 41 C.C.S. — which was in the direct line of fire. Everything in a turmoil and buzzing with excitement and patients everywhere.

Slept soundly, too tired to be disturbed by noise.

About 2 pm wakened to the sound of 'Girls! Get up quickly, you have to be dressed in 10 minutes, the train goes in 20. We have to leave everything.

Who said so? We must pack. What is the matter? But we tumble out.

Miss Baird comes in to our Nissen Hut, 'Are you up girls? The Germans are advancing, we have to leave everything. Train goes in 20 minutes, take what you can carry'. Exit all of us — with suit cases, boots, rugs, haversacks, dorothy bags in our arms, to see the rest of the Sisters in the same plight waiting at the Mess Hut. Dinner half eaten, I feel jolly hungry. Bright sunshine, clear sky. Troops, waggons, lorries, ambulances, gun-carriages, pack-mules all clattering down the road past the Hospital. One couldn't take in all that was happening. Guns still crashing and shells whistling. Two big naval guns of ours near by, made a good deal of noise.

Some of the M.O.'s meet us and we tramp down to the Station, feeling very disgusted at being sent off when we feel the patients need us. We ask what is to happen to them. Why can't we stay with them? Everyone asks questions and no one answers.

The line round to the Hospital is hit and 3 men killed where we went across from the Station. It is now nearly 6 pm and no sign of our train. The C.O. of 41 C.C.S. suddenly appears with the news that he has procured 5 lorries to take us all to Rosières. Once more we pick up our baggage helped by some men and take it across to the road.

6 Australian General Hospital Kangaroo Point, Brisbane, 1918.

Some Irish Officers of the Ulster Division talk to us as we board the lorries. They shout to the men to take cover as we see a fight between some of our 'planes and the Bosche close by.

Bright moonlight heard bombs not far away, it got very cold, we were covered in dust. We went through villages, nothing but a heap of bricks, and a few stone walls, a few feet high, trees that looked like sentinels, just the charred trunks and a limb or two standing. Got up to 47 C.C.S. Rosières at 10 pm. Equipment of other C.C.S.' piled about all round it. Patients pouring in. Received very kindly and given tea and bread and butter. Found an old friend there on Night Duty and camped in her bed.

Was told I would go on Night Duty, so was off to bed when the call to be 'Ready in 20 minutes to go' once more reached me. Miss Baird was to take 7 Sisters to a Railhead. Four of us were Australians and one South African and a Scotch and an English Sister. We thought the Colonies were well represented! The Colonel of 61 C.C.S. Miss Baird and 5 of us departed in an Ambulance, the other 2 followed in a lorry with our belongings. We found Villers-Bretonneux was our destination. Got out at the Railhead Siding.

24th. Got to the shed about 7 am to find about 500 patients, stretchers and walking cases. We fed them all and dressed the worst wounds. Abdominal and Chest cases we brought into the hut, the rest had to stay in the open. The Orderlies were awfully good, mostly Irish boys. They helped to cut up bread, open bully beef tins by the hundreds, biscuit tins; and made up the fires and had boiling water and helped with dressings. Others were busy making boilers of tea and opening tins of milk. Wounded kept coming in all day. One train went out at night taking walking wounded. Got to bed late!

Found we were to be an Entraining Centre and look after the wounded till they got on the trains. We were given a wooden hut which was being used by the sentries of a huge Petrol Dump which was a few yards away. The shed had some forms and 2 stoves in it. Major Grenfell paid us a visit; he had been in charge of the advance surgical supply depot in Ham. He sent us some Panniers and wool and gauze which we proceeded to cut up and get ready. We were supplied with plenty of wool, gauze bandages, safety-pins and splints, but very few instruments and lotions. Pot. Permang. and Iodine we used for everything.

25th. Got to the Shed to find about 8,000 wounded. Fed them all and went on dressing as hard as we could, more kept coming in. The space on either side of the hut and facing the line was covered with stretcher cases for hundreds of yards; the back bit, the hut and field kitchen were for the walking wounded. Col. Turner and Capt. Marshall were the only M.O.'s, they had some tents about 12 put up, two double dressing Tents. So three of us dressed the patients there, having the worst stretchers carried there to be dressed. The worst cases we kept in the Hut. Two of us were kept busy dressing outside and feeding the new arrivals. Later — More M.O.'s and Orderlies came. We had a double Tent full of Officers some very badly wounded. About 10,000 had been through our hands and still they were pouring in in lorries, Ambulances, etc. Very few trains getting them away.

At 12.30 we were sent up to the school to pack, had a hurried lunch at the Corner Café, finished packing the things in the Dressing Tents, heard a shout that the train was going out, ran down the hill, passed through the Hut where our very bad cases were as we had left our haversacks there. We were too hurried to think of the effect our leaving would have on them. I will never forget the expression on their faces when they saw we were going 'Oh they are leaving us', 'They are going'. I heard one man say. I went back to tell him we were going on a truck train and they would be going as soon as the Hospital train arrived. It didn't seem to comfort him much. They looked as if they thought their last hope had gone, poor things, we hated leaving them, and it made us realize our being there meant more than the actual work we did.

The evacuation of the hospitals to the hospital ships operating out of Boulogne and the servicing of the hospital ships while they were in the harbour demanded tireless service from the medical and nursing staffs, 16 hours a day being a normal tour of duty. Often the hospitals would fill and empty three times in twenty-four hours. Matron Gould in her report commented favourably on the work of the Australian nurses.

> The staff was indefatigable — it was only the thorough practical training they had received in Australia that enabled them to cope with the amount of work. Day in and day out they worked unselfishly and cheerfully. The constant rumble of heavy guns, the occasional air raid scare, the train loads of coo-ees passing up and down near the hospital, all contributed to the general efficiency.

Sister Gertrude Ada (Trudy) Nye was one who joined the QAIMNSR from the Brisbane General on 6 April 1915. It was initially an expensive operation as she had to borrow £20 to purchase various effects for the journey to England. In her diary she related how her patients 'clubbed together and bought a travelling rug and ladies companion comprising scissors, needles, thimble, button hook, nail files, all with pearl handles mounted in green velvet and in a very nice case'. As it was some weeks before they received their first army pay, the Australian Sisters had to have independent means in order to pay their way in the QAIMNS.

For Sister Nye, as with other Australian nurses, the first contact with battle casualties came as a terrible shock. 'Had my eyes opened at the sight of such ghastly wounds', she wrote in her diary. 'Never seen anything like it before — mostly shell wounds, very difficult to dress'. However, she soon accommodated herself to the situation, occasionally working in the Officers' Ward, occasionally with the 'Tommies'. Only rarely did she come across any Australians, which threw the Australian and Queensland nurses at No 3 British General and No 16 British General Hospitals together. Her diary continues with a very vivid account of the work at the hospital.

> Sept 27. Another big convoy in the night and another in the morning. Some work! The wounds are ghastly messes. Such heart breaking work seeing what war does to humanity.
>
> Sept 28. Still another convoy. Stretchers now in corridors, Officers mixed up with 'Tommies'. Capt Woods is on the verge of collapse, operating all day and well into the night. Everywhere poor battered humanity. 16 hours duty-must get some sleep.

These Sisters had to become used to tragedy, as gas gangrene and femoral haemorrhages carried off so many young lives. But it never hardened them to the point where they lost their femininity, for it was often this factor that brought some sense of normality to the lives of the sick and wounded. Sister Nye had frequently to decline gifts from grateful patients for it was 'against the rules'. On one occasion she relented.

> Lt B....... with tears in his eyes produced a little box containing a pretty blue enamel powder compact and begged me to accept this tiny token of gratitude for all my care and attention. He was so upset when I said I did not accept presents just because I had done my duty. In the end I simply had to accept it. It is a beautiful thing.

Then it was back to reality. Sister Nye and 5 other members of the QAIMNS were sent forward to No 4 British CCS forward from Abbeville, close to the front line. Their work never stopped, as they received the wounded from the muddy trenches.

> They were so caked in mud you would have thought they were niggers. Their wounds are septic, some have maggots in them. A good many amputations through gas gangrene setting in. Plenty of deaths. It's heart breaking work. Very tired.

Sister Nye commented on the rules made for nurses that they were not to go out with males, especially at night. There were relatively few women in what was essentially a man's world and army rules were designed to protect them from possible moral and physical danger. The rules were inconsistent, as Sister Nye pointed out in her diary. She noted the casual way in which the army effected transfers.

> When orders for transfer came through the Matron sends you off, sometimes alone and sometimes in the middle of the night and nobody seems to care what happens to you as long as you turn up for duty within a week!

On another occasion she was sent off from Harefield to London on her own to buy her overseas kit for France. That was bad enough for her but when she arrived back at the village station that night there was no conveyance waiting for her. So, in her own words, 'I set out and walked the 3½ miles to the hospital. As I was carrying 24 pounds on me I was glad when I arrived safely.'

In so far as treatment was concerned the Sisters were under the direct control of the medical officer in charge of the ward. While it was expected of them that they would carry out all the nursing duties expected of them, they had to carrry out without question orders given to them by their superiors and in the case of medical officers, this related specifically to the treatment of patients. Occasionally some may have forgotten — or in the pressure of time just did not get round to it, but they never questioned it or disobeyed it. Sister Nye, being of independent mind, was an exception. On one occasion she was rostered for duty on a hospital ship at Boulogne. The Sisters were responsible for bedding down the patients and giving them the preliminary treatments prior to the voyage to Britain. Since the *Anglia* had been torpedoed in the channel, the Sisters did not accompany them.

Sister Nye was very critical of one medical officer who ordered patients aperient pills prior to departure. She knew there were few facilities on board and few staff for the short journey, so she decided not to give this treatment. Eventually the doctor discovered she had not given the pills. 'He glared at me, turned on his heels and walked off', she wrote. When she discussed this with the Matron, the Matron agreed with her. No more aperient pills were ordered by that doctor!

Sister Nye was with a group of QA's sent to Genoa, as part of the medical support for the British campaign in Italy. Here for the first time she came across some American medical officers and she was not very impressed. She referred to one as the most ill-mannered doctor she had ever met. She clashed with him because he did his ward rounds chewing on a cigar! The account in her diary reads like some theatrical scene.

> Dec 19. Waited in patience for him to do his rounds. As he was entering the ward I very nicely asked him to leave his cigar in the ante room till he had finished his rounds. He stood and glared at me. I lifted my hand to remove it from his mouth and smiled as sweetly as I could, remarking that it would be quite safe and ready for him after he had seen all patients. Without a word he strode back into the ante room and laid the cigar on my desk — still glaring at me. I said, 'Thank you', and followed him into the ward. I had won that time but may not the next time.

One of the little known stories concerning the war in France and the service rendered by Queensland nurses has been uncovered in the documentation concerning Sister Susan Hughes whose family was well-known in the Warwick District. She trained at St Vincent's Hospital, Sydney 1910–1914 and served at the Randwick Hospital 1914–1916. When the war in France was at its most desperate stage, the French Red Cross (or the International Red Cross) is believed to have contacted the Australian Red Cross Society for some 20 Australian nurses to volunteer to help nurse the French wounded. Sister Hughes was among a group of 20 who sailed from Sydney on the New Zealand ship *Rohue*, in July 1916 and who served for three years with the French. They were not members of the AANS. One official record discloses that during this time, their very low salary was supplemented by donations from the Australian Jockey Club of Sydney and the Lismore (NSW) branch of the Red Cross Society.[(64)] Unfortunately the bombing in World War II and the NSW floods of 1974 have destroyed valuable evidence.

In 1917 with the 'Paeschendale push' in full swing surgical teams, consisting of a surgeon, anaesthetist, Sister and orderly were sent forward into Flanders. This was the period of gas attacks and Sisters had to be trained for that emergency. They had to learn to put on their masks within 6 seconds, they had to experience life in a gas chamber and in a gas filled trench and they were disciplined to have tin hat and respirator at the ready at all times. They had to learn to nurse gas victims, many badly burned with mustard gas and with smarting eyes. They had to assist patients to don face masks each connected to a central oxygen pipe running along between rows of beds, head to head. It was an eerie and awesome sight. Few

Nurses of the 2nd Australian General Hospital, photographed outside their quarters at Boulogne, in France, on May 11th, 1918. AWM E2112

would have envisaged at the beginning of the war that these 'nice girls' whose role was to nurse the sick and wounded far behind the lines would be so involved in the front line action, but the nurses carried out their tasks courageously and unflinchingly.

England

The War Office in London had the difficult task in the early stages of the war of integrating the medical requirements of the troops from the dominions with the established and prescribed organization of the Army Medical Corps in Britain. This was not easy, as each dominion had derived its own system for the sick and wounded and they had the additional problem of invaliding 'home' the unfit over many thousands of miles of dangerous waters. When Surgeon-General Williams, DMS, AIF, arrived in Britain in December 1914, he found most Australian troops had been diverted to Egypt and the Dardenelles and subsequently medical units, including the AANS followed them. Nevertheless he established effective liaison with the British authorities, working from his base in the High Commissioner's Office. He passed on to the Australian authorities the request of the British for 200 Australian nurses to volunteer for service with the QAIMNS. In the beginning it was assumed the Australian sick and wounded would be nursed in British military hospitals. It so happened that a patriotic Australian offered his home *Harefield Park* in Middlesex as a convalescent home for Australian soldiers, setting in train a pattern of auxiliary hospitals which were to become a feature of Australian medical treatment in Britain. These became known as Australian Auxiliary Hospitals, some small, some for officers only (*Digswell Place* and *Digswell House, Welwyn* and *Moreton Gardens*) but those at *Harefield* (No 1 AAH), *Southall* (No 2 AAH) and *Dartford* (No 3 AAH) were much larger and well-known to Queensland members of the AANS.

Sisters' quarters at the 2nd Australian General Hospital at Boulogne, in France, on May 11th, 1918. AWM E2175

No 1 AAH at *Harefield* grew to 1,000 beds as casualties flowed in from France. Mostly these were surgical cases and it was possible for special attention to be given to amputees before the return to Australia. It also provided for serious medical cases and for convalescents. The latter were important because of the shortage of manpower and it was there that a decision was made whether the patient came within 'the six-month's rule'.

No 2 AAH at *Southall* soon specialized in the fitting of artificial limbs, an area which became of increasing importance after the Somme offensive. No 3 AAH at *Dartford*, in a fine act of cooperation and co-ordination between DMS, AIF and the War Office, acquired a fine hutted hospital, previously used for the treatment of infectious diseases. Before long it held 1,400 patients and although its role was auxiliary, it developed a special interest in the surgery of nerves and the treatment of psycho-neuroses. One of the advances in surgery at this time was the establishment in 1917 of an Australian facio-maxillary unit attached to the Queen Mary Hospital at *Sidcup* in Kent, for the treatment of war injuries to the jaw and face. While the Hospital was at first imperial in character, there developed an Australian section with its own surgical and nursing staff. While many Australian nurses served at *Sidcup*, the greatest distinction came to Sister Jessie Kennedy, subsequently Matron, who was awarded the RRC for her exceptional service to nursing. The award was bestowed on her by Queen Alexandra, widow of King Edward VII. Another who served with great distinction in these hospitals was Sister Eunice Paten, Acting Matron at *Southall* and Deputy Matron at *Dartford*. Others who are mentioned in these hospitals include Sisters Loosemore, Andrews, Toft (Edith) at *Harefield*, Sister McClelland at *Dartford* and Sisters Farquhar and Avenell at *Southall*[65].

Nurses of the third Australian Casualty Clearing Station waiting, at Coblenz, in Germany, for a return boat to Eustachen, further down the Rhine, where their Hospital was established. AWM E5239

Many of the AANS wrote home feelingly about their nursing experiences in these hospitals in England. One of the most comprehensive is contained in the letters of Sister Avenell, writing to her mother in Gympie.

> All my boys are either winged or legs off, shoulders blown away, big head wounds, but nearly all healed up and just little pieces of dead bone keeping them from healing up altogether. They are such fine fellows. Some have only had 12 operations.
>
> We are getting more stumps every day and now have about 300 without legs and arms.
>
> I have about 30 leg stumps to dress every morning and about 40 beds to make. The orderly helps me but it is an awful rush.
>
> One of the nurses is to be married in the morning from here and Matron is giving her a morning tea. All the boys are standing with an archway of crutches on their one leg!
>
> Last Saturday afternoon another Sister and myself took the real stumpy's (boys without legs at all) to a cricket match at Uxbridge.
>
> Very busy time in surgical wards, with very special operations, such as removal of bulbous nerves from stumps, artificial wrists, etc. That is most interesting when a boy has his hand taken off. The Drs can now operate on the wrist and cut the bone in half making a joint. Of course the skin has to be cut back to get at the bone but it heals up and then he can use the hand.
>
> I am sorry for Australia for it will be nothing but broken down men after the war.

What stands out was the bond of sympathy and compassion with the wounded — and the bravery of the Australian troops in nonchalantly making light of their injuries. Sister Avenell continued,

The last draft of Sisters leaving France for demobilization. Principal matron Grace Wilson, C.B.E., R.R.C., 3rd A.G.H., is shown sideways to the camera. AWM C4820

We do love our patients. They are such bricks with their awful wounds. One of my patients was hit with shrapnel in five different places, his eye blown out, left arm blown off and other wounds on the back and body. He is a brave fellow. He says he is not too bad but thinks he got more than his share.

. . . back to my soldier boys. They are such hard cases and do tease me. Tonight they were all imitating animals, cats, cows, calves etc. I said I thought I was nursing heroes, not animals at the zoo, which made them worse of course.

I have 156 dressings to do about 30 one-armed men. Fancy some don't need dressing every day and all are up walking about. The "legs" are in the next ward and they are so funny, show off how they can hop on the one leg and frighten the heart out of me going past them. If they should fall they come down on their stumps.(66)

Sister A.E. Cocking described how they attempted to cheer up the blinded soldiers and the amputees:

Many blind boys were in the wards and it was a novel experience for Sister Sheehan and myself to take them to the movies and afterwards to tea at the Mia Mia Cafe. We had to describe the pictures to them, they enjoyed the music and their tea, but we had to be back by 6 pm. The boys will remember the many musicians who came to cheer them up, especially the dear old ladies with the auto-harp. The Red Cross supplied comforts and how the boys enjoyed the hot soup and dainties but even that did not compensate for loss of limbs, health and careers.

Sister Gertrude Moberly wrote feelingly of her impressions of the Australian wounded in these English hospitals.

I thought my heart would break when visiting the Sidcup Hospital. There were 600 men and not one with a whole face. Some of them had had as many as thirteen operations. I shall never forget. I was shown photographs of before

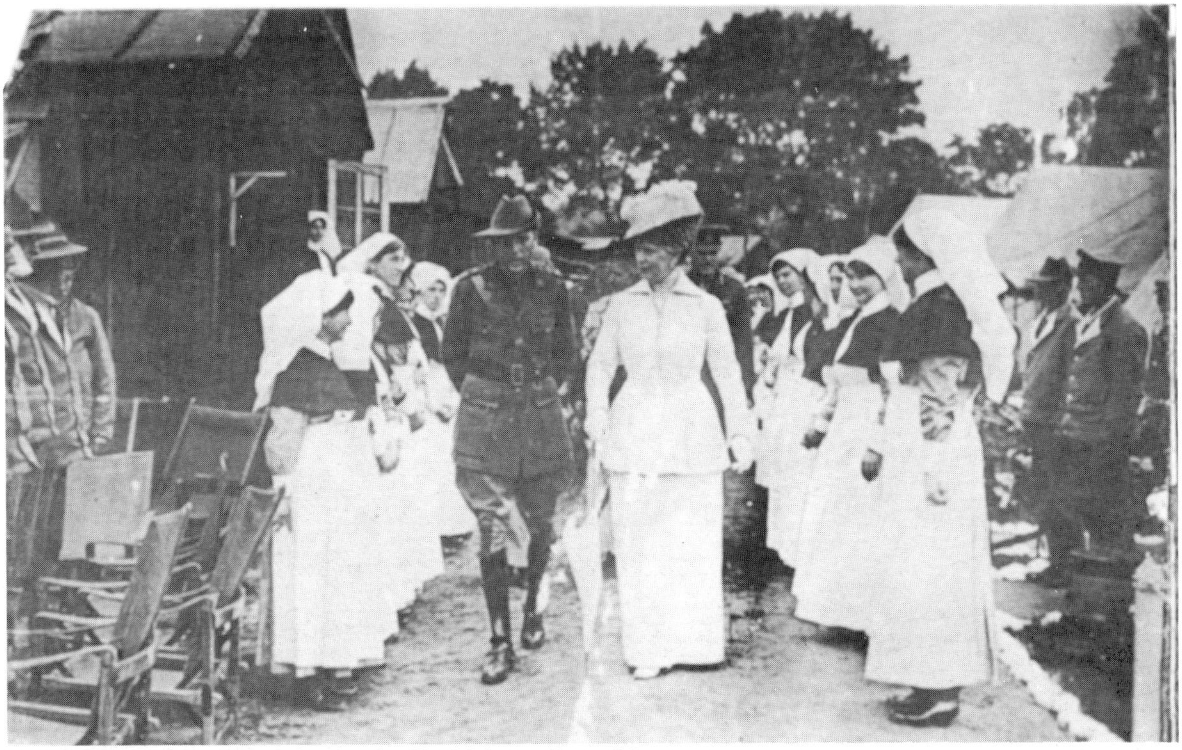

AANS in England 1915. Inspection by Her Majesty, Queen Mary.

and after the operations. My stomach turned sick and I left hurriedly. As soon as I was out of sight of the building I sat by the roadside and cried and cried.

Despite the demand for Australian nurses in other theatres of war, there was a continuing need for them in these Auxiliary Hospitals in England. In 1918 there were 215 members of the AANS still nursing Australian wounded in Britain. These nurses found that nursing limbless soldiers, in particular, was a demanding task, but the experience they gained in this field was to be of inestimable value in Australia in subsequent years.

Another group of Sisters who must be mentioned in this context were those Australian nurses who, impatient with the delay of being called up for overseas service with the AANS chose to enlist with the QAIMNS, even though the pay and conditions were not as good. The British War Office made the initial offer which was supported by the Australian authorities and at first 130 were sent overseas with the advice that another 200 were available. The situation changed both in Britain and Australia as the demand increased for more nurses in the light of heavy casualties in Gallipoli and France. Next time round, fewer Australian nurses elected service with QAIMNS and more with the AANS. In July 1916 a further request from the War Office for 100 nurses — 50 for Bombay and 50 for Egypt — found conditions for enlistment had improved. However the Australian Government insisted that they retain their AIF status and that they remain members of the AANS and not be members of the QAIMNS. Thus those serving with British units were in fact in either of the two categories. Beatrice Cheesman (Bundaberg) was one who joined the QA's and spent four years in France. She was mentioned in despatches 'for outstanding courage and heroism'.(67)

India

Members of the AANS, including a number of Queensland nurses, found themselves nursing in India, but the circumstances surrounding these appointments were curious and controversial. Of course, the British

AANS Sutton Veny (England) 1915. The Prime Minister of Australia, William ('Billy') Hughes has afternoon tea with the Sisters.

military presence in India dated back several centuries and after the Crimean War medical facilities had improved considerably. While most of the nursing work was originally performed by poorly trained soldier orderlies and assistant surgeons, the decision by the War Office in 1883 that nursing sisters should be appointed to all military hospitals with 100 or more beds had its impact on the forces in India. An Indian Army Nursing Service was established in 1888 and by 1891 fifty-two nurses had been despatched to India and Burma. Subsequently Princess Christian's Army Nursing Reserve was established in 1897 and Queen Alexandra's Imperial Military Nursing Service in 1902. Despite attempts at amalgamation the Indian nursing service remained as an independent unit, entitled Queen Alexandra's Military Nursing Service in India until 1926.

Australian Sisters found themselves in India from 1916 onwards as a result of a direct request from the Viceroy of India to the Australian Government. The Indian hospitals at this time were unable to cope with the flood of sick and wounded of the allied forces in Mesopotamia (including Indian troops and Turkish prisoners of war). Indeed many Indian nurses of the QAMNS in India had been sent to Mesopotamia to help the medical services there, which were inadequate to cope with the casualties. It so happened that following the end of the Gallipoli campaign and the movement of Australian troops from Egypt to France and England, the disposition of AANS was under review. On the recommendation of Surgeon-General Fetherston, some 50 members of the AANS were sent to India on a six months' engagement.

On arrival in India, AANS members were not posted to one major hospital, but they were split up into threes and fours and sent to station hospitals at *Lahore, Calcutta, Mooltan, Naoshera, Sialkot* and to

Sr Connie Frost.

Sr F. Myrtle Hutchison.

small medical establishments throughout India. The work was at times not military nursing, they were not always nursing military personnel but they were often filling in at civilian hospitals or nursing civilians in military hospitals. At the end of six months the first group of AANS, very dissatisfied with the conditions of service and their pay, were transferred back to England.

The need for help in India still remained, so after further negotiations, the Australian Government agreed to send one hundred nurses to India, paid for by Australia. The first fifty nurses under the control of Matron G.E. Davis (formerly of 3 AGH), the second group with Queenslander T.J. Dunne in charge and subsequent groups were sent until some 500 members of the AANS had seen service in India. In time their work extended into *Burma, Peshawar, Quetta, Naini Tal, Belgaum* and *Bangalore*. Others served on hospital ships operating from the Persian Gulf, Basra, Egypt and even round to Vladivostock.

The Indian experience presented many administrative problems for the AANS. In theory they were administered in part by the DMS, AIF in London and in part from Australia, but in fact they became an increasingly independent group with many decisions being made by the Principal Matron of the AANS in Bombay (Miss G.E. Davis). However their major concern was that they were being under-used in one of the backwaters of the war. Butler referred to one report in 1918 that many of the nurses resented 'the fact that for months they have been practically idle, and where work does exist it is not real nursing'. They believed they had enlisted to nurse the sick and wounded in battle, particularly Australian soldiers, but the Indian experience was a disillusionment for many of them. The nearest battlefront was in Mesopotamia where Australian and allied troops were fighting the Turks. The medical services in that campaign were

Sr Lillian Quinn (Bundaberg).

Sr Alice Leyland.

provided by British General Hospitals staffed by QA's. Members of the AANS believed (wrongly) that their service in India might not constitute active service and might affect their entitlements after the war.

Sister Rosine Derrer (Mackay) was one of those who left Sydney in September 1916 and who saw two years' service in various Indian hospitals. Her report on her experiences highlights the problems Australian nurses faced.(68)

> Leaving Brisbane on the 14 September 1916 with a number of Sisters we proceeded overland to Sydney where we boarded the SS *Kamala* en route for India. Landing at Bombay on 10th October I was sent to the Gerald Freeman Thomas War Hospital for duty. This was a very fine stone building in the centre of Bombay, it being intended for a College of Science, but was taken over as a Military Hospital. The patients were wounded British troops from Mesopotamia, having come by hospital ship from Basra. The staff, in charge of a British Matron of the Indian Nursing Service consisted of about 50 Sisters 12 of whom were Australian. After some months I was transferred to another British Hospital on the Hill Station at Poona, staffed by an Australian Matron and Sisters. This camp had about 14,000 troops so our cases were mostly local, with cholera, dysentery and plague being the most common diseases.

The problems confronting Australian nurses in their assignments in India eventually came to the notice of the Secretary for Defence in Melbourne in a curious roundabout way. The AIF Paymaster, Captain F.H. Wickham had been instructed to proceed to India and Mesopotamia (January 1918) to deal with pay and other matters concerning the Australian troops in those areas. After discussion, he was also given authority to act generally in all matters concerning Australia in this theatre of operations. One of his reports was related to the AANS and AAMC in India.(69)

> Since I have been in India I have been bitterly assailed by the Nurses with the complaint that they enlisted for six

Sr Grace Homewood. Mt Morgan.

months' service in India only and were then to proceed to another sphere of activities. They contend that the fact that they have been retained for longer periods in India has amounted to a breach of contract. The Indian climate is an exceedingly severe one and has told in many cases, and is beginning to tell in many other instances, on the health of the nurses. Malaria has attacked them and consequance of the nature of the disease they are subject to frequent recurrences.

The method of living in India is extremely unnatural. The performance of everyday tasks is considered menial and simply lowers all white people in the eyes of the natives. Nurses are therefore obliged to forego the exercise and recreation which they might gain in this way, and also to pay servants to do all manner of jobs they wish to do themselves. This leaves them out of pocket and inclined to indolence. They complain of the effect this is having on them and petition for a transfer to effect a rejuvenation. They state they do not mind making sacrifices if they are doing something for their country but they resent the fact that for months they have been practically idle and where work does exist it is not real nursing.

Captain Wickham went on to give examples of the under-employment of Australian nurses. Twenty Australian Nurses at the Victoria War Hospital darned socks and sewed on buttons and tapes for a period of two months while their wards had no patients. At the 44th British General Hospital, *Deolali* there were 23 Sisters to do duty in two wards containing 60 convalescent patients. As the stock was new there was no mending even to keep the 23 nurses employed.

The real reasons were financial. All Indian nurses, when attached to the War Organization became a charge on the War Office and the Australian Nurses who replaced them a charge on the Australian Government. Australian nurses were sent by the Australian Government to India and finished up doing Indian work paid for by Australia!

It would have been better if enlisted Indian Army nurses nursed Indian personnel. Captain Wickham pointed out this was the first time white women nursed Indian troops. Formerly this had been done by Indian orderlies.

Sr Martha Homewood, Rockhampton.

Some other interesting observations on the Indian experience are contained in a book by Sister Gertrude Moberly, RRC, who also served in numerous hospitals and on hospital ships at this time.[70]

> Cumballa War Hospital is a huge place. There are 11 AAN sisters here and I do not know how many English Army Sisters. The Matron is a "Regular" and doesn't understand our Australian freedom. (She is Irish) and awfully nice but oh, so strict, and the rules here are too absurd for words. Fancy trained women having to be in at 8.15 pm — that is the hour we dine! One must have special permission to stay out later than that.

Sister Moberly commented on the lack of responsibility given to nurses in Indian Hospitals. She said 'the usual procedure over here is for the assistant surgeon, (Indian or partly so) to give all the hypo-injections. You can imagine what our Sisters think of this rule. They, who I suppose have given hundreds of hypodermics in Aussie!

Sister Edith Avenell and Sister May Tilton.

Table 1

Staffing of Indian Hospitals by AANS, 1918

Hospital	No of beds	No of occupied beds	No of Australian nurses	No of other nurses
Victoria War Hosp. Bombay	600	100	40	—
Alexandra War Hosp. Bombay	250	30	26	—
Colaba War Hosp. Bombay	550	400	15	20
34th Welsh General, Deolali	3,000	300	42	20
44th British General, Deolali	1,200		23	10
Deccan Brit. War, Poona	1,200	300	46	4
King George's Hosp. Poona	600	60	27	1
Hislop Hosp. Secunderabad	750	200	10	12
Station, Bangalore	500	busy	18	2
Freeman Thomas Hosp.	500	100	27	10

Source: Butler, Vol 3, p 750

Sister Elsie Grant.

Sister Agnes Isambert.

Anyway, the rule now is that I, as Night Super, am to administer all the wards, the needle, which is a jolly bother in a huge place as this Colaba Hospital is, with its hundreds of patients.'

Matron G.E. Davis, writing from the Victoria War Hospital, Bombay (in December 1916) to the DGMS, Melbourne, summed up the problems confronting the nurses in India by stating that they seemed to be 'on active service', but were operating under 'home service' regulations.

The first grievance was that Sisters who came from Egypt expected to remain as a unit, but were sent in ones and twos all over India, as far as the Chinese and Baluchistan borders. They felt very lonely and too much was expected of them. Then there was the question of pay and allowances. Each district had its own pay office staffed by Indian or Eurasian clerks and it was some time before procedures were clarified so that nurses could collect their pay. However, Matron Davis had high praise for the Australian nurses and the way they coped with very difficult situations.

> I am very proud of our nurses out here and the way they work. It shows immediately we go into any hospital. If you saw the hospitals we take over and saw them again in three months you would be proud of their work — flower

Sister Gladys Wilsher.

gardens everywhere and flowers in the wards. The whole general appearance of the place, wards, dining halls, kitchens, was greatly improved. Sir Charles Munro GOC India has the highest opinion of Australian nurses.

A noted historian of this period Eric Keast Burke, author of *With Horse* and *Morse in Mesopotamia*, has given a discerning account of this Indian experience and the role played by Australian nurses.(71) When the *Mooltan* arrived with the first 50 Australian nurses in September 1916 an outbreak of cholera was raging in Bombay. Two Australian nurses from Egypt had died, (A.V. O'Grady and K. Power). Then the British nurses at the Victoria War Hospital were ordered to Mesopotamia, leaving the hospital in the hands of forty newly arrived Australian nurses, none of whom could speak Hindustani, and ten other nurses, mostly Eurasians with very little training. The AANS Sisters threw themselves into the work in a highly dedicated professional manner, learning to cope for the first time with 'tropical diseases, including cholera, dysentery and the plague'. Hours of work were from 7.30 am to 8 pm, with three hours off every other day. One compensating factor was that after 12 month's service they were entitled to one month's leave which enabled them to see something of the historic sights of India.

Eric Burke Keast made the following observations about the hospitals in which the AANS were located.

Victoria War Hospital, Bombay

This was a new building for the railways, of four floors, with 200 beds on each of the lower floors and the nurses' quarters on the top. Their first patients were 300 British prisoners of war, kept by the Turks in Baghdad and released after the fall of Kut. They suffered from dysentery, beriberi, old septic wounds, malnutrition and starvation. Christmas 1916 was remembered by the admission of 50 surgical cases, following heavy fighting. By March 1917 came a flood of wounded Turks, all in a filthy condition. Still,

Sister McClelland AANS with her brother James Ferguson McClelland. 1 FD AMB France. Girls are two sisters of late Walter Boyd of St Lucia.

this was good nursing experience — so septic were the wounds that on one floor of 240 beds there were 60 amputations. This posed some cross-cultural problems as many of the Turks preferred to die rather than lose a limb, as they believed an imperfect many could not enter Paradise! In the middle of 1917 came a flood of British cases with heat-stroke — nearly 1,000 per week for three months. By October 1917, German prisoners from East Africa began arriving. It was certainly a varied programme until the hospital closed in August, 1918. Among Queensland Nurses who served in Bombay were Sisters Ida Pearce, Elsie Pollock, A.G. Sim, Nonie Monckton, Marianne Dowling, Helen Lawson (who wrote about nursing 'British Tommies and Turkish prisoners'), Mary Derrer and Winifred Payne.(72)

Cumballa War Hospital

Situated on Malabar Hill, overlooking Bombay Harbour it had 600 beds, with mostly British patients. At first the staff was a mixed one with several Australian Sisters under a British Matron. Later, it was Australian policy that Australian Sisters, where possible, should serve under an Australian Matron.

Colaba War Hospital

In old bungalows outside Bombay city, this hospital proved to be a happy place for Australian Sisters, who worked well with the English Matron. Patients were mainly British soldiers suffering from cholera, smallpox and the plague.

Gerrard Freeman Thomas Hospital

A 1,000 bed hospital in Bombay with 12 of the 50 nursing staff Australian. Most of the patients were British. In September and October 1918 the influenza epidemic hit the city of Bombay. Deaths were 700 a day and Burke reported, 'at night the glare from the cremation grounds resembled that of a city in flames.' Rosine Derrer was one of the Queenslanders at this hospital.(73)

Sister Beatrice Cheesman, QAIMNS, MID. Front row, far right, British CCS France c.1917.

No 34 Welsh Hospital, Deolali

This hospital was in a hill station, four hours' journey from Bombay. The patients were a mixed lot — British Tommies, French Algerians, Mauritius Labour Corps and Turks. It was an enormous collection of bungalows, one and a half miles long by three quarters of a mile wide, a huge area for Australian matron Davis to cover. Much of the nursing involved treatment of joints and massage and fortunately one of the AANS, Sister A. Scott had a massage diploma, a rare thing in those days. By September 1916 there were over 2,000 patients in the hospital, when there was a serious outbreak of plague. The Sisters were kept very busy but none of them caught the disease. However, towards the end of the war the Spanish influenza epidemic struck the hospital and Sister E. Clare became a victim. No sooner was this epidemic under control than cholera struck again. It was a trying time for the nurses in the heat, dust and flies, coping with strange diseases in an unfamiliar environment. Sister Beatrice Leyland (Barcaldine) was one of the Queensland nurses at this hospital.(74)

No 44 British General Hospital, Deolali

Opened at the end of 1917, with Australian nurses under Matron A.E. Dowsley. It functioned for a year or so as an Isolation Hospital. Sister Janet McIlwraith was a prominent member of the AANS here.(75)

Officers' Hospital, Nasik

A small hospital, 7 miles from Deolali, with three Australian Sisters, including Sister E.A. Burke. It was a quiet, restful existence although snakes were an ever-present danger.

Deccan War Hospital, Poona

Set in the beautiful highland district of the Western Ghats, 140 miles from Bombay, with Queenslander Matron T.J. Dunne in charge. It soon expanded to 1,200 beds, with 50 Australian nurses, including Sisters Rosine Derrer and Mary Keating. Again, dysentery, plague and cholera accounted for most of the patients. Among the Queenslanders at this hospital were Sisters Georgina Farquhar and Lillian Dennis.(76)

AANS cricket team, India 1918. Front row: Sister A. Sim, Sister W. Payne.

King George's War Hospital, Poona

A hospital of 600 beds, with a staff of Australian Sisters, including Sister Ethel Butler as assistant matron.

Trimulgherry Hospital, Secunderabad

This hospital expanded from 500 to 1,200, especially with the return of native troops from East Africa with a serious form of malaria. Sister E. Hoadley and later Sister Gertrude Moberly acted as Matrons to the staff of Australian nurses.

Station Hospital, Bangalore

A large station hospital, two days and nights by train from Bombay. Taken over by Australian nurses with Sister Dowsley as matron. It was the time of the influenza epidemic which struck staff and patients alike. Another Station Hospital was at *Belgaum*, staffed by four Australian Sisters. *Maymyo*, a Station Hospital in Burma, also staffed by Australian Sisters, with Sister Agnes Ferguson in charge, received some publicity when the British Matron-in-Chief for India said the hospital was the cleanest she had seen in her thirty year's experience in India!

Frontier Wars

When fighting broke out in 1917, No 18 British General Hospital opened at Rawalpindi and 5 members of the AANS were sent there. The seriously wounded were kept at *Tank*, where they were nursed by Sisters Browne, E.G., Steel, V., McAllister, C.J., Furness, D., and Jack, E.K. Later in that year when malaria and smallpox broke out Australian Sisters were again sent to the frontier, where the temperature ranged from 116 to 124 in the coolest part of the hospital. Sister Victoria Christiansen (Brisbane) after a varied career in army nursing also found herself at Rawalpindi.[77]

Tank, on the Baluchistan frontier was a formidable place for Australian nurses. Matron Davis wrote, 'Here, where no woman had even been sent before — the last place God ever made — six of the AANS worked in the most appalling heat one could imagine'. Their bungalow was always guarded by a picket of 12 Ghurkas! Conditions at *Dera Ghazi Khan* were almost as bad. Three Australian Sisters, E. Horne, Beryl Tucker and A. Hodson found themselves working in temperatures between 118 and 120 degrees.

Sisters Browne, Steel, McAllister, Furness and Jack, members of the nursing staff of the Hospital at Tank, on the north West Frontier. AWM H12556

Afghan War

Although the war finished in November 1918, there was a reluctance on the part of the Indian authorities to speed up the release and return of the valued AANS Sisters. As it happened the Afghan War broke out in May 1919 and a number of Australian Sisters were transferred to hospitals in the Himalayas. According to Matron Davis, it turned out to be a beautiful and restful experience. Sister Lily Campbell was in charge at *Khuldana Hospital*.

Southern Persia

The only hospital on the Mesopotamian front staffed by Australian nurses was at *Bushire*, with 30 beds for officers and 100 for other ranks. Sisters Lily Stewart, Wellard, E.L., Waterstrom, M.B., Purcell, E., Zita Lyons and Parnell, M.T. had this unusual assignment.[78]

Hospital Ships

Bringing the sick and the wounded from Basra to Bombay and linking up at other intermediate ports was no picnic for Australian nurses. Sister Horne and Sister Scanlan made numerous trips to the Persian Gulf with a shade temperature of 124 to 130 degrees on board. On one trip Sister Larkan reported 500 heat-stroke cases on board. Sisters Alma Bennett and Gertrude Moberly had similar experiences. Sister Fletcher went on an epic voyage from Bombay to Singapore, Hong Kong, Vancouver and back to Vladivostock where the temperature was 13 degrees below zero.

Burke noted that it was much easier to send the AANS to India then to get them back. Lack of shipping between India and Australia and the desire of the Indian Government to retain the services of Australian nurses for as long as possible, especially when they were being paid by the Australian authorities held up their return until the end of 1919, a year after the Armistice. It had been a remarkable and valuable experience for Australian nurses, involved in other cultures, learning about diseases rarely seen in Australia. They performed well under very difficult circumstances, not in physical danger from the enemy as in France, but in greater danger from disease. Four members of the AANS lost their lives in India. Those who came back to civilian nursing in Australia were the richer for the Indian experience.

Table 2
Service in India W.W.1

Some Queensland Sisters known to have served there

Bassett, M.E.V.	Leyland, B.
Chataway, M.E.	Low, N.
Christiansen, V.	Monckton, N.
Dennis, L.	Moreton, L.G.
Derrer, M.J.	McIlwraith, E.G.
Derrer, R.	McPherson, A.I.
Dowling, M.	Payne, W.
Dunne, T.J.	Pearce, I.
Farquhar, G.	Pollock, E.
Francis, S.	Scott, A.
Imison, H.E.	Sim, A.G.
Lawson, H.	

The contribution of the AANS in the First World War would not be complete without reference to the influenza pandemic of 1918–1919. While nurses had been involved in the treatment of influenza on all battlefronts (and some had died from this disease in Egypt and Salonica), this new virulent form of influenza was to sweep across the world, taking friend, foe and neutral in its path. Whereas there were 8 million deaths from the four years of the war, 15 million were to die within a year at the end of the war, from this so called 'flu', epidemic'. It was thought at first that Australia's isolation, with several weeks' journey by sea, might act as a quarantine barrier, but as returning troopships called at Colombo or Durban or New Zealand and in view of the normal movement of people to and from Australia from all parts of the world it was a seemingly hopeless task to keep out the disease. When the epidemic broke out on one of the convoys approaching Western Australia, the call went out for nurses travelling on the *Wyrema* to volunteer their services at the quarantine station, Woodman's Point, near Fremantle. A large number did so and several contracted the disease. Unhappily three died, including Sister Rosa O'Kane from Queensland (21 December 1918). She was well-known in Townsville and Charters Towers and the residents subsequently subscribed funds for a memorial tablet in the Woodman's Point cemetery at the quarantine station. Lt Col P.M. McFarlane, who was officer in charge of those on the *Wyrema* later wrote an account of these events.

> Volunteers were called for from the nurses and there was not only a ready response but so many offered it was necessary to put the names in a hat and to draw out the 20 required. They knew perfectly well the enormous risks they were taking. Three of them have already paid the supreme sacrifice and 12 others have contracted the disease.

Chaplain, the Rev J.A. Ford from the troopship *Boonah* spoke of the tenderness and attention of the nurses towards the stricken men.

> To me this striking case of courage and devotion to duty equals the action of a body of soldiers going over the top in trench warfare, the casualties being equivalent to those sustained in such an action — three killed and 12 wounded out of a detachment of 20. I count it an exceptional honour to have been associated with such a gallant band of Sisters.[79]

Sisters Mary Derrer and Mabel Wiseman in front of a giant Handley Page aircraft, piloted by Capt. Ross Smith, Sgts Bennett and Shiers, somewhere in India c. 1917–18.

Queensland. The Home Front

For those members of the AANS not yet posted overseas and for those civilian nurses waiting for the chance to enlist, the war at first was a distant war. True, this was the attitude of most of the population. It was a war inspired by hatred of all things German, in which German names of suburbs were altered and persons with German names tried to alter them. It was a war in which patriotic fervour for Britain, for the British Empire, brought thousands of young men rushing to the recruitment centres. There was no fear of invasion, no doubt about the rightness of the cause or of the final victory and a firm belief it would be a short war, that it would all be over by Christmas.

The administrative confusion over the role and status of the AANS in Egypt was an inevitable consequence of the inefficient medical administration in Australia. As an offshoot of the Adjutant-General's Department, the Australian Army Medical Service was away to a very poor start in 1914. Surgeon-General W.D.C. Williams and later Surgeon-General R.H.J. Fetherston were never given the staff, the status or the authority to develop an efficient army medical service. They were not helped by the fact that while overseas the AAMS was under the control of the War Office. It took the crisis over the resignation of Matron Jane Bell and the subsequent inquiry to alert the people and the politicians that all was not well in the system responsible for the care and nursing of the sick and wounded.

In Australia, a great deal of power still resided in the Military Districts and with the Principal Medical Officers of those Districts. In Queensland Col A. Sutton, Col D.G. Croll, LtCol A.M. McIntosh, Col J.E. Dods filled this post between 1914 and 1918. They were responsible for the provision of adequate medical and nursing facilities in Queensland, including such matters as establishment of camp hospitals where large numbers of troops were gathered together, the examination of recruits, the selection and training of medical officers for overseas and in conjunction with the Principal Matron, all matters concerning the AANS. It was an enormous task, with so few facilities. For example, transport arrangements for men sick in camp were facilitated by the gift of the residents of Richmond, North Queensland, of a 40 HP Studebaker Motor Ambulance.

Sister Rosa O'Kane, AANS. Died of illness 21 December 1918. Buried Woodman's Point Cemetery, Quarantine Station, Fremantle.

It is significant that no major military hospitals were opened until 1915 when the first of many thousands of casualties from Gallipoli and the Middle East and later from the Somme began to arrive home. No 6 AGH with 270 beds and 32 nurses opened on 19 July 1915 and soon afterwards No 13 AGH with 300 beds and 40 nurses. The former was set up at Kangaroo Point with Matron E.E. Bishop and the latter at the Enoggera Army Camp, with Matron Ralston. Both these hospitals were to play an important role in the nursing of war wounded during the war years. The first report on 13 AGH, in the PMO's Annual Report for 1916–17[(80)] indicates the initial problems.

> As a 300 bed hospital it boasted of modern amenities, a septic tank, electric lighting and a hot water system. An up-to-date operating theatre in which 289 operations were performed in the first year was highly regarded. Accommodation for the Sisters of the AANS was in two rented cottages near the hospital and later within the hospital itself. This was the least satisfactory part of the arrangements. The hospital was frequently taxed to its fullest capacity, temporarily accommodating the excess from 6 AGH at Kangaroo Point. Meningitis and pneumonia provided the majority of cases.

The records reveal that in addition to Matron Ralston the foundation staff of AANS at No 13 AGH included Sisters Lynch and Goodman and Staff Nurses Cairns, Phillips, Black, Caves, Barron, Birt, Skyring, Homewood, Rowe, Bryden, Lawson and Toft. Most of these were eventually posted overseas.[(81)] Over at No 6 AGH were Sisters Waller and O'Neill and Staff Nurses Sprague, Rosine, Wilson, Wilsher and Cairncross.[(82)]

As the number of hospitals increased, there were over 100 members of the AANS posted for duty all awaiting the opportunity to move, all noting the possibility of enlisting in the QA's or of accepting requests to join the British hospitals or to go to India. It was a period of constant movement. While there was a

AANS Staff. 6 Australian General Hospital, Kangaroo Point, Brisbane c.1915. Sister Cairncross is third from right, front row.

Matron-in-Chief (in fact two at one stage, one at Horseferry Road, London and one in Melbourne) there was a Principal Matron or Matron for 1 MD. On her return from overseas, Matron A.C. Isambert took over the administrative duties for Queensland.

The dramatic change in outlook towards the war came with the homecoming of the wounded and the realization that for many years to come hospital facilities on a large scale would be necessary.

The homecoming of the wounded to Brisbane was feelingly described by Sister Isambert who on one occasion was rostered for duty on the ambulance train from Sydney to Brisbane.

> At first the men were taken to a large hall and entertained by the ladies of the Red Cross Society, Sydney. Some of the men subsequently got very thirsty and brought with them on the train a supply of intoxicating liquor. Seeing one of them refreshing himself from a flask and a small glass, I asked what it was and a serious face looked up to me and said ''Lime juice, Sister''. Later on the supply of liquor on the train was prohibited.
>
> We were nearly deafened by the whistles in the station yards at Sydney and all along the journey flags were flying from nearly every home. Children were allowed out of school and with adults were seen waving flags and handkerchiefs and shouting and cheering. Food was supplied in abundance, with sit down meals at stations en route. Flowers, cigarettes, tobacco and Queensland papers and magazines were supplied to every man on the train by the Red Cross Society, both in Sydney and at all stopping places.
>
> Then there was the climax — the meeting with relatives at different stations and in Brisbane. The bands are playing as we steam in. The heads of men are strained out of the windows, trying to catch a glimpse of their loved ones. Some of the scenes are pathetic when wives and mothers meet limbless or blind husbands and sons. Worse still is when the expected one has died on the way out, the news quickly broken to them there by one of the deceased's comrades.
>
> After a few minutes greetings, the men were lined up to be addressed by the Governor, Commandant, Premier or Mayor and then taken in motor cars to the Military Hospital at Kangaroo Point.[83]

First Sisters of staff — 13th AGH Enoggera, Queensland, 4th September, 1916. From left to right: Back row; Staff nurses Cairns, Phillips, Black, Caves, Barron, Birt, Skyring, Homewood. Front row; Staff nurses Rowe, Bryden, Sister Lynch, Matron Ralston, Sister Goodman, Staff nurses Lawson and Toft. AWM A3240

The development of military hospitals in Queensland during the war years centred on a number of smaller specialized hospitals rather than on one large hospital for the treatment of all kinds of disease and disabilities. VD cases, for example, were sent to the Lytton Camp Compound, but in 1918 these patients were transferred to Enoggera. Infectious diseases were nursed at No 4 AID Hospital at Rosemount from 1 April 1916, but this later became No 27 Australian Auxiliary Hospital. TB cases went to 'Kyooma' at Stanthorpe, while other auxiliary hospitals were established at 'Finchley' (Toowoomba), 'Staghorn' (Southport) and at the Old Main Beach Hotel Southport. The latter was equipped but never occupied. By 1917 the major army hospitals had been reduced to 4-6 AGH at Kangaroo Point, the Rosemount Orthopaedic Hospital, the Camp Clearing Hospital at Enoggera and the Military Sanitorium at Stanthorpe.

Thus there remained a great deal of nursing still to be done for members of the AANS. However, arrangements had to be made for the discharge of army wounded and the responsibility for their care assumed by the Repatriation Commission and by organization such as the Red Cross Society. While some nurses elected to return to private nursing, others transferred to the Repatriation Commission and spent many more years still caring for the war wounded. Red Cross Homes were established at Corinda (Ardoyne, where Sadie Macdonald was Matron for many years), Toowoomba (Simla), Graceville (Rakeevan) and in Brisbane (Grange Hill).

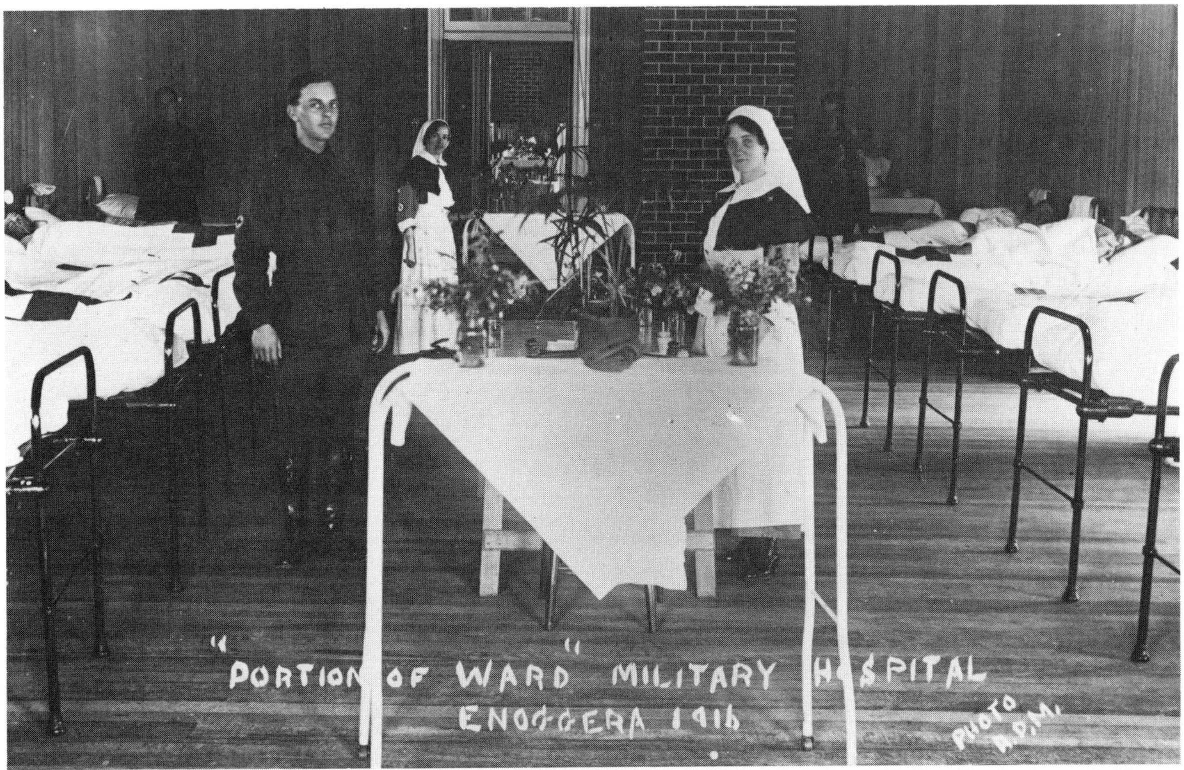

View of portion of ward in 13th Australian General Hospital Enoggera, Queensland. Left to Right. Private W. Temple, Sister Skyring (at back) and Sister Homewood. AWM A3239

Outside the sisters' quarters of the 1st Australian General Hospital at Rouen, in France, photographed on September 23rd, 1918. AWM E 3443

No. 2 Australian Casualty Clearing Station at Trois Arbres, France. AWM H 15349

The 34th Welsh General Hospital at Deolali, India. AWM H 12551

References

1. *Standing Orders*, Australian Imperial Force. Issued with Military Order, No 50 of 1918.
2. CHEESMAN, BEATRICE GRAHAM, MID. Trd Brisbane Gen Hosp 1910, midwifery at Lady Bowen Hosp; Nurse at Clermont Hosp, Matron Lady Chelmsford Hosp Bundaberg; Member QAIMNS; 4 yrs in France; Mentioned in despatches by Gen French for outstanding courage and heroism; 1921 Matron Clermont Hosp.
3. DODS, W.S. On Staff St Andrew's Hosp Bundaberg on enlistment 1915.
4. WILSON, M.E. See Fn 48. Sisters Madeline and Marjorie Wilson's names appear on the Nanango War Memorial, Sisters May and Myrtle Wilson's names appear on the Bundaberg War Memorial.
5. *Bundaberg Mail* 30 March 1915. Nurses for the Front.
6. BOWES, BEATRICE. b 22 July 1891; Trd Bris Gen Hosp and Chns Hosp; Staff Boonah Hosp; AANS 1915, Egypt 2 AGH (Mena House); Invalid home 1916. d 11 Jan 1970.
7. SMITH, ADA PRISCILLA, b Warwick, 5 Mar 1888; Trd Warwick Hosp; 1910–13 at Murwillumbah; AANS 1915. 2 AGH in Egypt, France, Eng, 2 AACS at Trois Arbres (B. of Messines). d Dec 1973.
8. AVENELL, EDITH FLORENCE ('QUEENIE'), b Gympie; AANS 1915, embarked Syd per *Mooltan* for Egypt; retd Aust with wounded 1916; Marseilles with 1 AGH, also 13 Stationary Hosp Boulogne; to Eng with 2 Aust Aux Hosp Southall; Retd Aust per HT *Euripides*; disch Bris Oct 1917.
9. HART, JULIA MARY ('BLUE'), MID, RRC, b Charters Towers; Trd Bris Gen Hosp; AANS 1914, sailed on *Omrah* Sept 1914 to ME. Service Egypt, France, Italy and Malta. Matron Rosemount Rep Hosp; Matron Prince Wales Hosp Randwick.
10. PATEN, EUNICE MURIEL HARRIET HUNT, (1883–1973) MBE, AARC, FCNA (Hon); Trd Bris Gen Hosp 1905, AANS 1914, sailed on *Omrah* 24 Sept; served ME and France; 1918 Actg Matron Aust Aux Hosp Southall, Eng and Deputy Matron 3 Aust Aux Hosp for Limbless Soldiers; 1920–36 owned and administered Holyrood Pte Hosp Brisbane; 1937 Mem Coronation Contingent; 1939–41 Princ Matron North Commd; Hon Life Mem Qld Bush Nurs Assoc, Fdn Mem Royal Flying Doctor Serv; Fdn Fell Coll of Nurs Aust; Hon Fell Coll Nurs Aust; Patron Centaur Mem Fd.
11. KEYS, CONSTANCE MABEL, MID (Twice), RRC, (1886–1964) b. 30 Oct 1886 Mt Perry, dau James Keys (teacher) and Margaret (nee Pelham); Trd Bris Gen Hosp, gold med for prac wk 1908–10; AANS 1914, embarked *Omrah* 24 Sept 1914 with first contingent AIF; service at Abbassia then to 1 AGH Heliopolis; retd Aust with wounded on *Thermistocles* and re-posted 3 AGH then to Kitchener Hosp Brighton, Eng 1916; 1917 to 3 AGH at Abbeville; 1918 Sister-in-Charge of 2 ACCS at Trois Arbres; 1919 to 1 AGH at Sutton Veny in Wiltshire. Retd Aust and disch AIF on 17 Feb 1920. Awarded MID and RRC, Medaille des Epidemies (en Vermeil); 1920 Matron Conv Home at Broadwater (Mt Gravatt); m 1921 Lionel Kemp-Pennefather; d Southport 17 Mar 1964.
12. WILLIAMS, D.M. One of four Qld Sisters who sailed on the *Omrah*. Later awarded the ARRC for her outstanding contribution to nursing.
13. BUTLER, *op cit* v 3, p 538, fn 19.
14. CROLL, WINIFRED MARION (nee Payne). b July 1886, trd at Bris Gen Hosp; m. Dr Gifford Croll, AAMC Col WWI and later closely associated with the then Brisbane Children's Hosp; Sr W. Croll, AANS 1914–1916. Killed in BOAC Air Crash Singapore, 7 Mar 1954. (Not to be confused with another Sister Winifred Payne, also WWI nurse from Qld.)
MOWBRAY NORMA VIOLET, dau of Thomas and Elizabeth Mowbray of Longville, Eagle Junction, Brisbane. Granddaughter of the Rev Thomas Mowbray, first resident Presbyterian Minister in Bris. Sister Mowbray was born at St George (Qld). Service AANS at Heliopolis with 1 AGH. Died of pneumonia 21 Jan 1916, aged 32 years. Buried in Cairo War Memorial Cemetery.

RALSTON, EMILY ANN VARDON. Enlisted 1914 — to ME on *Kyarra*, 21.11.14. 1 AGH Heliopolis. Mentioned in Isambert's Diary; Matron Lady Lamington Hosp, Bris: Dec 25 Dec 1939.

DUNNE, THERESA JOSEPHINE ('TRIXIE'), RRC. Matron St Helen's Private Hosp Bris for 12 yrs prior to enlistment; sailed on *Kyarra* 1914; served with No 1 AGH Heliopolis and No 2 AGH Ghezireh Palace; 1915 retd to Aust with wounded as sub-Matron; 1917 Matron; 1916 with AANS in India; second Matron RAMC at Rawalpindi, Matron British Deccan War Hosp, Poona d 19 May 1931. See obit ANJ 15 June 1931.

FARQUHAR, GEORGINA ANNE JESSIE, sailed on *Kyarra*, Nov 1914, served 2 AGH, prom Sister 1916; to France 1916; transport duties to Aust and then to 2 AAH, Southall, Eng; served with RAMC India 1918, Bombay, Deccan, Poona, Bangalore. Disch 10 Mar 1920.

GRANT, ELSIE ROSE. b. Clermont (Qld), dau David and Ellen Grant. AANS service with 1 AGH Egypt and 1 ACCS in France. Retd Aust Aug 1918. Photo in Wildman p 114. Referred to in Isambert's diary and in *The Grey Battalion*.

LANGFORD, ROSE JANE, MID. Born in Devonshire (Eng) and came to Qld at the age of four. Nursing training at Ipswich Hosp. Before outbreak of WWI Matron at Mt Perry Hosp and later at Mt Morgan and Mackay. 1914–1917 AANS. Left on first voyage of Hosp Ship *Kyarra* for Egypt. Served with No 1 AGH in Egypt and in Flanders. Mentioned in despatches for devotion to duty. After war on staff of Kangaroo Point Hospital and later Matron of Rosemount Repat Hosp. Married James Walker and lived in Brisbane until her death in 1935, aged 56 yrs.

LYONS, ZITA STELLA. b Rockhampton, Trd Bris Gen Hosp; Sailed *Kyarra* 21 Nov 1914; Service AANS, Persia, Palestine, Egypt and in Hosp ships off Gallipoli. After war in charge of State's first radium clinic.

SORENSEN, CHRISTENSE, MBE, MID, RRC, FCN. b Sandgate 5 Sept 1885, dau Conrad Sorensen, vet-surg and Hannah Jacobsen, second in family 11 chn; educ Sandgate State School; Trd at Bris Gen Hosp 1910–1913; AANS 1914–1919 with service in Egypt, India, Mesopotamia, Macedonia and England; 1915 attached duty H.S. *Guildford Castle*; 1917 Temp Head Sister AANS; 1918 Matron 60 British Gen Mil Hosp (Salonica); 1918 MID for 'Gallant Conduct' and Distinguished Service, 1917 and 1919 RRC for valuable service with British forces in Balkans; 1919 Médaille des Épidemies en Vermeill (France) for conspicuous service; 1921 Matron Repat Hosp Rosemount; 1922 Matron Hosp Sick Children; 1926 Dep Gen Matron of all Hosps under control of Brisbane and South Coast Hosps Bd and in 1928 Gen matron; 1920–1940 held many positions in prof nursing associations; 1949 Fell Coll Nurs; WW2 duties with Union Jack Club; 1952 MBE; d 1958 aged 72 yrs; Name on Honour Roll, Sandgate Baptist Church; Christense Sorensen Memorial Fund estab assist post-grad students at Colleges of Nursing; Christense Sorensen Ward for sick nurses est at Royal Bris Gen Hosp.

SCULLY, MARY (MAY). Before enlistment nursing Charters Towers and Matron Gladstone. AANS: Heliopolis, Egypt. After war, Matron Rosemount Hosp Brisbane.

WEBB, DOROTHY, RRC. Sailed on SS *Kyarra*. Served at Ghezirah Palace, Egypt, France and Belgium. Theatre Sister with AACS. Awarded RRC.

WILSHER, Gladys. b 1890; trd Hosp for Sick Chn, Bris, 1912–1915; Mat Sandgate Con Home; Staff Ips Hosp; Kuridale & St Martin's; AANS 1918–20; 6 AGH Kangaroo Pt, Rosemount & Stanthorpe. WWII inst in WRANS.

15. *Official History of Australia in the War* Vol X *The Australians at Rabaul* p 59. Capt Brian Colden Antill Pockley died of wounds, 11 Sept 1914.
16. Diary Sister M. Gibbon, AWM File.
17. BUTLER, *op cit*, Vol III, p 549.
18. *Report*, Court of Inquiry 4–7 Oct 1915, MP 133/1 Box 8 238/6/78, AA Melb.
19. *Parl Deb* (Comm), 27 Aug 1915 (p 6232), 10 Sept 1915 (p 6923)
20. General FETHERSTON: *Report* 8 Jan 1916. MP 729/4 Box 1 10/3/297 AA (Melb)
21. KENNY, *op cit*, p 33.
22. BUTLER, *op cit* Vol III p 548, fn 33.
23. KENNY, *op cit*, p 39.
24. *Ibid.*
25. DONNELL, Anne. *Letters of an Australian Army Sister*. Angus & Robertson, Sydney, 1920, pp 126–127.
26. BUTLER, *op cit*, Vol III, p 548.
27. TILTON, MAY, *The Grey Battalion*. Angus & Robertson, Sydney, 1934, p 19.
28. *Ibid.*
29. AWMF 509/2/399.
30. *Ibid.*
31. *Ibid.*
32. 'Reminiscences of the ANS 1915–1918' by One of Them. *Mufti*, 1939 pp 10–11.

CONYERS, EVELYN AUGUSTA, (1870–1944), OBE, CBE, RRC (and bar); FNM. b N.Z. dau Wm and Fanny Conyers; educ privately; early 1890s moved to Vic; Trd Chns Hosp 1894, then at Melb Hosp; Fdn Mem Vic TNA; Fdn Matron Queen's Memorial Infect Diseases Hosp Fairfield; AANS Reserve 1903; AANS AIF 1914, to Egypt Army Hosp at Abbassia nurse NZ wounded; July 1915 1 AGH, Actg Matron 3 AGH; Dec 1915 Matron-in-Chief AANS at AIF HQ Horseferry Rd, London, Rtd Vic 1918, RRC, MID; 1919 OBE, CBE, Bar to RRC, 1921 FNM; d 6 Sept 1944 Melb.

FETHERSTON, RICHARD HERBERT JOSEPH (1864–1943); b 2 May 1864 Melb; educ Wesley Coll; MB,CH.M Univ Edinb 1886; MD Melb 1889; Capt Vic Militia 1887; 1914 DGMS AIF; MajGen 1916; RMO leading Melb Hosps, practice in Prahran; Memb Legis Assembly Vic Parlt 1921, many posts in BMA; Fell Coll Surgs (Aust); Butler, in Official History, paid tribute to his absolute impartiality and wholehearted devotion to duty. d Melb 3 June 1943.

BELL, JANE (1873–1959) OBE. b Scotland 16 Mar 1873, dau William and Helen Bell; educ small school Dumfries; emigrated Australia 1886; Trd RPA Sydney 1898; 1903 succeeded sister Euphemia as Matron Bundaberg Dist Hosp; 1904 Matron Bris Gen Hosp; 1906 midwifery course Queen Charlotte's Hosp, London; 1907 Sen Asst Super Nurse Edinb Royal Infirmary; 1910–34 Lady Super Melb Hosp; 1913 Lady Super No 3 MD; recruited Vic nurses for 1 AGH; 1914 Princ Matron on first voyage SS *Kyarra*; 1915 Matron 1 AGH; June 1915 Matron Inspectoress; July returned to Aust for comm of inquiry 1 AGH; October 1915 resigned and retd to position Melb Hosp; Foundation memb ATNA 1899; Memb RVTNA 1910, Pres 1931–34, 1938–46; OBE 1944; d 6 Aug 1959 Melb, aged 86 yrs.

FIASCHI, THOMAS HENRY, (1853–1927), b Florence (Italy), 1874 to the goldfields in Nth Qld; 1877 grad MD and ChD at Florence; retd to Sydney to practice in 1879; 1891 Hon Surg (Capt) NSW Lancers; 1896 served with Ital Army in Abyssinia and was decorated by Ital; 1899 as Maj in Sth African War, Commd 1 NSW Fd Hosp, awarded DSO and MID (Twice), prom LtCol and in 1911 PMO 2 MD; 19 Aug 1914 married Sister Amy Curtis at Christ Church, Bundaberg; May 1915 CO 3 AGH at Lemnos; resigned Comm AIF 1916 to be surg in Ital Mil Hosp; wife worked with Ital Red Cross Soc; retd to Aust 1917 and rejoined AAMC (Res) as Col; retd 1921 as Hon Brig Gen; d Sydney 17 April 1927. Sister Anne Donnell in her *Letters of An Australian Army Sister* (1920) relates how she had to remain in Alexandria to nurse Col Fiaschi's wife who gave birth to a daughter (Alexandra Elisa) when 3 AGH went to Lemnos.

33. Nursing orderlies: Butler *op cit* Vol III, pp 537, 585, 1009–10.
34. BUTLER, *op cit*, Vol III, p 552.
35. *Ibid.*
36. *Ibid*, p 586
37. *Ibid.*
38. From letters to her mother in Brisbane. Sister Selwyn Smith sailed on the '*Mooltan*' in 1915 and saw service in England, Egypt, Lemnos and France, mostly with 3 AGH. After war completed a dietitian's course at Edinburgh and continued nursing career at Fairfield Hosp (Melb), Matron Toorak College Mt Eliza and as Matron, Glennie School, Toowoomba (Qld).
39. CORFIELD, A.B. Buried Chatby Military Cemetery, Egypt.
40. BUTLER, *op cit*, Vol III, pp 554–556.
41. BUTLER, *op cit*, VOL II, p 832.
42. BUTLER, *op cit*, Vol III, p 565.
43. BERRIE, CHARLOTTE. Believed to have come from Mareeba. While working at Belah in Palestine contracted 'Spanish influenza'. Admitted to an American hospital in Jerusalem and died there, 8 Jan 1919. Buried in the Light Horse Cemetery on the Mount of Olives. Mentioned by H.V. Morton in his book '*In the Steps of the Master*; p 323 but he mistakenly confuses her with her sister. 'She was a marvellous nurse and little did we dream that she would be the only woman buried with full military honours among many of the troops she looked after so faithfully. She truly gave her life on active service''.
A friend's letter.

BLACK, CATHERINE REID, b Rockhampton, Trd Diamantina Hosp, Brisbane. AANS service, 14 AGH Abbassia, H.T. 'Nile'; retd Aust on *Shropshire* 1920. Name appears on Bundaberg War Memorial. Photo in Wildman p 27.

PARKER, JANE, Trd Hosp for Sick Children (Bris); Midwifery Cert at Women's Hosp (Melb); Matron at Charleville and Tweed Hosp (Murwillumbah); Served with AANS in Salonica, Egypt and England.

MCLAUGHLIN, ELSIE, b Mt Morgan, dau Joseph and Annie McLaughlin of Rockhampton. Educ Girls' Convent School, R'ton; Trd R'ton Gen Hosp; Head Nurse Longreach. AANS Egypt 31 Brit Gen Hosp, Brit Citadel Hosp and Nazareth Hosp. Photo Wildman p 182.

COOTE, EVA FRANCES, b and educ Brisbane, dau Samuel and Fanny Coote, Ithaca. AANS, service at Enoggera, 14 AGH Abbassia, Port Said and Cairo. Retd Aust 1919 in *Orara*. Photo in Wildman, p 59.

MONCKTON, CATHRINE, b Rockhampton, dau William and Catherine Monckton of Mt Morgan. Trd R'ton Gen Hosp and on staff Mater Hosp (Bris). AANS Egypt 1917 14 AGH Abbassia, Port Said, Cairo, England. (Dartford). Retd Aust 1919. Photo Wildman, p 140.

44. CREAL, ROSE (1865–1921), b Young NSW. At 16 began working in Parkes Hospital; Trd Sydney Hosp 1899 Matron; Fdn memb TNA (NSW); 1914 Princ Matron 2 MD; 1916 Matron 14 AGH, Abbassia, 1919 RRC; Rose Creal Medal highest award by Sydney Hosp to students of Lucy Osburn School of Nursing. See also, *Gullett, Sir Henry, The AIF in Sinai and Palestine*, Sydney 1923.

AXELSEN, IDA MARIE, b. Tiaro and educ Qld; Trd Maryborough Gen Hosp and Women's Hosp Melb, staff Child Welfare Centre Bris and Matron St Mary's Priv Hosp Mary; 8 yrs Nursing Sister Central Meth Miss Bris; AANS No 2 Hosp Ship *Kanowna*, 9 voyages ME and Aust. After war Matron St Helen's Priv Hosp Bris, Westwood Sanitorium. Five of her sisters also trd nurses. her sister Clara 22 yrs at Diamantina Hosp. She and Ida opened Westwood in 1919.

LIMPUS, ETHEL, b 4 April 1889 at Capella (Qld) dau Mr and Mrs C.M. Limpus of Bundaberg; Trd Bundaberg Dist Hosp; AANS, served on medical transports ME and Aust (3 trips); No 6 AGH Kangaroo Pt; also at The Lady Chelmsford Lying-In Hosp, Bundaberg: d 1977.

45. KENNY, ELIZABETH (1886–1952), see Cohn, Victor, *Sister Kenny, The Woman Who Challenged the Doctors*. Kenny, E. *And They Shall Walk*. also Central Army Records Office, Melb; *Toowoomba Chronicle* 13 June 59; 4 Feb 65; Brisbane Telegraph 4 June 82.

46. ANDREWS, LINDA, dau of George and Ellen Andrews, 'East Kyogle'; Trd Bris Gen; lived at Southport; AANS, embarked *Mooltan*, 7 June 1917 for Egypt; service at Salonica, att 60 Brit Gen Hosp; retd Australia per *Oxfordshire*, 1919; after war Dep Matron Diamantina Hosp; dec 26 July 1940. Photo in Wildman, p 19.

DOWLING, MARIANNE, b and educ Rockhampton, dau of Vincent and Mary Dowling; Trd Brisbane Gen; Rosemount Hospital on enlistment. AANS, India 1916, Victoria War Hosp, Bombay, then to Brit Gen Hosp at Salonica; 1919 No 1 AGH Sutton Veny; returned to Aust after Armistice; Matron Yeppoon Hosp in post-war period; d 14 Oct 1939. Photo Wildman, p 84.

LAWSON, HELEN, b and educ Rockhampton, dau of Walter and Margaret Lawson, The Range, Rockhampton; AANS, 2 yrs 8 mths in India and Salonica. Photo Wildman, p 154.

WALPOLE, FRANCES GRACE, b Vic, 27 Aug 1876 educ Qld; Trd Bris Gen Hosp 1902–5; Certs Gen Obstets, Childwelfare. Matron South Warrego Dist Hosp, Cunnamulla, Nanango and at Priv Hosp Laidley. AANS 1914–18, Lemnos, Eng, France, Salonica, Egypt; after war Matron Nanango Dist Hosp and Wilson Ophthalmic School Hostel; d 30 May 1974.

MCLENNAN, JANE, b Bundaberg 27 Dec 1889; Trd Bundaberg Hosp 1914; Sister at Warwick and Maryborough Hosp. AANS 1917, service in Salonica with 3 AGH. After war Sister at Rosemount Hosp. d 21 Nov 1958. Name appears on Bundaberg War Memorial.

47. COWEN, KATHLEEN AMY, b 18 Sept 1884 at Glenelg (near Stanthorpe), 8th child of Mr and Mrs Alex Cowen. Course of pupil teacher trg at Inglewood and Marburg. Early nursing trg at the Bris Gen Hosp for Adults, followed by trg course at Bris Hosp Trg School for Nurses (1916); also completed Invalid Cookery Course at Central Tech Coll. AANS 1917–1919, service Macedonia, Salonica and England. Transport sunk on way to Port Said. While in Eng completed courses at Nat Trg School of Cookery and in Household Management. 1921 Midwifery course at Lady Bowen Lying-In Hosp Bris. 1922 worked in NZ Health Depart; 1926 Matron Harrisville Dist Hosp. 1952–49 owned and managed 'St Kilda' Nursing Home. d in Bris 1976 aged 92 yrs.

48. CHATAWAY, MARY ESME, b 20.7.1890 Cooktown. Trd Mackay District Hosp 1908–9 later Midwifery Course at Lady Musgrave Hosp Maryborough. Enlisted from Emerald. AANS 1915, service 2 AGH, Egypt, France, Rouen, India and England. After war Matron Mundubbera and Grafton Hosps. d Feb 1980.

HOMEWOOD, GRACE. One of seven sisters, all trd nurses. Trd Mt Morgan Gen Hosp, AANS 1917, embarked on *Ayrshire*, arrived Egypt and served at the Citadel, Cairo, 31 Gen Hosp Abbassia, 14 and 18 Brit Gen Hosps, and returned Aust 1919. On staff Rosemount Repat Hosp and later nursed privately in Melb. 1937 went with brother, Dr John Homewood to India. Joined The Lady Minto Nursing Association and nursed in India 8 years. Then in charge Air Cadets in Burma. 1947 retd Australia and joined YMCA. Again sent to India and also to Germany on nursing duties. Retd Melb active in care and management Homes for Elderly. Led a most colourful, interesting and full life. d Qld 1983.

HOMEWOOD, MARTHA. Trd Mt Morgan Gen Hosp, AANS 1916. Embarked on *Thermistocles*, arrived Eng March 1917, service in France, various Brit and Aust hospitals. Early in 1919 granted 6 weeks leave to attend course in Motor Driving in London and motor maintenance, a rare course for ladies at that time. After discharge married and lived on land near Rockhampton.

NORTON, ANNIE, b Somerset (Eng), trd at Mt Morgan Hosp, Matron Proserpine and Emerald Hosps, AANS Oct 1915. Embarked *Orsova*, Dec 9th at 1 AGH Heliopolis, also at Luna Park, Sporting Club, Choubra and in 1917 to England; Nursed at No 7 Stationary Hosp Boulogne, No 25 Brit Gen Hosp Hardelot, 49 CCS Achiet le Grand, No 3 Aust Gen Hosp Abbeville, then to Dartford (Eng). Retd Aust 1919, with her sister Ellen Agnes Norton on the *Zealandia*.

NORTON, ELLEN AGNES, b and educ at Capella (between Clermont and Emerald), dau of Richard and Mary Norton. Sister of Annie Norton who also served in AANS. Trd Rockhampton Gen Hosp, Act Matron Springsure, AANS Oct 1915. Embarked on *Orsova*, with her sister. Served No 3 AGH Heliopolis, 25 Brit Gen Hosp, Boulogne, 49 CCS Achiet le Grand, 3 AGH Abbeville, then to Dartford (Eng). Retd Aust 1919 with her sister, Annie Norton on the *Zealandia*.

SKYRING, GERTRUDE MAY, b Bundaberg 1890, educ there and trd Bundaberg Hosp, AANS 1915–1919, service Heliopolis, England, France, d 1981.

TOFT, ALICE MAY, b Bundaberg 9 June 1885, dau Joseph and Sarah Toft. Trd Bethesda Hosp Melb, also completed Midwifery course there. AANS 1915. Embarked *Orsova*. Service with various Brit and Aust hosps in France. Retd Aust per *Aeneas* 1919. d Bundaberg 1928.

TOFT, C.S. Embarked 5 May 1917 on *Ulysses* to Eng via Cape. Sent to Croydon War Hosp, Engram Rd Section (massage, hot air and electrical treatment). 1918 to France (Hardelot) and to 3 AGH (Abbeville). Later posted to 3 AGH at Dartford.

TOFT, EDITH MARY, b Rockhampton, dau Prudence and William Toft of Koongal. Trd Rockhampton Gen Hosp, private nursing Sydney. AANS, 1915. Service Enoggera Hosp, embarked for Eng late 1916, arrived Mar 1917. Sent to France and attached Brit forces Rouen and Boulogne. Served at Harefield and Sidcup. Retd Aust per *Aeneas*, 1919.

WILSON, EVELYN, WILSON, MADELINE, WILSON, MARJORIE. The names of these three sisters appear on the Nanango War Memorial.

WILSON, MAY, WILSON, MYRTLE, WILSON LILIAN. These three names appear on the Bundaberg War Memorial.

BIRT, MARY TRAVENEN, b and educ in Adelaide; Parents moved to property near Windorah; Trd Bris Hosp 1912. Nursed at mining town Hampton, (between Townsville and Cloncurry;) 1915 AANS, to Eng and France on *Thermistocles*. Nursed wounded at Rouen with No 12 Brit Gen Hosp; Ret Aust 1918. After war Sister Birt and Matron Macdonald ran Chelmer Conv Home. d aged 66 yrs.

49. Carrel Dakin method: Devised by Alexis Carrel, American surgeon, and Henry Dakin, English research chemist. 'The treatment of wounds by regular intermittent irrigation through surgically placed rubber tubes, using an antiseptic solution of chlorinated soda to obviate infection in contaminated wounds and to hasten asepsis in suppurating wounds.

50. NICHOLLS, RUBY b Ipswich, educ at Toowoomba; dau of Charles and Marie Nicholls, 'Trelawney' Stuart St Toowoomba. Embarked on *Orsova* in July 1915 for Egypt. Service 2 AGH, also on transport duty to Australia, also in Boulogne, France, with No 13 Brit Stationary Hosp and 83 Dublin General. Attached 1 ACCS. Retd to Aust and on duty at Stanthorpe Sanitorium. Photo in Wildman p 192.

LOOSEMORE, M.A. b and educ at Gympie; dau of William and Ann Loosemore of Inglewood Hill, Gympie; AANS — Mil Hosp Enoggera 1916. Emb overseas 1917. Service at Harefield and Southall Mil Hosps and then to Brit Hosps in France and back to Southall in Eng; Retd Aust 1919 on SS *Ulysses*, to Enoggera Mil Hosp. Photo in Wildman, p 157.

MCCLELLAND, MARIA ALEXANDRA, b and educ Rockhampton, dau of Lowry and Fanny McClelland, Mt Morgan; Trd Mt Morgan Hosp, private nursing Bris. AANS 1915, embarked on *Mooltan*, to No 1 AGH at Heliopolis, then transport duty to Aust No 6 AGH Kangaroo Point, then retd to Egypt to 14 AGH Abbassia. In 1917 'lent to British for the duration'. Service in France with 2 BGH Le Havre, 38 BGH Calais, 38 Brit Stat Hosp Genoa, 9 CCS Italy and at Dartford Hosp, London. Retd Aust 1919. Photo in Wildman p 179.

CLERKE, EVA. Referred to as a friend of Sister Mary Birt. Both trd together at Bris Gen Hosp 1912 and after trg both went to the small mining town of Hampton near Cloncurry. Both joined the AANS and were at Enoggera together. Both served in Brit and France and after war remained firm friends in Brisbane.

51. As stated in a letter home to her mother in Brisbane.
52. BUTLER, *op cit* V, III, p 557.
53. BUTLER, *op cit* VII, p 382.
54. *Ibid.* pp 382–383.
55. GOGGINS, MARGARET ELLEN, dau Michael and M. Goggins of Tannymorel, Warwick; Trd Warwick Gen Hosp, staff of Gympie Hosp and Matron Longreach Hosp; Head Sister Cue-Day Dawn Hosp (W Aust); AANS 1915–19, service 2 AGH Egypt and 2 ACCS France; wounded in bombing attack and invalided to Aust, continued nursing in NZ and in New York. d Mater Hosp Bris 7 Oct 1926.
56. DERRER, MARY JANE, M.M. b 3 Feb 1898, Rosehill near Mackay; educ at State Schools and at the Misses O'Connor School, 'Duporth' at Oxley; Trd at Mackay Gen Hosp, 1911–14; AANS 1915, embarked on *Orsova*, service with 2 AGH in Egypt, with 14 Stat Hosp, 3 AGH and 2 ACCS in France; awarded Military Medal during B of Messines 1917, also at Dartford, England. Retd to Aust 6 AGH Kangaroo Point; embarked for India 1918 on *Wiltshire*. In India nursed 34 Welsh Gen Hosp Deololi, Brit Gen Hosp Rawalpindi, discharged Jan 1920.

KING, EILEEN, MM, RRC, b 1884, dau T.M. King, ISO, Auditor-General of Qld; believed to have trd in England; QAIMNS 1915–1918; served in France, Egypt, Lemnos & hospital ships. Sister of R.M. King, MLA, Deputy Premier in the Moore Gov; retd to England 1937 and nursed there during war years 1939–43. She was presumed lost with other passengers on the *Melbourne Star* when it was sunk returning to Aust in 1943.

KING AMY, RRC. Sister of Eileen King, above, b 1882; believed trd in England; served QAIMNS France, Egypt, Lemnos & hospital ships.

57. Report. War Diary No 2 ACCS, AWM.
58. PARKER, ELIZABETH. Trd Rockhampton, AANS in Egypt, No 3 AGH Lemnos, also in Salonica, France, Belgium and England. Private nursing in England after the war. 1938 attended unveiling of Villers — Bretonneux Memorial by His Majesty King George VI. d London 1965.

FISHER, MARY ELLEN, RRC. b Gladstone (Qld) 8 Sept 1889; AANS 1915–1919. Served in Eng and France. Awarded RRC and ARRC. Served on *HT Ypiranga* as Sister-in-Charge, with other Queensland Sisters Nea Low, Mary Matheson, Florence Auld. d Caloundra (Qld) 5 Oct 1981, aged 92 yrs. Photo in Wildman p 90.

JESSUP, ELIZABETH, AANS 1915, service with 6 AGH at Kangaroo Point. Embarked Feb 1917 for Plymouth via South Africa. Posted to 3 AGH, then to 3 AGH, Abbeville, 14 Brit Gen Hosp at Wimereux and 3 ACCS. With 2nd Army advance into Belgium and retd to 3 AGH.

PARNELL, LOUISA S, b Port Douglas, dau of Hon Arthur Parnell and Barbara Parnell of Rockhampton, educ at Bris Girls' Grammar. Trd at Rockhampton Children's Hosp: AANS Aug 1915 — sailed on *Morea* for Egypt, attached No 1 AGH, Heliopolis. Then to Mena House and Luna Park, and No 3 AGH Abbassia; Dec 1916 to England — went to Netley Hosp, 1917–18 Southall Hosp. Then to Harefield and Rouen. Prom from S/Nurse to Sister. 1919 retd Aust per *SS Orsova*.

KENNEDY, JESSIE VIOLET MARION, RRC. Trd as a nurse at Rockhampton Hosp; AANS 1915–1918, service in Egypt, France, Great Britain. Matron Sidcup Military Hosp, Kent. Matron HMT *Somali* Awarded RRC for exceptional service to nursing. Bestowed on her by Queen Alexandra, widow of King Edward VII.

THE MACDONALD SISTERS. The six daughters of Mr and Mrs Donald Macdonald of Raglan, south of Rockhampton became a nursing legend in Queensland. All became nurses. *Sadie* served in Heliopolis, Egypt, France and Eng in First World War. Between wars she was Matron Ardoyne Hosp Corinda. In World War II she was Matron Red Cross Home Chelmer, and Matron Pres-Meth Girls' Hostel, Archibald Hse New Farm for 12 yrs. Awarded RRC and MBE and first Qldr awarded FN Medal. Died Repat Hosp Greenslopes, aged 95 yrs.

Flora also served in First World War. Later she opened a Baby Clinic and Health Centre in Brisbane.

Mary organized Bush Nursing Association and opened a Nurses Home in Brisbane.

May and *Lena* opened Brisbane's first maternity hosp at Highgate Hill 1920–1945. May died at age of 100.

Grace was Matron of Hillcrest Private Hosp Rockhampton for many years.

59. BUTLER, *op cit*, Vol II, pp 383–384.
60. From letter home. The town was Ypres!
61. BUTLER, *op cit*, Vol II, p 384.
62. *Ibid*, Vol III, p 564.
63. *Ibid*, pp 558–563.

JAMES-WALLACE, FLORENCE E. Trd Bris Gen Hosp; AANS 1914; Served Egypt, Lemnos and France, 1918 att 61 Brit CCS.

64. HUGHES, SUSAN. From Central Army Records Office, Melb.
65. ECHLIN, GLADYS. Trd Childs Hosp Bris. Sailed on *Kyarra* 1915. Service in Egypt, England and France.
HERBERTSON, FLORENCE, b Clunes Vic 21/3/86, educ Gladstone S.S. and Maryb. Girls' Gr Sch. Trd Gladstone Hosp and Bris Hosp. Father pd £5. Nurses prov own uniform and laundry. At end trg each nurse pd for certificate! She and Matron Marion Dowling first staff Kingaroy Hosp. 1913–14 Longreach Hosp. (Matron Henry). Relieving at Peak Downs (Clermont) and Roma. 1915 AANS Enoggera. Emb on *Orontes* to ME to No 1 AGH, Heliopolis, Ismalia and Abbassia (No 1 Aux Hosp); 1916 to Dartford in Kent, to No 3 Aux Hosp; 1917 retd on *Ulysses*. Then Stanthorpe and Rosemount. d 1962 aged 76.
66. From letters to her mother in Gympie.
67. See Fn (2)
68. DERRER, ROSINE, b Mackay and trd Mackay Hosp, enlisted 25.8.16 aged 29. AANS War Service Record: 16.9.16 emb Sydney *Kermala*, 24.11.16 Gerard Freeman Thomas Hosp Deccan, 5.3.17 Deccan War Hosp. 16.9.18 Emb for UK. 8.8.19 RTA Katoomba. 10.11.19 Disch 1 MD.
69. There was also some difficulty in settling accounts between the Indian and Australian Governments. Capt F.H. Wickham was sent from France to act as AIF representative in India and Mesopotamia, attached to A-G's Dept at Indian Army HQ. The quotation is from his Report AWM Files 509/7.
70. MOBERLY, GERTRUDE F., RRC. *Experiences of a 'Dinky Di' RRC Nurse*. Aust Med Pub Co Sydney, 1933, p 46. Gertrude Moberly was the daughter of the Rev E.C. Moberly. She became Matron of No 6 AAH, Moreton Gardens. Later she served in India, in Bombay and on Hospital Ships. She was Matron of Hislop War Hospital, Deccan.
71. BURKE, ERIC KEAST. *With Horse and Morse in Mesopotamia. The Story of Anzacs in Asia*. Arthur McQuitty & Co. Syd, 1927. pp 124–130: The Australian Nurses in India 1916–1919. pp 198–200: The Australian Army Nursing Service (India).
72. PEARCE, IDA ELIZABETH HORTON, b Crow's Nest Qld, educ and trd in Qld; 1917 AANS, staff of Rosemount Hosp, sailed for India SS *Wiltshire*, 12 mths Bombay and Quetta. After war St Martin's Hosp Bris and Matron St Denis Priv Hosp, Toowoomba.
PAYNE, WINIFRED. Trd Chns Hosp Bris 1913. AANS India 1917, nursing at Bombay, 34 Welsh Gen Hosp. After war at Chns Hosp. One of the founders of the Union Jack Club, Brisbane.
POLLOCK, ELSIE JANE, b and educ MaMa Creek, Grantham; AANS — sailed July 12 1915 — to Lemnos and various hospitals in Egypt 1916; on transport duty to Aust; then to India 1916–18. Returned to Bris May 1919. Photo in Wildman, p 216.
MONCKTON, NONIE, b Rockhampton. Educ at Presentation Convent, Wagga Wagga. Trd St Vincent's Hosp Sydney. Nursed at Garrison Hosp 1915-16. Then to India — nursing there 1916-19. Sailed on *Wiltshire* for Eng. Duty at Dartford and Weymouth. 15 April 1919 m Capt R. Woodside RAMC. ANJ Vol XVII, No 7 July 15 1919, p 237 + Photo in Wildman, p 140.
73. See fn 68.
74. LEYLAND, BEATRICE ALICE, b Barcaldine, educ Barcaldine and Cambridge College, Stanthorpe; dau of James and Susan Leyland of Barcaldine. AANS — Enoggera 1917, India 1917 at Deolali (34 Welsh Gen Hosp), then to Cairo (71 Brit Gen Hosp Eaza), 3 AAH Dartford, Eng; retd Aust 1919. Photo Wildman, p 155.
75. MCILWRAITH, JANET ELLE GORDON. Trd Mt Morgan, AANS — 6 AGH Bris, service with 44 Brit Gen Hosp, Deolali, India. 1919 Rosemount Hosp Bris.
76. DENNIS, LILLIAN BEATRICE, b Bris 1886, educ Normal School Bris; dau of Alfred and Elizabeth Dennis of 'Greylands', Harts Rd, Indooroopilly. AANS: 2½ yrs in India (Bombay and Poona), 6 mths in Eng. Matron Anzac Hostel, Kangaroo Point.
77. CHRISTIANSEN, VICTORIA. Trd at Chns Hospital, Bris. Enlisted 1914, sailed on the *Orsova* 1915; AANS service: Convalescent Home Sth Moreton Gardens, Sth Kensington; (Col Symes wanted someone 'attractive but not frivolous'). Then to Coulter, the Duke of Manchester's Home where she nursed Canadian wounded. Next to 3 Brit Gen Hosp at Tooting, where she was made an Hon Member of QA's. Also transferred to hospital at Rawalpindi, India. 1919 rtd to Kangaroo Point.
78. BASSETT, M.E.V.
Name appears on Nanango War Memorial. Served in India.
FRANCIS, STELLA. One of a family of 7 daughters, 6 of whom were trd nurses. Trd at Bris Gen Hosp and with 2 of her sisters served with AANS in WWI. Service in India. On return Matron St Mary's Priv Hosp, Maryborough. With 3 of her sisters bought home in early 1920s and 1945 sold to Presbyterian Church. Active in affairs of Nurses Rest Home at Bowen Hills. d June, 1969.
IMISON, A.E. Enlisted Brisbane 1916. Served at Kangaroo Point with Miss Bishop (Matron). Served in Britain and India. Served with QA's in WWII. Evacuated from Dunkirk. Awarded RRC for work in Britain. Also served in ME three years at Heliopolis. Assistant Matron in Algiers.
79. O'KANE, ROSA. Dau Mr and Mrs J.E. O'Kane of Charters Towers, Qld; prior to enlistment service at Charters Towers and Townsville Hosps; died of illness 21 Dec 1918, aged 28 yrs. Buried at Woodman's Point Cemetery at the Quarantine Station. Also commemorated in the Western Australian Garden of Remembrance.

HOGG, JESSIE, b. 19 April 1888, dau. Dr James Hogg, Med Sup Goodna Mental Hosp; educ Ipswich Girls' Grammar School; nurse trg Gen Hosp 1909–13; served Brit Red Cross 1915; enlisted AANS 1918, embarked on *Wyrema* for Salonica; nursed flu victims at Woodman's Point Quarantine Station and contracted disease. d. Toowoomba 1943.

ROBSON, JANE SELINA, b 16 Nov 1892. Trd Ipswich Gen Hosp. 1913–1917. AANS — S/N 7/10/1918. Also nursed at Woodman's Point, d 11 Oct 1962.

MORRIS, STELLA. 1918 on *Wyrema*. Armistice. Ship diverted at South Africa. Served at Quarantine Station, Woodman's Point.

80. *Annual Report*, PMO, 1MD, 1916–17. AWM, File 399/17.

81. AWM File 399/92.

PHILLIPS, VIOLET — married Maj N. Hamilton Fairley of Vic, Feb 12, 1919.

O'BRIEN, MARY ANN. Longreach Hosp Probationer 1909, Trd 1913, Staff 1913–16. (Matron Henry in charge). Muttaburra Hosp 1916–17. AANS 1917–19. She and Matron Henry went to Cairo on same troopship; Citadel Hosp Cairo, Enoggera Hosp 1919–20; Boulia Hosp 1920–25.

82. CAIRNCROSS, SARAH LEA. Trd at Mt Morgan Hosp, 4 yr trg course; completed 1916. Matron Nanango Hosp and accepted for AIF. Attached to 6 AGH at Kangaroo Point. (Matron was E. Bishop). Subsequently married H.S. Rowe, Kingaroy. d 1957 aged 72 yrs. Her name and names of other Sisters appear on the Nanango War Memorial.

HUTCHISON, FANNIE MYRTLE. Trd Bris Gen Hosp. Gold Medal for Pract Nursing, pre-enlistment at Wattlebrae; AANS: 6 AGH, Kangaroo Pt, 17 AGH Enoggera until 1920.

83. Diary of Sister Isambert.

Other AANS Sisters known to have served in WWI include the following:

CHAPMAN, MINNIE LOGIE. b 1885 Orkney Is. Scotland; educ Gympie, father Pres Mins there; trd Bris Hosp 1911–14; Army nursing 1914–18 with QAIMNS, Eng, France, Belgium; retd Aust 1920. d 1968 Brisbane.

GILL, HONORA JEANNE. b 1892; trd Warwick Gen Hosp; Staff Nurse Grafton; AANS 1917–1919; Served Citadel, Heliopolis and Abbassia; also 14 AGH Port Said.

GEDDES, MARION. b Rockhampton 1882. AANS WWI; Matron Goondiwindi Hosp some 20 yrs.

JONES, MARGARET ELIZABETH. b 1891, trd Rockhampton Gen Hosp, obstet Launceston Queen Vic Hosp; AANS 1917–1920, service England, Sutton Veny, Salonica. After war, Matron Rockn Gen Hosp, Maternal & Child Welfare. d 1983.

MOORE, IDA ANNA. b 1889 Charleville; trd Bris Gen Hosp 1910–1913; Staff Warwick Hosp 1914–15; AANS 1915. Sailed *Kyarra* 1915, service 2 AGH Egypt and France. Staff Rosemount and country hospitals. d Brisbane 1971.

SEARCH, DOROTHY. AANS, WWI; Service in India. Matron Beaudesert Hosp.

TOLMIE, HELEN (ELLA)

TOLMIE, SARAH

TOLMIE, AGNES These three nursing Sisters were members of a well-known Toowoomba family. *Ella* was Matron of Toowoomba Gen Hosp 1897–1917, when at the age of 53 she transferred to military nursing and was posted to 6 AGH, Kangaroo Point. In 1893 she was present at the first appendicetomy operation performed in Australia. This was at Drayton. After WWI she opened a private hospital 'St Andrews'. d 1945, aged 79 yrs. *Sarah* served on the staff of Toowoomba Gen Hosp while *Agnes* was Matron of Maryborough Gen Hosp in 1905 when there was an outbreak of the pneumonic plague. For her services to nursing Agnes Tolmie was awarded the OBE. Neil Yeates, *Stone on Stone, A Pioneeer Family Saga*. pp 75–77.

WARNER, ISABEL ANNABEL. b 1876 Gympie, one of 13 chn; trd Toowoomba Gen Hosp; AANS 1914–1918, England.

WINNING, MARGARET Y. Trd Ipswich Gen Hosp 1903–1908; AANS 1914–1918. Service France & ME. College of Nursing

CUSKELLY, ANNIE. b. 29.9.1887; trd Toowoomba Gen Hosp; served AANS France & England, 1915–18; in 1985, aged 97 the oldest surviving member of AANS in Queensland.

Chapter Four

Continuing Traditions (i):

World War II (1939–1941). Middle East, Greece, Crete, Far East and Britain.

Members of the nursing staff were all called upon to undergo considerable strain and hardship under conditions of no little danger, under which, they without exception displayed considerable courage. Their bearing and cheerfulness did them the highest credit.

LtCol Coppleson to MajGen Burston on the evacuation from Greece and Crete.

Between the Wars

With the end of the war and the eventual return and demobilization of all servicemen and women, including members of the AANS, consideration was given to the future requirements of the army medical corps, both by the establishment of a reserve and the continuation of a small permanent army. With regard to the nursing service, the review considered the status of the service at the outbreak of the war, its contribution during the war, its position in 1920 and plans for its reorganization in the years of peace ahead.(1) In 1914 the total number on the reserve was 108 (1MD — 16). The reason for this was partly economic, as the budget allowed 1 pound ($2) per annum for each 'efficient', i.e. one who had qualified in first aid and had attended 3 out of 4 lectures on organizing Military Hospitals, Hygiene and Military Surgery. This organization was so revitalized that during the war 2,582 nurses were enlisted; in 1 MD — 258 served abroad, 169 served within Australia and 16 served with Imperial organizations.

In 1921 a decision was made to review the AANS lists in all Military Districts, to remove the names of deceased members, to retire others on account of age, marriage or for other reasons. After this an attempt was made to bring up to date a seniority list for each MD. In 1 MD (Queensland) the names of 8 members who had reached the retired list, 55 who had married, 12 deceased and 42 for other reasons were removed. The DDMS 1 MD proposed that as the number of troops in 1 MD exceeded one Division, the establishment for medical purposes should be based at that level, i.e. 2 General Hospitals, 2 Stationary Hospitals and 1 Casualty Clearing Station. He recommended that the number of nurses required for 1 MD be a total of 132 to cater for the needs of the above medical units. To keep within budgeting requirements, vacancies due to natural wastage would not be filled, Matrons and above to receive £1 p.a., Sisters and below 15/-

An earlier generation of Sisters relax at the Pyramids 1918. From left: Sisters McLaughlin, Grace Homewood and Rowe.

per annum, 'as the duties to be performed under present peace conditions do not warrant additional expenditure'.[2]

Between the wars the Australian Army Nursing Service was kept alive by the maintenance of the 'reserve'. Records were kept in all States of those trained nurses who had volunteered to join the 'Reserve' and who were willing to serve in a national emergency. However, there was no organized attempt to give them any army training and no attempt was made to use their services in the militia or in army camps. In Queensland, responsibility for the administration of the 'reserve' was vested in Miss E.M. Paten, a veteran of the First World War, who had been appointed Principal Matron 1 Military District, from 1 July, 1923. This was a part-time appointment while she was engaged in civilian nursing. It is no reflection on her ability that so little was done, but Australia went through a period of bitter reaction from the horrors of the world war, believing in the idealism of the League of Nations that there would be no more wars. It must have been very frustrating for a dedicated nurse administrator to find so little government support for what she believed were essential defence requirements. As a token gesture from a grateful government Miss Paten and other Principal Matrons were sent with the official contingent to the coronation of King George VI in 1937. Queenslander Grace Wilson who had been Principal Matron 3 MD since 1923 and subsequently Matron-in-Chief led the contingent.

War Memorials

This was a period when throughout the land war memorials were erected in memory of those who had enlisted in the Great War and for those who paid the supreme sacrifice. Some were large and imposing memorials in the capital cities, but all cities and towns, large and small, built monuments on which were

Another generation of sisters at the Pyramids, 1941. Sisters Palmer, Carmichael, Soorley, Vad, Gordon.

inscribed the names of the young men (and in some cases, women) of the district who had volunteered. Sometimes it was a plain obelisk, with a soldier standing at the ready with his rifle, sometimes it was a grander edifice. At the entrance to the lovely city park in Warwick, memorial gates were erected and an imposing memorial listing the names of several hundred who enlisted from the district was formally dedicated. Among the many names were Sisters Holmes, P.L.C., Goggins, M.E., McLean, C.E., Smith, A. and Wood M.[3] A simpler memorial in Nanango recorded the names of Sisters Bassett, M.E.V., Cairncross, S.L., Wilson, E., Marjorie Wilson and Madeline Wilson.[4] In Bundaberg, while a major memorial was erected in the main street, a special plaque was set in the small park near the hospital, on which were inscribed the names of those nurses from the district who served in the war — Sisters Edwards, Limpus, E., Mitchell, Harte, M., McLennan, J., Quinn, L., Relf, G., Skyring, M., May Wilson, Myrtle Wilson, Black, -, and Lilian Wilson.[5] So it was in Gympie, Rockhampton, Mackay, Townsville, Cairns and Toowoomba. In Maryborough there was a curious twist to a beautiful piece of statuary reflecting the armed services. As it was executed in Italy by an Italian sculptor who had no uniform to copy, the figure of the nurse is modelled on that of an Italian nurse![6]

In towns and cities throughout Queensland, grateful Councils presented imposing scrolls and certificates to those who had volunteered, including members of the AANS. The Gladstone Town Council and Calliope Shire Council presented a magnificent scroll to Sister Mary Ellen Fisher of Gladstone, but it was a replica of one given to all soldiers, even addressed to 'Dear Sir', and concluding, 'We have the honour to be, Dear Sir!' Sister Daisy W. Squire was more fortunate, as she received from the Mayor of Rockhampton a scroll obviously composed specifically for nurses.

War Memorial, Maryborough (Queensland). Note scissors and crucifix on left side.

Churches throughout Queensland also remembered those in their congregations who served their country. Honour Boards were dedicated, stained glass windows were a prominent feature, while other physical reminders such as gates, chapels and plaques adorned the churches. Christense Sorensen's name appears on the Honour Board at Sandgate Baptist Church, Nurse Dorothy Brown and Ruby Browns' names appear on one at St Paul's Presbyterian Church, Brisbane. Each city built a major memorial, the one in Anzac Square, Brisbane containing a crypt commemorating the fallen nurses. One of the most interesting memorials was the Nurses' Honour Board unveiled on 18 June 1917 in the War Chapel of St Luke's Church of England, Brisbane. The event was recorded in *The Church Chronicle* of 1 August 1917 by David Garland, Resident Chaplain, Director, Soldiers' Church of England Help Society.

Underneath the cross which surmounts the board the name of Florence Nightingale is carved, a tribute to the Mother of the British Army of today, as well as of the Red Cross movement. The inscription reads: ''In honor of those

AANS Memorial Plaque, The Crypt, Anzac Square, Brisbane.

St Luke's Honour Board, now at Anzac House, Brisbane.

Memorial to Allied Nurses, Reims, France.

members of the Australian Army Nursing Service who went from Queensland to serve in the Great War 1914–." "Women which followed Jesus ministering unto Him"; and then come the names of 200 of the noblest of our womanhood. The service, which was brief, was very memorable, and for many reasons, is one which will stand out in our memories among all the services we have ever attended. We had the hymn for absent friends, and then the board was unveiled. We said the "Our Father" and the 20th Psalm, a prayer for the absent nurses and a prayer for those who had died on service. After the address we sang a hymn for the departed, and then the National Anthem kneeling, ending with the blessing. In the front of the Church were all the army sisters who could get off duty, among whom were some who themselves had been to the Front and endured hardship and illness for the sake of our boys. Very sweet they all looked in their red capes, worn in memory of Florence Nightingale, over their uniform and with their white veils. Military officers and other high military officials honoured us with their presence, and all of us of different denominations prayed together for those noble women.[7]

On the 8 June 1933 a meeting of '26 nurses and masseurs' met at Anzac House and after an address by Mr Huish, State President of the RSSILA, decided to form a Returned Sisters Sub-Branch with Miss Paten as the first President. In 1936 an 'Honour Board' was handed to the Returned Sisters Sub-Branch, by Canon Garland, but it apparently was not the original Board. This second Board is now at Anzac House, Brisbane. The fate of the original Board is not known. St Lukes remained the Mission Headquarters until 1950 and the Synod Hall until 1977. It is today no longer a church building but has become well-known as the Pancake Manor.

Bundaberg War Memorial to Sisters of World Wars I and II.

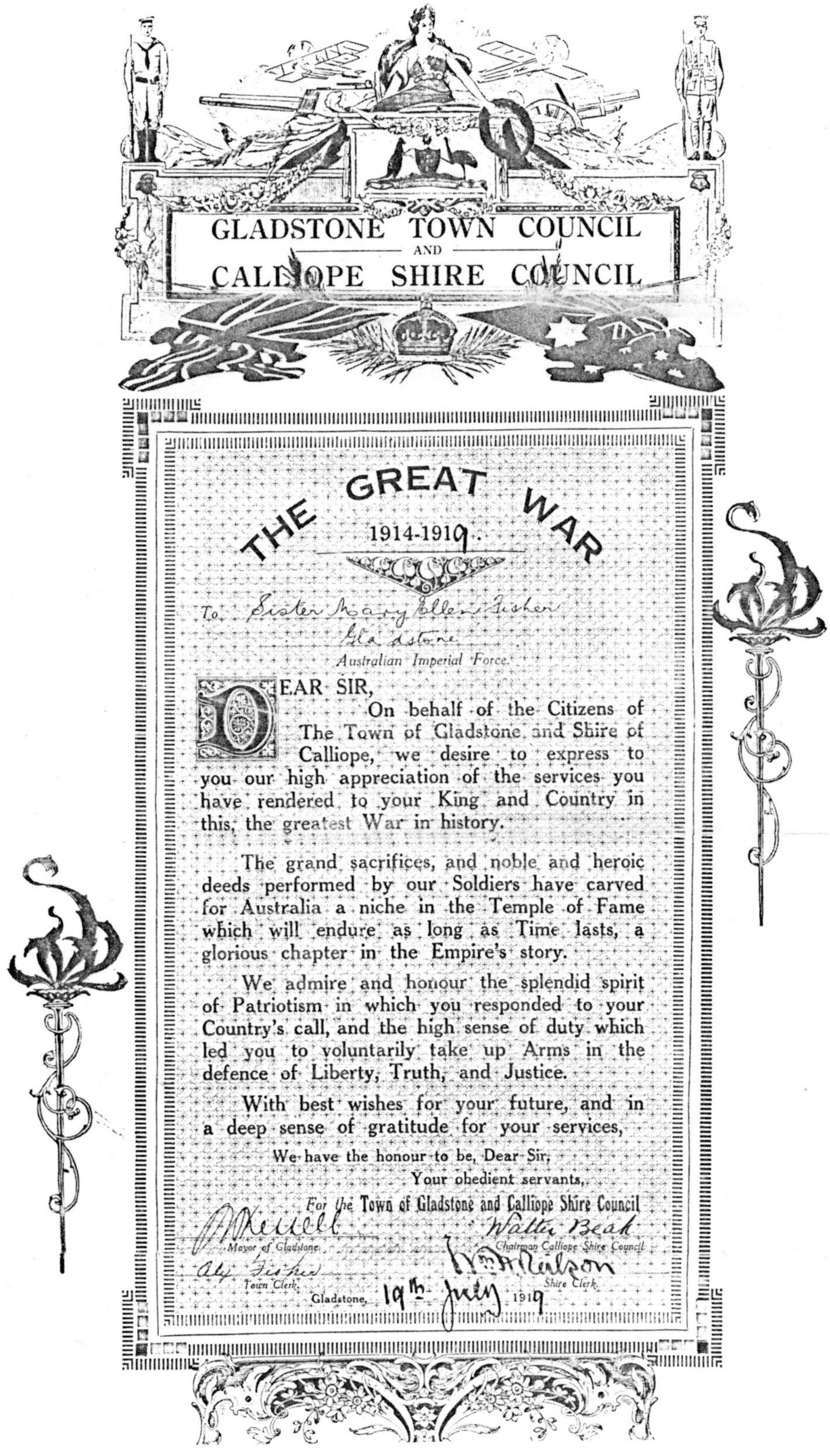

GLADSTONE TOWN COUNCIL
AND
CALLIOPE SHIRE COUNCIL

THE GREAT WAR

1914-1919.

To: Sister Mary Ellen Fisher
Gladstone
Australian Imperial Force.

DEAR SIR,

On behalf of the Citizens of The Town of Gladstone and Shire of Calliope, we desire to express to you our high appreciation of the services you have rendered to your King and Country in this, the greatest War in history.

The grand sacrifices, and noble and heroic deeds performed by our Soldiers have carved for Australia a niche in the Temple of Fame which will endure as long as Time lasts, a glorious chapter in the Empire's story.

We admire and honour the splendid spirit of Patriotism in which you responded to your Country's call, and the high sense of duty which led you to voluntarily take up Arms in the defence of Liberty, Truth, and Justice.

With best wishes for your future, and in a deep sense of gratitude for your services,

We have the honour to be, Dear Sir,
Your obedient servants,

For the Town of Gladstone and Calliope Shire Council

Mayor of Gladstone.

Walter Beak
Chairman Calliope Shire Council.

Town Clerk.

Shire Clerk.

Gladstone, 19th July 1919

To

SISTER DAISY W. SQUIRE

A. A. N. S.

On Behalf of the Residents of **Rockhampton** and District we desire to express to you our high Appreciation of the Services you have so nobly rendered in this, the most cruel of all Wars.

We Admire the Patriotic Heroism you displayed in Voluntarily going Overseas and facing the dread perils of War and Disease. The Tender Care with which you Nursed our Sick and Wounded Sailors and Soldiers has restored to us many who would otherwise have been lost, and your Loving Devotion soothed the last hours of those whose lives could not be saved.

With a deep sense of gratitude for your services, we wish for you the Peace and Happiness you have so amply earned.

R C Hartley
Mayor of Rockhampton.

Thos Smith
Chairman of Livingstone Shire.

Jas Armstrong
Chairman of Fitzroy Shire.

The Great War

Another war memorial had an international connection for Australian and Queensland nurses. In June 1923 the Australian Red Cross Society received a letter from the President of the League of Red Cross Societies, Paris with the information that a subscription had been opened by Madame Juliette Adam for the purpose of raising a monument to the glory of French and Allied Nurses. A site was selected on the Esplanade Ceres in the town of Reims and on it were inscribed the names of those nurses who gave their lives in the Great War. The names of Queensland Sisters, Norma Mowbray, Agnes Corfield, Charlotte Berrie and Rosa O'Kane were forwarded to Paris.[8]

On the outbreak of war in 1939 there were an estimated 13,000 trained nurses in Australia, but only 600 on the AANS reserve throughout all States. In Queensland Miss Paten was called up for full-time duty as Principal Matron of Northern Command, a position she held through the initial but important early years of the war until her retirement in February, 1941. In September, 1939 recruitment and training centres were set up at strategic places around Queensland. Even though the war seemed far away, the response to the call to arms was spontaneous. Of immediate concern was the provision of adequate medical facilities to service the needs of many thousands of Queenslanders flocking to the training centres for medical examination. Despite the need for nurses, entry requirements remained high.

Conditions for enlistment

(i) a trained nurse, registered by the Nurses Board in a State of the Commonwealth of Australia, preferably with several years' nursing experience as well as an additional nursing certificate;
(ii) a British subject, domiciled in Australia;
(iii) single or a widow or a divorcee without dependents;
(iv) between the ages of 25 and 35 for overseas service in the case of sisters and staff-nurses and below the age of 40 in the case of matrons;
(v) passed fit A1 by medical examination;
(vi) of good character with personal attributes essential to the making of an efficient army nurse.

Sister E. Hanrahan who was in charge of the first group of Queensland nurses was reported in the local press as saying that so rigid were the fitness tests for nurses enlisting for service abroad that a bunion was sufficient to disqualify an applicant!

Rates of pay for members of the AANS were announced by the War Cabinet.

Matron-in-Chief	19/6' per day 3/- per day deferred pay
Principal Matron	17/6 per day 2/6 per day deferred pay
Matron	15/- per day 2/6 per day deferred pay
Sister	10/6 per day 2/6 per day deferred pay
Staff nurse	8/6 per day 2/- per day deferred pay.

There was immediate newspaper criticism of the conditions laid down for the nursing service, their rates of pay and their uniform grants.

The Principal Matron of Northern Command, Sister E.M. Paten, said that nurses going overseas would have to spend 'little if anything' on their equipment. They received a grant of £30 ($60) each from the Commonwealth Government to cover cost of uniforms, another of £10 ($20) from a private fund while all nursing equipment would be provided by the military hospitals to which they were attached. There was some public criticism of this procedure as most nurses found their basic outlay was more than £50 ($100). One newspaper described this system as 'ridiculous and illogical'. As an integral part of Australia's overseas service, their position was no whit different from that of the men. For them the Government provided the uniform and clothing and all their necessary equipment. Why should nurses be treated differently?

The senior officer of the Australian Army Nursing Service at this time was Queenslander Grace Wilson, who had had such a distinguished nursing record during World War I. Between the wars she had continued her nursing career as Matron of numerous hospitals, while still serving on a part-time basis as Matron-in-Chief of the AANS. On the outbreak of war Grace Wilson was immediately called up for full-time duty, a challenge she willingly accepted, although at sixty years of age she might have been considered too old for active service. The army wisely decided to make full use of her years of nursing experience, both in military and civilian hospitals. The Army Medical Directorate was reorganized and plans drawn up to cover the medical requirements of the armed forces, both militia and those enlisting in the Sixth Division AIF. As Matron-in-Chief, Grace Wilson's responsibility was to advise the Director-General of Medical Services, Major-General R.M. Downes, of the nursing requirements for the great variety of medical units being established, to which nurses might be expected to be posted. Within Australia there were camp hospitals, which at first were staffed by male nursing orderlies, base hospitals which were planned in each State, while other units were being raised for overseas service. These included General Hospitals, and Special Hospitals, Casualty Clearing Stations, Hospital Ships and Sea Transports, Ambulance trains and Convalescent Depots.

AANS 1940. Some of first Queensland nurses to enlist. Photo taken at Hotel Australia, in Sydney, at a party given by NSW Returned Sisters, World War I.
Back Row L. to R: K. Stewart, M. Glasgow, J. Dickson, D. Coleman, M.V. Reid, U. Hely-Wilson, I. Thompson.
Centre Row L. to R: E. Hanrahan, Miss Carey (Princ. Matron NSW) F. Petersen.
Front Row L. to R: L. Keys, E. Doig, V. Harland, F. Harvey, I. Hoey, P. Miller, R. Stanmore (NSW), R.E. Davis.

There were, of course, other medical units, such as Field Ambulances, to which nurses were not normally posted. It was still laid down as a basic principle that nurses should not be placed in positions of danger or where they might be taken prisoner of war. This principle related to the mostly static war of 1914–18 but as subsequent events proved in the Second World War, this policy had to be modified considerably.

Early in 1940 an advanced medical centre had been established in Jerusalem to plan for the arrival of the Sixth Division, estimated to be about 20,000. For normal sickness at an accepted rate of .03 per cent, almost 1,000 beds would be necessary. The 2/1 AGH was to be established at Gaza, well-known to an earlier generation of medical and nursing staff. The first group of Australian nurses embarked for overseas in the *Empress of Japan* on 9 January, 1940, the destination known to all to be the Middle East. These members of the AANS were to form the 2/1 Australian General Hospital at Gaza under Matron C.A. Fall and the CO Colonel J. Steigrad.

On January 19, 1940 the first group of Queensland Sisters left South Brisbane Railway Station for the Camp Field Dresing Station at Ingleburn before proceeding overseas. *The Courier Mail* report stated that 'cheers smote the air as the Kyogle Mail drew out from the interstate station'. Long before the departure time trim figures in their grey uniforms with brown shoulder tabs and ties began to arrive accompanied by their quota of luggage. In charge was Sister Hanrahan, E.F., while the group consisted of Staff Nurses Keys, L.E., Marks, I.A., Hely-Wilson, U., Dean, E.M., Reid, M.V., Harvey, F.E., Harland, V., Doig, E.N., and Hoey, I.[9] A second group was to follow later — Sisters Petersen, F.M., and Knudsen, V.M., Staff Nurses Coleman, D.M., Dickson, J.C., Stewart, H.C.D., Thompson I., Miller P.B., Davis, R.E. and Glasgow M.M.[10]

At the civic reception in their honour prior to their departure the Lord Mayor, Alderman A.J. Jones, said,

Nurses leaving the city hall after the farewell reception by the Lord Mayor 1940. E. Dean, J. Dickson, A. Marks, I. Hoey, ..., F. Harvey, V. Harland, Miss Paten, E. Doig, D. Coleman, E. Hanrahan, F. Petersen.

Scene at South Brisbane Railway Station 17 Jan 1940 as the first detachment of Queensland members of the AANS left for the south. Sisters Hely-Wilson, Petersen, Harland, Harvey, Doig, Dickson, Stewart, Miller, Keys, Dean, Hoey, Hanrahan, Reid, Princ. Matron Paten. (Telegraph photo).

> On behalf of Brisbane and of Queensland I wish you good health and happiness overseas and a safe and speedy return to your native State. If you are called upon to minister to the wants of Australian soldiers wounded in battle, we feel sure that like your sisters of 20 years ago and more, you will rise to the occasion nobly.[11]

The Telegraph reporter referred to them as looking 'very dapper and bonnie', laden down with gifts of books, sweets, flowers and cigarettes. Many of those farewelling this first group of Queensland war nurses had themselves memories of similar farewells in the First World War. Matron E.M. Paten, Principal Matron of Northern Command and Mrs L.H. Pennefather (Sister C.M. Keys) were in the first group of Queensland nurses who sailed in the *Omrah* in 1914. Mrs Pennefather was farewelling her niece, Staff Nurse L.E. Keys. Also in the group was Staff Nurse Iris Hoey whose grandfather fought in the Crimean War. Rev Hely-Wilson, a Padre in World War I was there to farewell his daughter Staff Nurse Una Hely-Wilson. Another interesting fareweller was Miss Jean Barron, Matron of Atherton Hospital, who saw service in the First World War and Miss Sadie Macdonald, well known nurse of the previous war and latterly Matron of Ardoyne Hospital.

The Middle East

The mobilization of the Sixth Division and its transport to the Middle East by February 1940 was characterized by great haste and inevitably there were deficiencies in training. Likewise the provision of medical services for such a large body of troops could not be satisfactorily arranged in a matter of weeks.

AANS. Queensland 1940. Sisters Hely-Wilson, Harvey, Marks, Reid, Dean, Hanrahan, Hoey and Doig. (Photo Consolidated Press).

Even a modest sickness rate of 0.3 per cent over 20,000 troops would require up to two general hospitals, or about 1,000 beds. In the beginning the medical units comprised the 2/1 Australian General Hospital, the 2/1 Australian Field Ambulance, the 2/1 Field Hygiene Section and the 2/1 Convalescent Depot. Each of these had an important role to play in the overall medical plan for the division. As it happened, members of the AANS were appointed only to the 2/1 AGH in the first instance. While the advance party which left a few weeks ahead of the main body had, with the assistance of the British authorities arranged for the basic pattern of the hospital site, with main roads, kitchens, messes and ancillary buildings to be completed, much remained to be done. The nurses found only three showers, three latrines, no hot water and no tents erected for wards. As well, there was a deal of ill-feeling between the Arabs and Jews, so that no one could leave the hospital area without protection. Moreover, the arrival of the unit coincided with a cold, wet season, making things very uncomfortable for everyone. The CO, Col J. Steigrad, found the whole area a

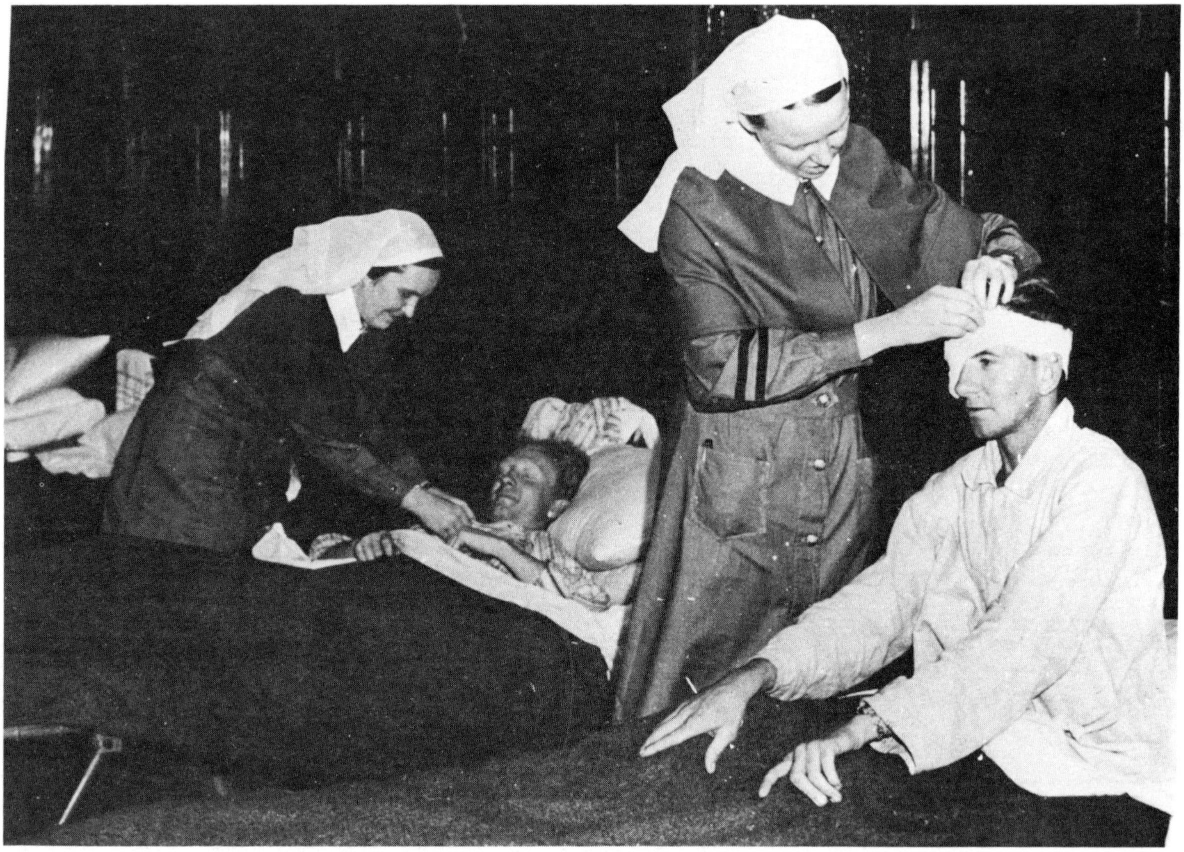

AANS tending patients at the CCS at exhibition grounds 1940. Sisters M. Wallace and V. Paterson.

quagmire, 160 acres of ploughed, sodden land upon which to erect the hospital wards. Tentage had not been brought with the hospital from Australia and unit personnel had to learn the difficult art of pitching EPIP tents, hospital marquees, hospital tents extending and of 'brigading' them. It was soon obvious that floors should be concreted, which meant that tents already erected had to be 'struck' and 're-pitched' over the concrete floors. Eventually more permanent hutted wards were erected, of light local brick construction, but it was many months before 2/1 AGH gave the appearance of a 'permanent' base hospital.(12)

Meanwhile men did get sick and had to be admitted for medical treatment — and nursed. Dysentery was an immediate problem. While the Jewish settlers had improved local hygiene standards considerably, the habits of the local Arab population encouraged the breeding of flies and inevitably of bowel-borne diseases. It took some time for health standards to be set and accepted among the troops, but even in the hospital itself, where flywire was almost unheard of and unprocurable, protection against flies, particularly in the kitchens, the messes and the wards was a major problem. Fruit and vegetables, as well as eating utensils, were methodically washed in a weak solution of permanganate of potash. The building of appropriate sanitary blocks, the disposal of faeces and provision of sullage disposal were problems for the Field Hygiene Section which did a remarkable job in preventive medicine.

By 18 May there were 228 patients in 2/1 AGH and within a few days the figure jumped to 430 when the second convoy arrived from Australia. At this stage there was an establishment of only 80 nurses to a 1,200 bed hospital and 50 to 600 bed unit. It was some two years before a more liberal staffing scale was adopted. By October the bed state was 2,200 which increased enormously the work load of the medical staff, originally posted on the scale of a 600 bed hospital. Another basic problem for the hospital was an adequate supply

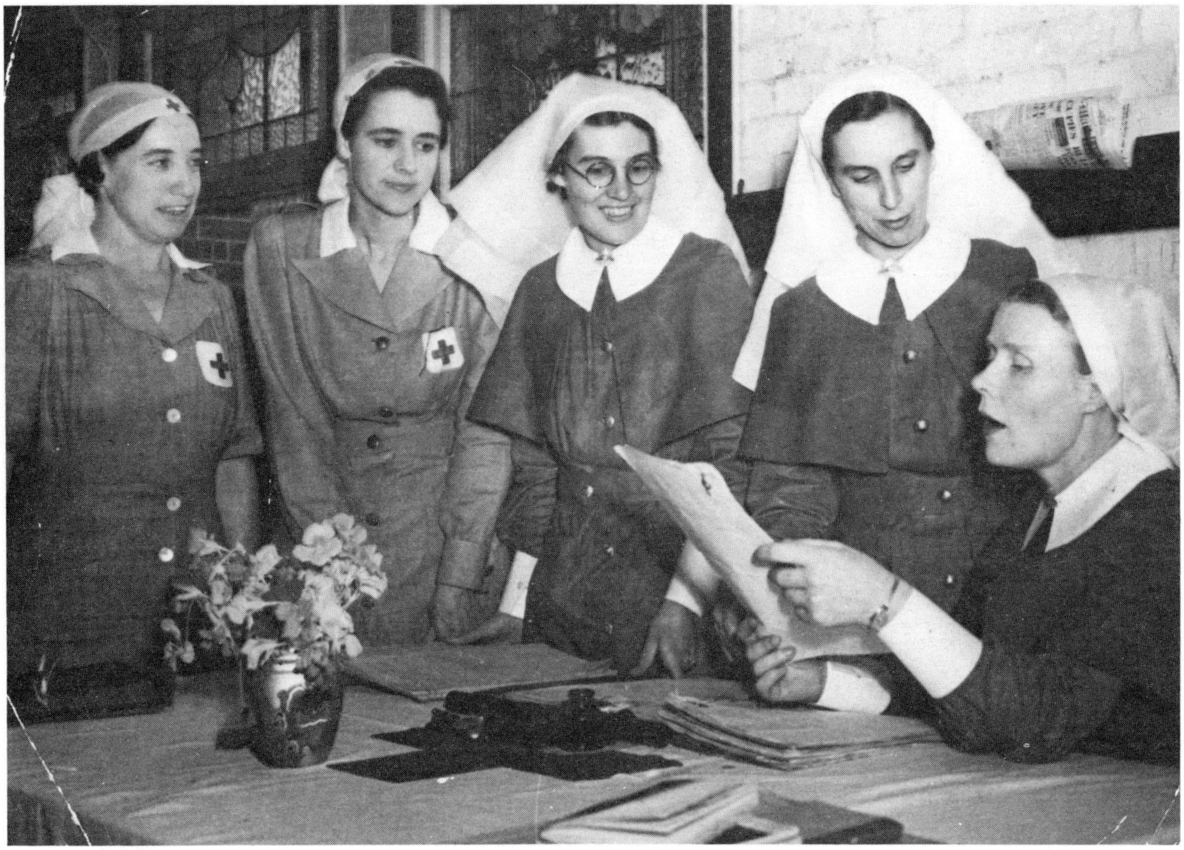

AANS at exhibition grounds Brisbane 1940 Sister Mary Luddy (seated). (Telegraph photo). On left, two VAD's.

of heating and steam, for sterilization, for washing, for cooking. A small generator was sufficient for the autoclave, while gas cylinders from Egypt provided help in the kitchens. Wood was unprocurable in Palestine, so coppers were impracticable. The major heating device was the temperamental primus stove. Many a nurse and orderly fumed over these obstinate devices and despaired at the scarcity of methylated spirits and kerosene.

Sister Amy McConnel was one of the fortunate ones moved from the tents in the dusty ploughed field to hastily prepared huts. As nursing sisters have done everywhere, she set about making her room 'homely', using her initiative to improvise with the limited materials available. Writing home to her mother in Clayfield, Brisbane, she described her new quarters:[13]

> It has a primitive wardrobe with shelves in it and a camp bed while an orderly procured a make-shift table and a mirror to stand on it. I tactfully did not enquire where these came from. The floors are of very soft bitumen, not nearly as good as ant-bed, and the bed legs sink into it. So I bought some reed mats from an old Arab called Frowzy, a most apt name, also a waste paper basket and some old stools, so the room looks quite habitable. Of course I was swindled over these but things are frightfully expensive, whether bought from the Arabs or Jews.

Originally all members of the AANS posted to this unit in May 1940 were from NSW but when the 2/1 AGH was established at Gaza it became a major staging centre for all major hospitals arriving in the Middle East or awaiting re-posting. It became well-known to many Queensland nurses. One of the first matters of concern was the health of the nurses, for diseases were common in the new country and replacements for those incapacitated were many weeks away. Fortunately the high medical standards required on enlistment

AANS leaving Australia for Middle East. Sisters: R. Davis, J. Hoey, E. Bowe, L. Keys, Una Hely-Wilson, V. Reid, E. Doig, V. Harland.

stood them in good stead. In his first review of their health the OC (Med) LtCol A.S. Walker commented on their good health, but recommended a review every three months and that provision should be made for regular leave whenever possible.

The second group of nurses left for the Middle East on 15 April, 1940, as members of the 2/2 AGH, with Matron A.M. Sage in charge. Sister Doherty (NSW) was placed in charge of the Queensland contingent of AANS on these transports. Her instructions from the Principal Matron with regard to dress are worthy of recording.

> As to uniform, you must all dress for dinner at night — this means caps, red capes, white cuffs. If you have mess dresses, you can wear them. For going ashore at ports, or any out-door wear, you will of course wear your grey suits but no parcels, purses or anything of the kind should be carried in the hands. If it is hot going ashore at any port, you may wear your cotton dresses, white collars and cuffs and your ship board hats. No red capes should be worn out of doors which means in your case, off the ship. During the day you can be lax with uniform. They can wear ship board hats or not, cotton dresses without cuffs, cardigans if they are cold, ship-board shoes and great coats if they wish. Unless they are on duty caps and red capes need not be worn except for dinner at night. You should see that everyone comes punctually to dinner, waits for you and all sit down together.[14]

On arrival in the Middle East they found little in readiness for them, indeed it was not clear where they would be stationed. In the interim, they quartered with 2/1 AGH at Gaza Ridge, but many of the nurses were attached temporarily to various British hospitals in Palestine and Egypt. Not until December 1940, some six months later, were they able to re-assemble at the new hospital site, El Kantara, a desolate sandy place close to the Suez Canal. On the 29 December, 1940, they were able to receive their first patients. As the war intensified they found themselves in the centre of enemy air raids on the Suez Canal Zone, so blackouts had to be strictly enforced. Then they suffered the inconvenience of the khamseen, as the desert

2/5 AGH at Rehovot, Palestine, early 1941. In front: Matron K. Best. Others in photo include Sisters Skyring, Short, Burnett, Patrick, Roe, Tomlins, Hayes, Zielke, Newman, Machon, Exton, Taylor and Casey.

winds brought the blinding, choking dust that forced the cessation of many medical and surgical procedures. Queenslanders who were attached to 2/2 AGH at this time included Sisters (or Staff Nurses) Blain, E.M., Marks, I.A., Dickson, J.C., Glasgow, M.M. Stewart, H.M., Davis, R.E., Hanrahan, E.F., Harland, V.M., Harvey, F.E., Thompson, I., Hoey, I., Keys, L.E., Hely-Wilson, U., Miller, P.B., Reid, M.V., Dean, E.M., Doig, E.N., Coleman, D.M., Petersen, F.M., and Sinclair, L.[15]

When Matron Grace Wilson returned to Australia in May, 1941, Matron Sage from 2 AGH became Principal Matron, Middle East and Sister E. Bowe took over as Matron of 2/2 AGH.

Another group of nine nurses departed from South Brisbane interstate station on the 11 October, 1940 in charge of Sister Dorothy Roe, who had trained at St Martin's Hospital. The party included Staff Nurse Hooke, J.S., from Rockhampton and Maryborough and Hooke, H.D., who had trained at the General, Staff Nurse Margaret Exton from Toowoomba (trained at Hillcrest Private Hospital, Rockhampton), Staff Nurse Zielke, J.B. who had trained at the General and was previously on the staff of Sandown Private Hospital, Southport. Another trainee of the Brisbane General was Staff Nurse Dorothy Burnett from Buderim. St Martin's Hospital was well represented among the Queensland members of the AANS, an appropriate training centre, as it was dedicated as the War Memorial Hospital of the Church of England after World War I. Staff Nurse Taylor, M.J. was another St Martin's trainee who had served on the staff

2/5 AGH on train from Suez to Gaza 1941.

of Atherton Hospital. Staff Nurse Patrick, E.E. also from the General had been engaged in private nursing, while Staff Nurse Casey, E.L. had been a member of staff at the Brisbane General Hospital.[16]

Matron Sage very quickly got down to the details inherent in the administration of the army nursing service. She made an initial tour of inspection of all medical units where nurses had been appointed and arranged for a conference of all Matrons in the Middle East in August 1941. Some of the problems which emerged were dealt with speedily — the matter of adequate accommodation and overnight leave arrangements — but others required long term decision by army authorities. The abolition of the Staff Nurse rank, seniority and promotion in the AANS, staffing and war establishment for hospitals and CCS's, badges of rank and the perennial question of an appropriate summer uniform were all to be resolved in the fulness of army time.

Life at El Kantara

For more than a year the 2/2 AGH found itself in a key role for the reception, treatment and evacuation of the sick and wounded from the various campagins in the Middle East. As the hospital bed state increased from 1,200 to 1,500 and at times to around 2,000, as AANS staff from other medical units serving 'up the desert', especially the 2/4 AGH and the 2/2 CCS, were temporarily posted to them, the centre grew to the size of a small town. While life was always busy and sometimes hectic, it was never dull. The following excerpts from letters home by Sister Jean Dickson and from her diary give a realistic impression of what life was like at El Kantara from Christmas 1940 to Christmas 1941.

The Wards

> We have only very basic equipment. Half our beds were standard iron hospital beds, the remainder were folding tubular steel stretchers with wire mesh mattresses. Most wards contained 60 beds, 30 each side of a central service area. When extra beds were requuired EPIP tents were quickly erected close to the hutted wards. Each hospital bed had a 'tester', with chain and ring.

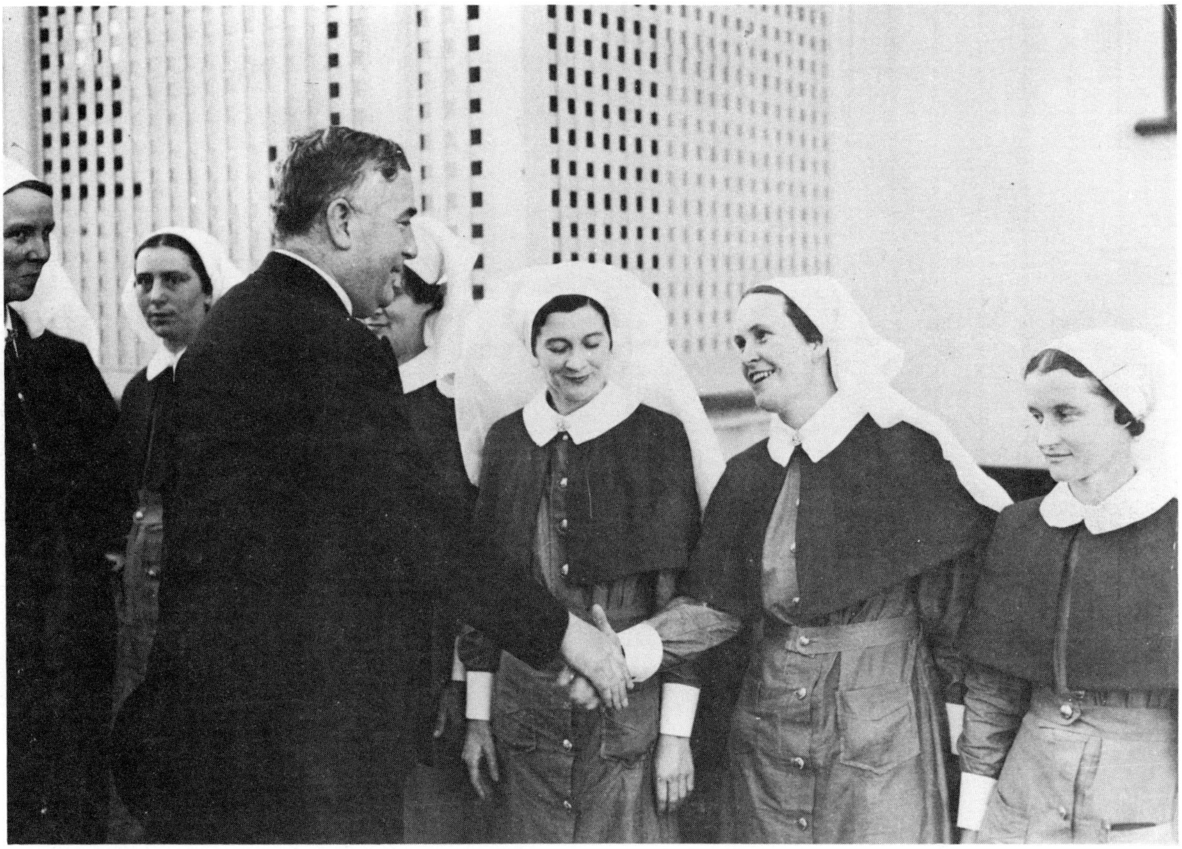

AANS greeted by P.M., Mr Menzies. Sisters Paterson, Patrick, Palmer (hidden), Sinclair, Wallace, Zielke.

Light and heating

Lighting was usually by hurricane lamps, about 3 or 4 to each ward. later, wards and messes were supplied with carbide and pressure lamps, and very much later with electricity. Heating for boiling water, using sterilizers, making tea, etc was by primus stoves. Coppers were useless as no wood was available. We learned to live with the temperamental primus stoves and became quite expert at 'fixing them'. (We all carried spare washers and vaseline for the pumps). Our boys made the sterilizers — a 4 gallon kero or petrol tin opened on the side, with a handle soldered on to each end and a tight fitting lid made from another tin, with another handle soldered on top.

Australian men and women proved to be very adaptable to all circumstances. It must be our experience of living in the country.

Wireless

One of our most prized possessions in the mess was a battery operated wireless set, "a present from a 4BC radio station, Brisbane". When we could we gathered round to hear the chimes of Big Ben, the latest news from the BBC and occasionaly one of the stirring speeches of Winston Churchill.

AANS Mess

This was really our "home", for apart from the dining area, one part of the hut was partitioned off as a lounge, including another small section for "brothers and others". As only commissioned officers were allowed in our mess, we needed an area to meet, our brothers, uncles, cousins — and "others"!

Recreation

Sisters worked a 6 day week, usually 12 hours a day, sometimes longer. They could expect 6 days leave every 6 months, circumstances permitting. On days off, after signing out in the leave book, they could go, if fortunate, in

2/6 AGH under snow, Jerusalem 1943.

the "ration truck" to Ismailia. Alternatively, a horse drawn gharry could take them a mile down the road to the Canal ferry, with the railway for Palestine on one side and Cairo and Port Said on the other. The Canal was crossed by a punt on a continuous cable, carrying cars, donkeys, trucks, people of all nationalities, run by a "Gyppo", with British "Red Caps" or Australian MP's on either side of the Canal.

Sometimes we purchased a second class railway ticket and travelled first class per favour of the Egyptian conductor, for a small tip. He would usher us in, dust the seat, swirling dust everywhere and make us comfortable! We always travelled in groups and made frequent use of the YWCA's at Port Said and Ismailia.

Patients and staff had to improvize their own amusements and recreation. They staged their own impromptu concerts, they organised "donkey races", polo contests on donkeys — and then there was the hilarious "picture show" in the open air outside the hospital grounds. It was put on by an enterprising Egyptian entrepreneur who had fenced off an area, divided it into "dress circle" and "stalls" and charged a small fee for admission. Sometimes there was sound without pictures, sometimes pictures without sound, sometimes neither. Breakdowns were common — and the audience reacted as they did in matinee peformances throughout Australia. On one occasion it was so bad, the Aussies demanded their money back. When the "Gyppo" refused, that was the end of the picture show!

Recreation for Patients

A canteen was available for patients, for small items and the possibility of one small bottle of beer per day. The Red Cross also provided many extras in food, clothing and writing material. Ward patients appointed a "canteen runner" whose job it was to collect orders and subsequently to deliver the goods. Cards and Chinese checkers helped pass many a boring hour.

Work in the Hospital

There were periods of intense activity when convoys of wounded arrived. Working conditions were then "pretty tough", especially when the attached nurses were returned to their own units. Night duty was always strenuous, with one Sister to cover 4 wards and a nursing orderly in each. Her task in moving around and giving the required treatments was demanding and tiring. Surgical nursing was particularly heavy, as there were frequently no walking patients. If there were "up" patients, they were a great help, even with their disabilities, in feeding and looking after their "mates"

Sister Dickson paid tribute to the untiring work of the orderlies, both general and nursing, who were an essential part of the team.

When the time came for the hospital to close down, ready for a move to an unknown destination in the Far East, the CO, Col K. Fraser, praised the work of the AANS, whose dedicated service had been a feature of life in the hospital at El Kantara.

The CCS's

After the experience of having nursing staff in casualty clearing stations in the First World War there was never any doubt that they would be posted to CCS's in the Second World War. The 2/1 CCS and the 2/2 CCS were also established as an integral part of the medical services in the Middle East. While it became evident they would have an important role to play in forward areas, there was no general agreement on how far forward the nurses would be permitted to serve. Nevertheless, nurses were posted to these units and took part in all the initial training, including the routine tasks of packing and unpacking medical supplies at speed, in keeping with the anticipated mobile role of the unit.

The 2/1CCS to which 8 Sisters were attached, soon found itself in action. Again there was some confusion on its arrival in Palestine, another medical unit to be aided by 2/1 AGH at Gaza Ridge. For a brief period they set up a camp hospital at Gaza and another at Qastina. With the entry of Italy into the war and the movement of the Sixth Division to Egypt, 2/1 CCS, with Lt Col J.K. Adey as CO and Sister M.J. Hanna in charge of the Sisters, moved to Amiriya, west from Alexandria. This was a tented section and the nurses found the desert nights bitterly cold. Then came torrential rain, forcing the abandonment of the dug-in, sandbagged tented wards. On the 31 December, 1940, the 2/1 CCS, including the nurses, moved closer to the front line at Mersa Matruh, taking over the quarters of British nurses who had been evacuated. There was an immediate conflict with British command, which felt that it was too dangerous for nurses.[(17)]The 2/2 Australian Casualty Clearing Station commanded by LtCol K.J.G. Wilson was noteworthy in that for most of its time, members of the AANS attached to it were all Queenslanders. Sisters Paterson, V.M., Thorpe, M.F., Wallace, M., Marshall, M.A., Finlay, M.E.F., Pym, P., Golden, R.J. and Wilson, H. were later joined by Sisters McDonald, M.M. and Hamilton, V.F. (the only non-Queenslander). These had joined the unit initially at the Brisbane Exhibition Grounds where it operated an eighty bed hospital in the basement of the Ernest Baynes stand. The Sisters were billeted at the 'Sylvia Moffatt' ward at the nearby hospital for sick children. They also staffed Camp Dressing Stations at Redbank and Enoggera. On 9 November, 1940 the unit entrained at South Brisbane where they were farewelled by Col Macartney, DDMS and Miss E.M. Paten, Principal Matron. Half of the unit embarked on the *Orion* in Sydney and the remainder on the *Stratheden* from Melbourne. While the *Stratheden* eventually disembarked four Sisters at Kantara, the *Orion* disembarked her passengers at Haifa. In the process of time the 2/2 CCS Sisters were gathered together by Matron K. Best of 2/5 AGH and taken to the transit camp at Gaza.[(18)]

It then moved from Dimra in Palestine to Amiriya, ready for a forward move, if necessary. Conditions in this area were never pleasant, with constant sand storms, insufficient water and a general uncertainty about their future role. In January 1941 the nurses were transferred back to Alexandria, while the men prepared to go 'up the desert', following the battles against the Italians. The success of that initial campaign pushed the Italians back through Bardia, Tobruk, Derna and beyond Benghazi. Medical units (minus the Sisters) were an integral part of that first desert campaign, with the 2/3 Fd Amb, 2/8 Fd Amb, 2/11 Fd Amb, the 2/2 CCS and the 2/4 AGH making a significant contribution. With the Italian army in disarray beyond Benghazi it seemed only a matter of time before Tripoli would be taken and the enemy removed from North Africa. By this time, March 1941, the 2/4 AGH was established at Barce and it was decided to bring the AANS from the 2/4 AGH and 2/2 CCS forward to these bases.

On the 27 March 1941, the British hospital ship *Dorsetshire*, arrived in Tobruk, bringing 63 members of the AANS for 2/4 AGH and 2/2CCS. However, the entry of Rommel's Afrika Korps into North Africa changed these plans. His successful counter attack soon made the whole area unstable. A hospital at Barce

2/2 CCS about to entrain at south Brisbane railway station 9 November 1940.
L. to R: M. Thorpe, M. Marshall, E. Finlay, H. Wilson, V. Paterson, P. Pym, M. Wallace and R. Golden.
(Telegraph photo).

was out of the question and urgent signals were sent to hold the AANS at Tobruk. The 2/2 CCS, minus the Sisters, was sent from Tobruk to relieve the 2/4 AGH, but their stay in Barce was also quite brief. On the 3 April they too had to join in the hasty retreat to Tobruk. There they set up a new site outside the town, while the 2/4 AGH took over the old Italian military barracks as a town hospital and established a 'beach section' for medical cases.

For ten days the nurses of the 2/4 AGH and 2/2 CCS literally slogged to put the hospital wards into shape. As many of the orderlies had not arrived back from Barce, the nurses set to work cleaning the floors, erecting beds, putting gauze over windows to keep out some of the persistent Libyan flies. Blackout requirements were strictly enforced. Meantime patients were admitted, the casualties from bombing and machine gunning increasing hourly. Patient numbers built up as no system of evacuation by land or air was possible and evacuation by sea uncertain.[19]

The Sisters of 2/4 AGH and 2/2 CCS were billeted at the 'Albergo', a derelict hotel, once occupied by Mussolini. It took much hard work by the Sisters before it was habitable! Each morning and evening duty parties had to march, complete with capes, to the hospital, some four kilometres distant, but for some curious reason, Italian POW's were driven in trucks. Water was rationed and clean linen soon ran out. When the supply of beds and mattresses ran out, patients were bedded down on their stretchers under the beds!

As the military situation in the desert deteriorated, the future of the nurses came in for serious consideration. By 3 April the fall of Benghazi was imminent and the rear parties of 2/4 AGH and 2/2 CCS were fortunate to make it safely through Derna back to Tobruk. Each day at the hospital became more eventful, with casualties increasing and the medical and nursing staff working round the clock. By Sunday, 6 April, German patrols were stopped only 40 kms from Tobruk. In view of the gravity of the situation, higher command made two vital decisions — to stand and fight at Tobruk and to evacuate all members of the AANS. As it happened similar situations arose in Greece, Crete and Malaya, requiring other decisions, with different consequences.

So on the evening of the 7 April, the British Hospital Ship *Vita* took on board 63 nurses and masseuses from 2/4 AGH and 2/2 CCS, as well as 323 patients, although its capacity was only 250. One of the nurses subsequently described the departure:

> We were told to pack all our belongings into our kit bags but by the time they were packed we could not lift them off the floor. At the wharf we embarked on a large open bellied naval launch. Our ship, a Hospital Ship with the usual white, green and red markings was about two miles out, just past the hulk of the sunken Italian cruiser, *San Gorgio*. We were all attired in our usual outdoor uniform — grey suit, hat, greatcoats, and we carried a small brown suitcase. Our respirators, Italian water bottles and steel helmets were slung across our shoulders.[20]

A recent writer summarized the dilemma of the evacuation of nurses.

The withdrawal of the entire female nursing staff from a hospital with almost 1,000 patients was a shattering blow. In his official report on the matter, the CO of 2/4 AGH had this comment to make:[21]

> Orders were received that all nurses should be evacuated, and when I interviewed Col Cook, who was the Fortress Commander, requesting some modification, he said they had to go, as it was more than likely that we should be under artillery fire, and it was no place for women. I did not think we could function with any degree of efficiency, but all protests on my part were unavailing. I begged for six to run the theatre and personally supervise, but only got an emphatic "No".
>
> The picture was dreadful to contemplate when one pictures the duties of the AANS. The course for an operating room assistant is two years, and that of a nurse four years. We had two ORs (male) on strength, trained in the unit. We had 25 Nursing (?) Orderlies and 20 ward orderlies, more than 900 patients and no nurses. I cannot but feel very strongly that the decision to evacuate the nurses was wrong. They left under protest, with a very deep feeling of injustice, and all stated they were willing to take their chance with the rest of the unit. What were these chances?
>
> (a) Killed?
> (b) Wounded?
> (c) Captured — raped?
>
> After all, we were fighting civilized nations, and I'm sure public opinion would have been all in favour of their retention, in view of the grand work they could have performed. Certainly they would have had to rough it, probably adopt male attire, slept in cubby holes, lost that school girl complexion, but I am sure they would have carried on the traditions of the AANS with fortitude and honour to themselves.

Of the last claim there can be no doubt. Whether it was a wise decision only history can judge.

Members of the AANS who were evacuated from Tobruk to Palestine and Egypt never let up in their desire to get back to the front in the Western Desert, where the action was and where they believed the nursing skills could be best used. The Matron-in-Chief, Grace Wilson wrote back to one of these Sisters in the following terms.

> I was very touched by your letter and very sorry for you too. I did all I could to send you back again, but the men will not listen to me. Don't be afraid you are not keeping up our tradition — you are — and some of you are making new ones for yourselves. I had hoped something might be arranged, but now the final decision has been made. All Sisters are to stay at the base.[22]

Perhaps the most touching and personal letter to one of the Sisters on this matter came from Senior Chaplain C.L. Riley (Bishop of Bendigo) who wrote as follows.

> It all boiled down to this. Much as the hospital missed you girls and your nursing skills, it was universally accepted that it was no place for you to be in. I know you will say you volunteered to serve and take the same risks as your menfolk, but there's a third risk in the case of capture which does not apply to the men and from which they have pledged themselves to keep you safe, if they can. When the hospital suffered its worst bombing raid and so many

were killed and injured I heard man after man say "Thank God, the girls aren't here". The men all know how willing you girls were to take risks and they all honour you for it, but they go about their work much more confidently when they know you are safe.[23]

While members of the AANS did not get to the front in the Western Desert, little did they realize that ahead of them was the near capture in Greece and the tragedies of Singapore, Sumatra, New Guinea, Rabaul and of the *Centaur*.

All the nurses, including the Queensland members of the 2/2 CCS begged to stay, but to no avail. On the Hospital Ship there was much work to be done. The ship arrived safely at Haifa and the nurses were dispersed to the 2/1 AGH at Gaza and to the 2/2 AGH at El Kantara. This was no relief as casualties flowed in from the desert, while enemy bombing raids increased, especially at night. Sister Paterson and her group of Queensland Sisters of the 2/2 CCS, having escaped from what might well have been an extremely dangerous situation were to find themselves in other equally dangerous situations in the months ahead.

Britain

Meanwhile other members of the AANS had been heavily involved in nursing on different battle fronts. The third party of 32 nurses under Sister Bowe, E.J. left for the Middle East on 5 May, 1940 in one of the largest convoys of the war — the *Queen Mary*, the *Aquitania, Andes, Mauretania, Empress of Japan, Empress of Canada* and *Empress of Britain*. However the entry of Italy into the war diverted the convoy

King George VI inspects the AANS at Salisbury Plains 5 July, 1940.
L. to R: Sisters Wilson, Harland, Harvey, Dickson, Dean, Keys, Reid, Hely-Wilson.
Front: H.M. King George VI, Sister Bowe and (behind) Brig. Morshead and Maj-Gen Wynter.

to Britain via South Africa. This contingent arrived in Scotland on 16 June, 1940, at a time when the Battle of Britain had begun and invasion was a distinct possibility. The Australian nurses first of all went to Salisbury Plains where they had the honour of being inspected by His Majesty, King George VI. A further group of 42 nurses which left Sydney on 27 May in the *Stratheden* reached England on 16 July and when 4 more arrived the total AANS strength in Britain rose to 78. These were subsequently attached to the 2/3 AGH which opened at Godalming in Surrey on 1 August, 1940. These nurses endured the horrors of the blitz, but fortunately no bombs fell on the hospital. Nevertheless, air-raids were frequent, requiring the nurses to wear steel helmets. It was an unnerving experience of slit trenches, air-raid shelters and anti-gas drill. Sleep was almost impossible as the bombing went on, hour after hour night after night.

When the Australian troops, known as Austral force, arrived in the UK on 16 June 1940 no Australian General Hospital accompanied it. However the 3 Australian Special Hospital was in the convoy and numerous members of the AAMC and AANS were on duty in the transports. Authority was given to the ADMS to establish an Australian General Hospital, which became 2/3 AGH, at the King George V Sanatorium at Hyde Stile in the Godalming district of the county of Surrey, some 40 miles from London. It was however 60–100 miles from the camps where Austral force was located. LtCol D.B. Loudon was appointed CO, with Sister E. Butler, as Matron. 18 of the AANS contingent were Queenslanders, the most senior being Sisters Florence Mary Petersen and Ethel Francis Hanrahan.

That same Queensland fellowship which was a common thread in the lives of Queensland members of the AANS during the First World War was again noticeable during the Second World War wherever they were located. Sister Jean Dickson recalled this feature of their life in England during the blitz.

> Many firm friendships were made during our stay in Britain. Most of us had been paired off as we had to share rooms, cabins, etc. In our group we used to have a "Queensland Family" of eleven of us — we used to say each of us had ten friends. We used to share everything, especially about home and Queensland. Those friendships are still as good today as when we first made them over 40 years ago.

The scarcity of mail was a matter of concern, as sometimes it took as long as five months to arrive, while at other times it did not arrive at all, having been diverted to the Middle East. Around the hospital there was an upsurge in interest in gardening as part of the 'dig for victory campaign'. While the crops of cabbages, sprouts and potatoes did not fill many cooking pots, the therapeutic and psychological value of 'doing something' was undoubted. Other Queensland Sisters who served at Godalming with 3 AGH included Sisters Dickson, J.C., Marks, I.A., Harvey, F.E., Hely-Wilson, U., Harland, V.M., Hoey, I., Keys, L.E., Doig, E.N., Dean, E.M., Reid, M.V., Coleman, D.M., Davis, R.E., Miller, P.B., Glasgow, M.M., Thompson, I. and Stewart, H.[24]

Accommodation for the nurses was provided in three huts, separated into small cubicles. They were always noisy and as winter came on, bitterly cold. The CO in his official report, commented on another simple problem which apparently was insoluble.

> I wish to comment here that there was a pigsty within 20 yards of all three huts. The biggest representations were made to have it removed, by me, by the ADMS, by the GOC, UK and by the High Commissioner for Australia, but without success. The stench at times was deplorable.[25]

Perhaps only Winston Churchill was aware of the military significance of the pigsty in the confines of a hospital!

During this period six members of the AANS were transferred temporarily to the Tidworth Military Hospital, to assist in the nursing of Australian patients. After their return from three months duty in this strict British hospital, the CO wrote to Matron Butler commending these Sisters for their excellent work. His letter is striking testimony to the efficiency and dedication of those members of the AANS.

'Digging for Victory' at 3AGH Godalming, rear of Ward 5, 1941.
L. to R: Sisters Miller (Qld), Metzner, Dean (Qld), Umpherstone, Langham, Davis (Qld). Mounds of dirt at rear hide slit-trenches.

> It has given us great pleasure in having them with us, and working in cooperation with them. Their services have been invaluable, always willing, bright and most efficient. Nothing has been too much trouble for any of them. We shall be very sorry at their departure and we very much regret that the time has come for them to return to their own hospital.[26]

One of the interesting cases at this time was a critical test of the nurses' status and authority. A patient returned to hospital AWL (and apparently drunk) at 0200 hours. The Night Sister noticed his shirt was wet and ordered him to remove it. Subsequently when he again refused, she placed him under arrest and laid a charge against him, in that he refused to obey the lawful command of a superior officer. On appeal, it was held that a nurse is not a 'superior officer' of a soldier. Although nurses had authority in army hospitals that authority did not confer rank on them. They were given honorary status and privileges but in fact they were not even soldiers, within the meaning of army regulations! While the soldier could have been charged with disobedience under another regulation, the whole case cast doubt on the nurse's status, rank and authority within the hospital. Until this was clarified the ability of the AANS to maintain discipline in the wards was always in question. It took the army more than two years to clarify this matter and to put the AANS on a proper military basis!

Later in the year it was decided to transfer all Australian troops, including the nurses to the Middle East. Despite the incessant bombing of Glasgow and Liverpool, their departure ports, they arrived safely in the

Two Sisters pose for gas drill. Sister Miller on left is in mustard gas uniform and Sister Doig on right in her decontamination rigout. Godalming October 1940.

Middle East. The nurses and other medical people were transferred to various hospitals in Palestine and to all intents and purposes 2/3 AGH, as such, ceased to function as a medical unit.

By November, 1940 the situation in the Middle East had changed dramatically, with large numbers of Australian troops arriving for the expected clash against the Italians in the Western Desert. So it was decided to despatch, among others, 4 Sisters and 24 Staff Nurses to the Middle East as the nucleus of a new general hospital there. Matron Butler, in her report, explained why all the Queenslanders were chosen to go.

> Colonel Fraser, ADMS, wished these 4 Sisters and 24 Staff Nurses to go. The Queensland Nurses were feeling the cold, so he thought it wiser for them to be the first to move.(27)

Queensland Nurses with the QAIMNS

By the middle of 1940, 72 Australian nurses had, for various reasons, enlisted in the Queen Alexandra's Imperial Military Nursing Service. Some were in England at the outbreak of war, some were impatient at the delay in being called up for service with the AANS, others thought the conditions might be better and the experience more varied. Sixty-nine per cent signed on for the duration, 12 per cent for two years and the remainder for one year. They saw active service on many battlefronts — Britain, France, Malta, the Middle East and on hospital ships. Six Queenslanders are known to have served with the QA's: Sisters Hannah Murray, Frances Elliot, Letitia Whittem, Marie Williams, Mavis Twine and Imison, A.E.[28]

Sister Hannah Murray trained at Toowoomba, was in Britain at the outbreak of war and promptly enlisted in the QAIMNS. She was sent to France with the BEF and was one of the survivors from Dunkirk.

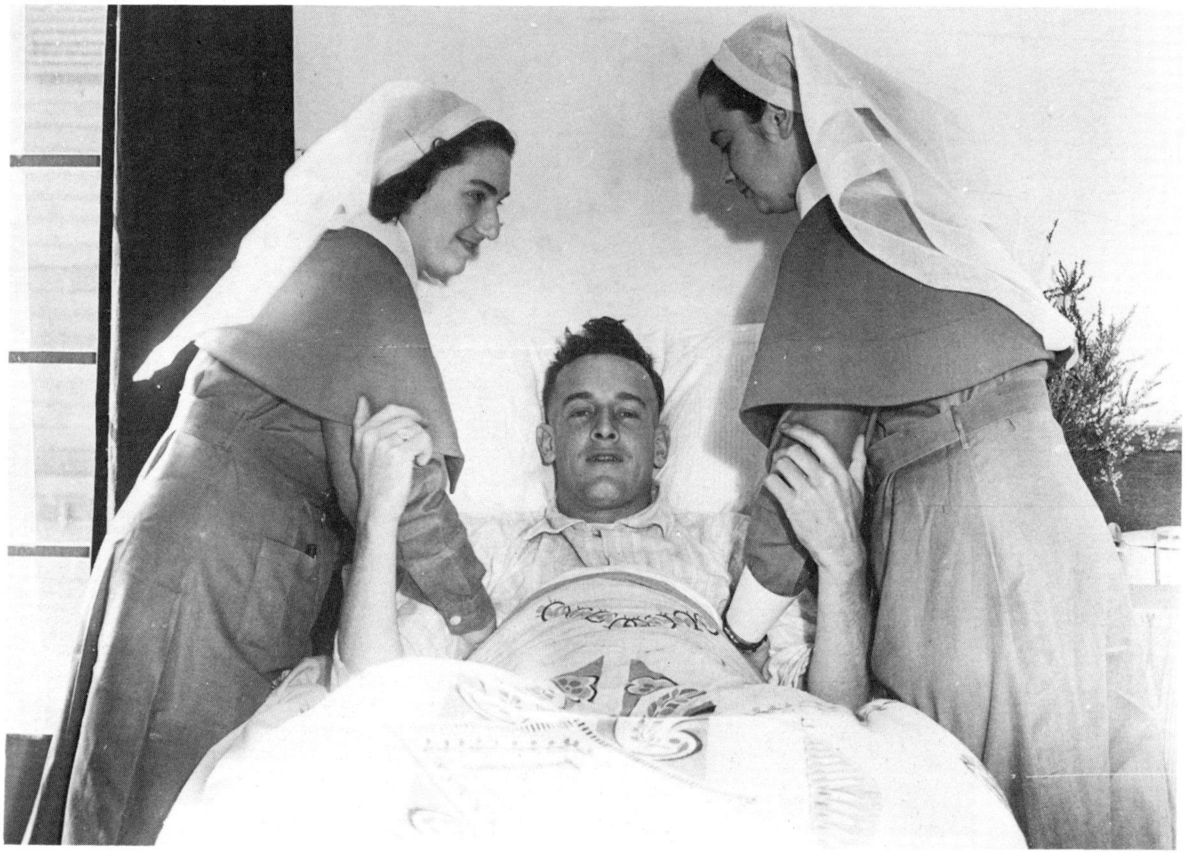

Sisters P. Miller and J. Harvey visiting a patient at Godalming November 1940.

Subsequently she served with the British forces in North Africa, Tobruk and the Middle East. While in Egypt she was nursing German POW's and in a letter home made these interesting observations:—

> At first I felt so hostile to them and thought, "Why should I have to treat them so well, especially when I thought of how I had to flee from them from the hospital in Northern France and the sufferings and slaughter I saw at Dunkirk.
>
> But after nursing these Jerries for a while one cannot help being struck by their great respect for us, as well as their discipline. As patients they are for all the world like children. They would not dream of disobeying an order. They have the greatest respect for our medical officers. When they come into the ward they stand like steel to attention, until the MO's leave the ward.

Sister Murray served at the hospital in Tobruk when it was occupied by British and South African forces and was evacuated before it fell to the Germans. In her letters she wrote of the very happy time she had with the British medical unit and recounted the interest they all had in Australia.

> They are all very interested in hearing of our Australian hospitals and marvel at the exploits of the Bush Nurses and the distances they travel on horseback in all weathers.[29]

Another Queensland nurse who served at Dunkirk was Sister Frances Elliot, who trained in Brisbane and nursed at Kingaroy prior to enlistment. In fact, she was in Britain at the outbreak of war and promptly enlisted with the QAIMNS. She accompanied the BEF to France and was one of that gallant band of nurses who performed such heroic work during the epic evacuation from Dunkirk.[30]

Another Queenslander who had a rich and varied career with the QA's was Sister Mavis Twine.[31] She had trained at 'The General' and went overseas for further experience prior to the war. On the outbreak of war she enlisted in the QA's and was sent to France with the 6th British General Hospital. When the

Photo taken at a reception in Queensland House, London, probably 1941. At rear, third from left is Sister Hannah Murray (Toowoomba) who served with The QAIMNS during World War Two.

Germans found a gap in the Maginot Line, the hospital was only a few miles away. It was ordered to the coast where thousands of allied troops were waiting to be taken off in small boats back to England.

After that dramatic rescue Sister Twine was sent to Egypt with 6 BGH for some 2½ years, nursing in hospitals in the Suez Canal Zone, occasionally catching up with Australian patients and former nursing friends. Then the unit was moved to Malta for another 2½ years, at a time when Malta was enduring heavy bombing. Again, she related her experiences to a friend.[32]

> Malta was interesting — some hospitals were up in the hills where the bombing was less severe than down in the waterfront area where patients with their nursing staffs, doctors etc often had to take shelter in the tunnels in the rocks which had been in use for many years, since the time of the Knights of Jerusalem.
>
> Life for the nurses on Malta was devoid of much besides routine work. They often strolled in their off duty hours and watched the village women making their beautiful lace, a craft that has almost gone into recess since the war. Our friend managed to be assigned to the midwifery section which looked after the wives of the men stationed on Malta on garrison duty.

Sister Whittem, who trained at Roma, also joined the QA's in 1940 and served most of the war in British Hospitals in the Middle East. After peace was declared in 1945 she continued on in the QA's, serving for some time with the British occupation forces in Germany. Of Sister Marie Williams little is known. She is believed to have trained at Rockhampton, served with the QA's and married an officer in the British army. It so happened that she was on the hospital ship *Dorsetshire* which took the AANS of the 2/4 AGH and 2/2 CCS to Tobruk in March 1941.[33]

Wherever they served Australian nurses proved to be competent and efficient, fitting well into the disciplined nursing tradition of the QA's. The Matron-in-Chief of the British army commented, 'The Australian women's loyalty and keenness delighted us.'

Syria

The protection of the Suez Canal and the defence of the entire Middle East region from the Western Desert through Egypt to Palestine and Syria was regarded as an essential part of the overall war stragegy. The axis powers for their part were seen to be operating a giant pincer movement to put pressure on the allied forces. From their base in Palestine, allied forces, especially Australian, faced the dilemma of bolstering up the defences of Egypt and the Western Desert to prevent Rommel's breakthrough to Cairo while at the same time dispersing troops to defend the northern flank. The abortive attempt to save Greece was part of this strategy. The allied war council was conscious of the instability of the Balkans, the uncertainty of Turkey's intentions and the danger of German infiltration of the pro Vichy French command in Syria. Adjoining these problem areas were the oil-rich regions of Iraq and Iran. As a first move it was decided to invade Syria to protect this flank. In theory it was thought to be a relatively easy task for a force comprising British, Australian, Indian and Free French forces. In fact it proved to be a bitter, short and nasty campaign of some six weeks, from 7 June to 12 July, 1941 before the Syrian and Vichy French surrendered. Australian casualties numbered 416 killed or died of wounds and wounded 1,136.

Members of the AANS were not attached to any of the medical units involved in the Syrian campaign. The 2/4 Fd Amb and the 2/6 Fd Amb were able in the beginning to bring Australian casualties direct through to Nazareth or Haifa; later they were supplemented by the 2/13 Fd Amb. Ambulance cars and ambulance trains coped successfully with the evacuation procedures to Palestine. If the nurses were not involved directly in the Syrian battles they were heavily committed in the medical units to the south. The 2/1 CCS had been set up in a monastery at Nazareth, where Sister Hanna and her small staff dealt with many convoys late at night. The 2/1 AGH (Matron Fall) at Gaza found its bedstate increased from the established 1,200 bed to more than 2,200. The 2/7 AGH (Matron Johns) was barely established at Rehovot when the Syrian campaign commenced in June but somehow managed to cope.

The second stage in this campaign, in so far as the nurses were concerned, came after the cessation of hostilities. It was decided that in order to protect the flank from possible German attacks that an army of occupation would remain, a task assigned to the 7 Aust Div after the 6 Aust Div was withdrawn for rest and regrouping. The situation in Syria was still regarded as too volatile for the establishment of large base hospitals, but it was thought that CCSs should be moved up to supplement the work of the Fd Ambs. The 2/3 CCS began work in school buildings at Beirut, while the 2/1 CCS took over a mental hospital nearby. The two CCS's worked as a team distributing cases between them. An innovation for members of the AANS was the decision to attach nurses to the 2/1 Fd Amb at Zebedani, the 2/2 Fd Amb at Homs, the 2/4 Fd Amb at Tripoli and the 2/8 Fd Amb at Aleppo. Meanwhile back in Palestine, the 2/9 AGH (Matron Marshall) which had been operating a small hospital near Alexandria moved up to Nazareth, adding to the hospital beds available.

Casualties from Syria, both in the fighting and during the occupation provided new experiences for the AANS. There were twice as many casualties due to sickness as due to wounds. The dangers of malaria in Syria were well known but despite all precautions and the work of the anti-malarial squads malarial cases increased. Dysentery was a constant problem, with the occasional typhoid. As the snows of winter in Syria increased so came cases of frost bite and conjunctivitis. The icy roads also increased the hazards of driving, with numerous accident cases. Perhaps the most difficult of all diseases to combat, especially in an occupation force, was venereal disease. VD rates among Australian troops were recognized as being too high, the rate being 48.46 per 1,000 in 1941. Initially these cases were handled within a wing of 2/1 AGH but it soon became necessary to establish the 2/8 Australian Special Hospital to deal with these cases. When that unit left the Middle East in October, 1942, Lt Col Loudon reported that 5,774 cases of VD had passed through the hospital. The increased incidence of VD in Syria led to the establishment of the 14 Aust Spec Hosp at Bhamdoun, near Beirut.

Greece

As the war lurched from crisis to crisis, the allies had to move fast in an attempt to hold the Nazis on another front. When in October 1940 Italy invaded Greece through Albania it was feared that Germany might add support and threaten the Middle East theatre, including the Suez Canal. As the situation worsened a strong force was sent to Greece, including the 6 Australian Division and the medical units 2/5 AGH,[34] 2/6 AGH,[35] and 2/3 CCS.[36] Nothing seemed to go right for this expedition. Sisters of the 2/5 AGH recalled the days of delay and indecision, of orders given and then cancelled, of the 'orderly confusion' at the docks before they embarked for Greece. Members of the AANS from 2/6 AGH (Matron Joan Abbott) and of 2/3 CCS (Sister Deane, F.J.) arrived in Greece on 3 April and of 2/5 AGH (Matron Kathleen Best) on 12 April. The Grecian campaign was a disaster from the beginning, doomed to be another allied reverse, despite the heroic fighting of the Commonwealth troops. Nurses of the 2/6 AGH were unable to join their unit at Volos and were instructed to work with a British RAP near the docks at Piraeus. Enemy bombing here was so ferocious the nurses had to be withdrawn for safety to Kephissia. The Sisters of the 2/3 CCS fared no better. Their unit at Elassan came under heavy enemy attack so on 9 April they were withdrawn to Pharsala. As the situation deteriorated further they pulled back to Athens on 15 April. All this was happening over Easter but there was little time to think about it. Sister Dorothy Roe remembered a Greek cook making a special treat for the Sisters on Easter Sunday — a caramel custard!

Time was running out for the allies and on the 21st April the decision was made to evacuate Greece with all speed. But to get the 6 Division and all the ancillary units as well as other Commonwealth troops safely out of Greece, without control of the air, was an optimistic undertaking.

Again as in Tobruk, there was a critical decision to be taken with respect to the nurses. Brigadier Large, DDMS, British troops in Greece, was of the opinion that the 26 British General and 2/5 AGH, including nursing staff should remain behind to look after the sick and wounded. The DMS of the Allied force MajGen Burston took the opposite view. On April 20 the hospital ship *Aba* arrived at Piraeus but before patients or nurses could embark heavy bombing caused a postponement. The nurses returned to Kephissia. They were prepared to stay but General Blamey intervened to order their immediate evacuation. Their escape from possible capture is one of the epic stories of the Second World War.

On 23 April one half of the 2/5 AGH nurses set out from Ekali in charge of Sister Cook, J.L. They were joined by 31 nurses of the 2/6 AGH from Kephissia under Sister McAlpine, H.B. Their objective was to reach Navplion. Enemy attacks including machine gunning from the air were so fierce that at times they had to hide in fields of barley and once in a cemetery. When they reached Navplion they were taken out in Greek Caiques to the destroyer *Voyager* waiting two miles out in the bay. It is reported that the sailors were so astonished to be hauling up women, they dropped the first batch into the sea! Despite numerous attacks by enemy aircraft the *Voyager* brought its human cargo safely to Crete, little knowing what lay ahead. Meanwhile the remaining nurses from the 2/5 AGH and 2/6 AGH were ordered to leave on 25 April. They made their way to the beach at Magara and in the darkness they were taken off in small boats to the *Thurland Castle*, which despite enemy air attacks, also reached Crete safely on 26 April.

Conditions on Crete were chaotic and it was obviously only a temporary staging place on the way to safety in Egypt. Australian nurses were attached for a time to British hospitals, occasionally sheltering in the olive groves. Again a decision was taken to evacuate all the Australian nurses. On the 29 April the *Voyager*, the *Corinthia* and the *Ionia*, successfully made the trip from Suda Bay to Alexandria.

The evacuation of the 2/6 AGH from Greece is well recorded in the diary of Sister Muriel Gordon. In it she refers to 'Mike' (Sister Carmichael) and 'Joan' (Sister Joan Soorley).[37]

Saturday, 19 April (1941)

Though we had heard rumours that we may be evacuated from Greece, the sudden order to pack up and be ready to leave on 2 minutes notice was quite unexpected. Forthwith we flung our things together and made ready to leave when called upon. Midnight found us, weary from sitting on our luggage since early afternoon, unrolling our bedding and making ourselves as comfortable as possible — in semi undress — on the floor.

Sunday, 20 April

Still waiting — down to Fatty Jimmys for breakfast and lunch and back to our luggage to wait. After 1 pm we

Sisters taking shelter in a Greek cemetery, awaiting transports for Crete, 1941.

dragged our luggage down to the gate and waited some more. About 4.30 pm along came two 20 ton army trucks and we — 31 of us — piled aboard — all standing or sitting on piles of kit-bags or valises — and off. Twenty-five of our number, including Matron had left earlier from another house further up the road — down to the docks through streets where Greeks smiled or waved. We hated that — driving off and leaving them to their fate without having done a thing. We wondered guiltily if they thought we were not evacuating after all, but going up to the front lines. Then there was an air-raid on — traffic was at a standstill and when we got to the wharf it was only to find our ship had up anchor and gone, out of the dangerous waters of the harbour and there was nothing for us to do but to turn straight around and back the way we had come, through Athens and its crowd of wondering faces to Kephissia. There we all tumbled out, outside the 26 BGH, tired and dishevelled and wondering where we would be landed for the night. None of us will forget the old digger from the hospital who scrounged a bucket of tea from the hospital kitchen and served out cupfuls to us all. Anyway we landed in the turretted house where some of the girls had been billeted before — bedding on the floor again.

Monday, 21 April

Still on short notice with luggage ready!

Tuesday, 22 April

In the afternoon volunteers were called to work at the 26th. We had been clamouring for this for ages since coming to Kephissia but only a very few of us had been on duty. The whole 31 of us turned out and worked till after 10 pm sponging, washing, shaving and doing dressings. I shall never forget the dirt and discomfort of those lads.

Wednesday, 23 April

Still waiting! Nothing to do but make tea on our little primuses and run up on to the roof each time there was an air-raid. There was an air of tension everywhere — the Greeks going about with unsmiling faces — doubtful glances cast at us as we go up and down the road to "Fatty Jimmys". Rumours rife that the Germans were getting closer each day. More bombing of the harbour — troops retreating from the front line. — On 2 minutes notice again and told we can only take what luggage we could carry — more packing and unpacking — but hopeful to the last we had all our luggage at the front gate. At about 3 pm 2 20 tonners arrived — the drivers, Tommy boys, agreed to put our luggage aboard and off we went keeping to the outskirts of Athens this time and out onto the road heading south — past the little cottages from which war seemed remote. — stopped by a military guard who refused to let us continue and so back again the way we had come, to a little railway station not far out of Athens, where we were to catch a train to continue our southward journey. No sooner had we arrived there than there was an air-raid warning, the trucks were driven into the trees and we took cover in the fields. There we stayed and ate our biscuits and bully in the barley and learned that the line had been bombed and our train could not get through. By this time there were quite a number of nurses in the fields — English, Australian and New Zealand, almost 150 of them. Finally about midnight we were told there was a convoy of 8 trucks lined up and by the light of dimmed torches we farewelled our heavy luggage and tottered down to road to load up with our essential military gear and what personal stuff we could carry. None of us could forget that night. There were 22 in our truck, including a NZ MO and a Tommy soldier with a .303 who watched the road behind us. We were too crowded to move more than an inch all night. We travelled with dimmed lights, under instructions that we were to take cover off the road should there be an air-raid. Our drivers were truly marvellous — though they had been driving for days, they were ever watchful. At our place there was a huge bomb crater in the road, around which we made a detour. Our OC was most anxious that we should reach the canal into Corinth before dawn, with the fear that it might already have been bombed. Before that we had to negotiate a pass, a really horrible place, with steep hills rising sharply from the road with not a skerrick of cover on them. We took 2 hours to do the 4 miles as the Jerries had bombed and machine gunned the road, with trucks and cars tipped over and dead horses and mules lying alongside which blocked the road.

Thursday, 24 April

At dawn we crossed the canal, had a brief snack and then on again. About 9 am we had our first air-raid and the speed with which we left the trucks and into the barley fields was pretty good. We lay on our faces there for an hour and the whine of bullets and the plonk of shrapnel was as close as ever I want to get again. Later we learned that one truck with NZ girls tipped over with quite a number of casualties. We walked on for about half a mile to a cemetery where we sheltered for the day among the tombstones. About 8 pm the trucks lined up again and we clambered aboard — just as an air-raid alert sounded. So it was out again into the cemetery, but nothing happened. On again, with instructions to make for the wharf with all speed. It was a slow and tortuous ride to the docks in complete darkness. Even then we had to walk the last ¾ mile, with our baggage, much of which was dumped by the wayside.

There we tumbled aboard a waiting barge, past a burning ship and out to an Australian destroyer — never was there a more comforting sight! Those Aussie sailors who hadn't seen home for two years were kindness itself. They gave us their beds and their blankets and couldn't do enough for us. Despite a couple of raids we arrived safe and sound at Suda Bay, Crete on Anzac Day, Friday 25 April.

Here we were landed as quickly as possible, put aboard trucks and rushed across the island — such a pretty island, with green hills, red poppies and snow-capped mountains. At the camp we were taken to the officers' mess where colonels, majors, captains and lesser ranks ran around to give us food and drink. — and so to bed, the first night between sheets for a week.

Saturday, 26 April

We heard a convoy of wounded was expected — so in haste a hospital was prepared. Our tents were taken over (we slept in the open) — mattresses and bedding were found and in no time we had a hundred patients bedded down on the ground. I was on night duty that night and over 200 more patients came in. The British boys of 7 BGH were wonderful in establishing a "hospital" and coping with the ever increasing stream of evacuees.

Sunday, 27 April

We were given an empty house up on the hillside and moved in with our few possessions. We had to sweep it out to get rid of the bird lice. Sleep was difficult with dozens of people walking about.

Monday, 28 April

Our ward was full — 80 beds. I was awakened with the news of another move. We were to be packed and ready to move.

Tuesday, 29 April

After boiled eggs and black tea we stumbled down the hillside in the dark, into trucks and down to the wharf again. The wharf was crowded with Tommies, Aussie, NZ troops — Air Force, all dirty and bedraggled, tattered uniforms, hospital wear, civilian clothes, even one in a straw boater! — A retreating army, but not beaten!

We boarded a dirty little tub called the *Ionia* with a doubtful and motley crew. An air-raid sounded and the crew took off to the hills. We waited for hours until the Captain collected another crew — with Aussie volunteers as stokers, we finally crept out of port. The tub was filthy and greasy from bow to stern, and our "cabin" was disgusting. We slept most of the time on the decks, talking with the men, occasionally singing. Wednesday and Thursday passed as we slowly made our way.

Thursday, 1 May

About 11 am and we made Alex. Good to see Egypt again — and smell it, altho' we had been glad to leave. From the wharf we were put in buses and taken to 64 BGH. We were glad to be back in civilized surroundings again.

Matron Abbott was very proud of the conduct of the AANS during the perilous days of the Greek campaign, but her leadership was also a factor in ensuring that all members came through safely.[38] The CO, Col R.A. Money was highly critical of the whole plan and of the arrangements in Greece. In a subsequent report, he commented:

> In brief, the Unit arrived in Greece on March 8; the final decision to establish at Velos was not made until March 18; the actual site was only approved on March 24; and apart from CRS equipment, the first big batch of our equipment only arrived at Velos on April 12. The whole story of the campaign is indeed a sad one and it would appear that the relatively small amount of surgical and medical work done, not only by this hospital, but by the other hospitals in Greece, scarcely justified the tremendous loss of valuable personnel and equipment that occurred.[39]

Another Queenslander who featured in the Greek withdrawal was Sister Dorothea Burnett of 2/5 AGH.[40] Following training at the Brisbane General and a period on the staff of that hospital she enlisted in September, 1940 and subsequently sailed to the Middle East on the *Aquitania* on the 20 October, 1940. After disembarking at El Kantara, Sister Burnett was one of a small group of AANS, ostensibly allocated to 2/5 AGH but sent on to 2/1 AGH at Gaza to await further instructions. After a period of nursing at Rehovot, Sister Burnett and other AANS members of 2/5 AGH became part of the force to go to Greece. On Thursday, 10 April, 1941, they embarked at Alexandria and two days later they were in the beautiful harbour of Piraeus.

Strangely they were returning to the area where Florence Nightingale and her nurses operated during the Crimean War. Not far away was Salonica where Queenslander Christense Sorensen had been Matron of the 60 British General Hospital in 1918. Another generation of Australian nurses followed in the footsteps of their mothers, aunts and near relatives.

Sister Burnett worked in the operating theatre of the 2/5 AGH when it set up at Ekali where for ten days they were subject to enemy bombing and strafing. It was the familiar story of rumour, confusion, order and counter-order, with no one sure what would happen next, except retreat from the advancing Germans. Sister Burnett was one of those who volunteered to remain with the patients.

> From all reports and contrary to what we hear, Germans do respect the Red Cross and we are perfectly safe. It seemed clearly my duty to stay and do what I could for our lads who have been through a most terrific experience.(41)

In the event, a top decision by General Blamey required all to evacuate. Sister Burnett's diary takes up the story of her escape.

Friday, 25 April

New orders to get dressed immediately. Take only one haversack, and pack it as lightly as possible, blanket, rug and respirator, 3 days iron rations. So with tin hat on and haversack on shoulder, plus tucker bag and water bottle we packed into two buses . . . it was a very dark night, lit by the fires of ships in the harbour. About 1 am we were told to get out and lie in the gutter to await further orders . . . eventually we went out to the harbour in small boats, expecting a battleship, but it turned out to be the *Thurland Castle*. We climbed aboard as it was already on the move.

Saturday, 26 April

Jerry came to visit us at 7.40 am. Quite a lot of them, in fact. The convoy set up a most terrific barrage. There was a dreadful whiz as a bomb went past and dropped into the sea. Then came another mighty bang and splinter glass splashed over all of us. A piece of shrapnel came through the cabin and hit one of the lads outside. Throughout the day we had continuous bombing but were not hit. We reached Crete with three feet of water in the ship. What a day! On arrival, we were taken to a British Hospital to help with hundreds of casualties pouring in.

Sunday, 27 April

A count showed 84 Sisters of 2/5 AGH, 30 2/6 AGH and 8 2/3 CCS.

Monday, 28 April

Again hold to be ready to move at a moment's notice.

Tuesday, 29 April

Up 4.00 am — taken in trucks to the wharf. Taken on board small Greek ship, carrying wounded. Another air-raid. Ship so crowded we slept on the dining room floor — but a very disturbed night with so many air-raid alarms. Up early as room required for breakfast.

Wednesday, 30 April

Awoke to find about half the British navy escorting the convoy — destroyers, cruisers, an aircraft carrier, 8 troop ships, about 23 in all. Thank God the navy is efficiency plus.

Thursday, 1 May

Arrived safely in Alexandria Harbour. Loss of luggage the least of our worries.

Similar observations on the nerve wracking experiences in Greece and Crete was made by Sister Ivy Machon.(42) After training at the Brisbane General and serving at the Gregory Hospital (Qld), she enlisted in the AANS and was sent to Darwin in 1940. Subsequently she was sent overseas and joined the 2/5 AGH at Rehovot, with a warm welcome from Matron Best and the CO Col Kay. She too was sent to Ekali, 18 miles from Athens, right alongside an airfield — the bombing and the noise of the ack ack was frightening, she wrote. Sister Machon was one of those who volunteered to stay behind to nurse the wounded, knowing full well the probable consequences. Among other Queenslanders she recalled in this group were Sisters Casey, E., Tomlins, D., Hayes, B.E., Taylor, M.J., Patrick, E., Exton, A.M. and Zielke, J.B. They were saddened by the news that their CO Col Kay had been killed while supervising the loading of the wounded at the wharf. Most of the AANS escaped from Greece on 25 April, on 27 April the Germans took over the hospital at Ekali. It was certainly a near thing for the nurses. Medical staff of the 2/5 and 2/6 AGH who remained behind became POW's.

In her report on the evacuation of the AANS from Greece and Crete, Matron Kathleen Best gave a graphic account of the events and in the process paid a glowing tribute to the nurses.

> About 11 am on Wednesday, 23 April Colonel Kay came to see me. He was very anxious about the situation as far as the AANS were concerned. He told me that the Sisters of the 6 AGH and 3rd CCS were to be evacuated that afternoon together with 24 Sisters and Masseuses of the 5 AGH. The other 40 were to remain with the Hospital. He told me there was no chance of these being evacuated and asked me how I would select them. I said I would call for volunteers. I called a parade of the Sisters and explained the situation to them. I told them that those who volunteered to stay behind would almost certainly be taken prisoner. I asked them to write their name on a slip of paper, together with the word "Stay" or "Go". Not one Sister wrote "Go" on her paper. I then selected 39 Sisters to remain. I asked them to respect my decision as final and to make it easier for me by not questioning it.[43]

It so happened that these Sisters were evacuated, under fire, first to Crete and then to Alexandria. Summing up the whole eventful exercise, Matron Best said:

> I am afraid it is beyond me to describe fully those last few days in Greece and the loyalty and courage of the Sisters. Their one thought was for the patients and the Hospital. Their loyalty to me was the finest thing I have known, never once did they question any decision made for them, and they stood behind me and helped me in every way they knew . . .
>
> No one wanted to leave, but they all realized the common sense of it. Then the long journey through the night and the nightmare voyage from Greece. Yet their self-control remained perfect, their sense of humour always quick to respond to any funny incidents — and they never complained.[44]

In a confidential memo to DMS, AIF (MajGen Burston) Lt Col Coppleson expressed appreciation of the service of the AANS involved in the evacuation of Greece in the following terms:

> Members of the nursing staff were all called upon to undergo considerable strain and hardship under conditions of no little danger, under which, they, without exception, displayed considerable courage. Their bearing and cheerfulness did them the highest credit. It is difficult to single out any particular persons, when all acted so well, but I commend the following: Sister MacAlpine, Sister Beattie, Sister Sherwood, Sister Cook, Sister Dean and Sister Moir.[45]

The narrow escape of the nurses from Greece and Crete and their safe evacuation to Egypt., without casualty, was indeed a fortunate occurrence. It did however raise the same questions which Col Speirs had raised at Tobruk, only a few weeks before. Should members of the nursing service be exposed to possible injury or death by serving in advanced areas? The rapid changes in modern warfare provided the answer to that question. The bombing of cities in Europe and Britain with heavy military and civilian casualties meant that there was no longer any front line. Nurses and medical staff everywhere were at risk — and they were prepared to accept the risk as part of their duty in caring for the sick and wounded. Nevertheless, in the battle order, nurses were protected. In the main they served in base hospitals and field hospitals and in some circumstances in Casualty Clearing Stations. They were not as a rule posted to Field Ambulances except in special circumstances, as in Syria.

The more difficult question arose when it seemed clear that the medical unit would be overrun by the enemy. Should medical and nursing staff remain behind with the patients and become POW's with all the risks entailed in that, or should they escape to have their skilled services available to many more troops for the duration of the war.

While nurses believed they had an obligation to remain with their wounded and were prepared to do so, Higher Command in Tobruk, Greece and Crete ruled otherwise. The Sisters argued that the Germans were a civilized nation and that they would be properly treated as POW's. Overall, this proved to be the case, although there were lapses when clearly marked hospitals were bombed in Tobruk and Crete and with the wanton killing of medical personnel in Crete. Similar arguments were raised in Malaya but there the allies were dealing with a brutal, savage, ruthless enemy with different values with respect to human life, the Red Cross and the treatment of prisoners of war.

When the Sisters returned to Gaza they were welcomed by their colleagues at the 2/1 AGH. Sister McConnel recalled in her letters meeting Beatrice Hayes, Dorothy Roe, Janet Taylor, Phyllis Hawken and other Queenslanders, 'all of whom have safely returned to the fold after the most harrowing adventures in Greece. They are very cheerful and one would have thought they were sight-seeing tourists!'(46)

2/11 AGH Alexandria

During these desperate months of 1941, the 2/11 AGH was also part of the action in the Middle East. As a hospital originally established in Hobart in August, 1940, it found itself in April, 1941 established in a modern Greek hospital in Alexandria. It provided a first port of call for sick and wounded being evacuated from the Western Desert and from Greece and Crete. It was never a large hospital, but it was always extremely busy and always operating within an area which was a constant target for enemy bombers. At first Matron Bowe and members of the AANS had to be transported each night 9 miles to the 2/9 AGH at Amiriya, but later they were accommodated in a building some 3 miles from the hospital.

Among the Queenslanders who served with the 2/11 AGH at this time were Sisters O'Keefe, E.M., Riddell, G., Freeman, E.L., Couche, J.W., Hall, E.M., and Netterfield, M.M. When the HMS *Orion* was hit while evacuating troops from Crete (28–29 May) casualties were heavy and the Sisters found themselves working round the clock, treating many of those suffering from burns and shock. They often recalled the horrifying task of nursing 'burns cases' whose bodies were soaked in black oil. While work at the hospital had to go on the bombing seldom stopped, but the Sisters were fortunate that there were no casualties among them from this cause.

When things quietened down later in the year they were moved briefly into a pleasant home at 11 Rue Peake, within comfortable distance from the hospital. With the final evacuation of Australian troops from Tobruk and with Greece, Crete and Syria behind them, the AANS moved back to Beit Jirja, in Palestine, only to be caught up again in the Stepsister Movement.

Asmara

Time and again medical personnel, including AANS were put at risk and the work of the AAMC rendered less efficient than it might have been by movement policies which to all concerned were extremely difficult to fathom. Senior officers of medical units were not always consulted or advised about the reasons for impending moves, while members of the AANS and other ranks were left guessing. This was extremely bad for morale. The move of the 2/5 AGH from Gaza to Asmara was a case in point. It is true that the unit had to be refitted and reorganized after the Greek tragedy and it needed a period of rest but why in Abyssinia, where there were no Australian troops and in a place most inaccessible for evacuation purposes? The movement from Gaza to Asmara was a nightmare to all concerned — by train to Port Tewfik, thence by the *President Doumer* to Port Soudan and Massawa, followed by a long truck drive to Guru camp site — an expedition to test the fitness of all. Everything that could go wrong went wrong. At Suez the AANS had to sleep on concrete floors, with no washing and inadequate toilet facilities.

Queensland nurses on this primitive safari were Hayes, B., Roe, D., Tomlins, D., Machon, I.F., Patrick, E. and Zielke, J. It was the end of September before the hospital was functioning, to receive 'invalids' from Palestine and Egypt. No sooner had it begun functioning than it was closed down and all the personnel were returned at some danger and great inconvenience to Gaza. Troops of all ranks, including AANS members could not fathom the inscrutable wisdom of higher army authorities.(47)

Meanwhile back in Egypt and Palestine the 2/1 AGH, 2/2 AGH, 2/6 AGH and 2/7 and 2/9 AGH continued their work of treating and nursing Australian and allied casualties from the various battlefronts. By this time there were many Queenslanders in these units.(48) Members of the AANS proved adaptable to all the changing circumstances, to the physical hardships, to the rumours and counter rumours, to plans and cancellations of plans to to hasty, ill-considered moves. Five nurses from 2/7 AGH were sent to Aleppo to join the 2/5 Fd Amb; 2/5 AGH had a signal to move to Rhodesia but this was cancelled. The 2/6 AGH also had moments of excitement.

This hospital functioned for five months in the Kaiser's Palace on Mt Scopus, Jerusalem — an address rivalled only by the 2/4 AGH nearby the Street of the Prophets, Damascus Gate! The AANS, along with other medical groups had the unique experience of working in this historical city, moreover at Christmas time. Added to this it snowed in Jerusalem for the first time in many years, giving a traditional Christmas card atmosphere to the whole scene. Many of the AANS were fortunate enough to trudge through the snow to Bethlehem for the Christmas service. The 2/9 AGH also had its moments when the changing fortunes of war appeared to save it from disaster, as though it were being preserved for a greater contribution elsewhere. After arriving in Egypt it took over the 2/4 AGH site at Abd el Kader, very unsuitable for a hospital. It was on the battle order for Greece but the sudden collapse of that campaign caused it to be sent to Nazareth, again in unsuitable buildings. Sister Carmody, later to have a distinguished career in the AANS gained her first experience in these ups and downs of 2/9 AGH in Egypt and Palestine.

Dr Allan Walker summed up the varied experience of the Australian nurses in the Middle East:

> In the Middle East the AANS worked under very varied conditions — in a workhouse in Nazareth, an Italian and German Hospital at Jerusalem, the Kaiser's Palace on Mount Scopus, underground at Tobruk, in an Egyptian barracks at Mersa Matruh, a marble and brick edifice at Alexandria, on an aerodrome in Eritrea, in stone houses in Syria, and in tented wards in many places. Their living quarters were often primitive: in some cases four nurses shared an EPIP tent and at Mersa Matruh they lived in an underground shelter infested with bugs. Climatic conditions varied from the dry heat and dust storms of the desert, through the semi-tropical humid atmosphere of the Nile Valley, to the winter cold of Palestine and Syria.(49)

As 1941 drew to its close the tempo of the war increased. Australia's 6th, 7th and 9th Divisions were still in the Middle East. There was still the threat from German forces as their initial onslaught against Russia carried them to the shores of the Black Sea. Turkey was an unknown factor. The Syrian border had still to be protected. In the Western Desert the war had not gone well for the allied forces. Attempts to raise the Siege of Tobruk failed until December 8. Rommel's Afrika Korps was still a force to be destroyed and it appeared that the allies did not have the strength of armour nor the technique to destroy him.

All this changed on December 7, 1941 when Japanese planes attacked Pearl Harbour and the USA entered the war. Suddenly the Pacific and the Far East became the focus of the world war. Within weeks the Japanese had occupied the Philippines and advanced down the Malayan coast. The threat to Australia became very real. Immediately the Australian Government sought the return of the 6th, 7th and 9th Divisions from the Middle East, but Churchill recognized that in the overall strategy of the world war, this would have left the Middle East dangerously vulnerable to onslaughts by the Axis powers. In this dilemma, the 'Stepsister Movement' was born — the 6th and 7th Divisions would depart from the Middle East to places unknown, the 9th Division would remain. Inevitably, this meant dramatic changes in the distribution of medical services, including the AANS.

The following medical units and members of the AANS departed from the Middle East in this period February–March, 1942 — 2/1 AGH, 2/2 AGH, 2/4 AGH, 2/5 AGH, 2/9 AGH, 2/11 AGH, the 14th Special Hospital, 2/1 CCS and 2/2 CCS as well as related Field Ambulances and ancillary units. Their story unfolds a little later. At this time there was much elation among the AANS with the announcement of a promotion list. Those nurses who had given valuable services in the Middle East received some recognition for their dedicated service. Among them were 41 Queenslanders.(50)

What then remained? the 2/6 AGH took over from 2/1 AGH at Gaza and remained there until February, 1943. The 2/7 AGH had a more mixed career as Queensland Sister Broomfield found.(51) Early in 1942 the policy that no Australian General Hospital be set up in Syria was reviewed and plans were made to establish the 2/7 AGH at Sidon, as well as the 2/4 Australian Convalescent Depot. The 2/7 AGH was orderd to close its hospital at Rehovot and to move to Sidon where the first ward was opened on 4 July, 1942. As they moved up to Syria the 2/7 AGH members were surprised to find the 9th Division going flat out in the opposite direction — to the Egyptian frontier. Clearly something 'big' was on. Within a few days the 2/7 AGH was ordered to pack up, hand over to the 3 British General Hospital and to move to Egypt.

Some members of the AANS found themselves mixed up again in the ebb and flow of battle movements. Nurses from units involved in the Stepsister Movement were pulled out of the Field Ambulance Units in Syria to return to their 'parent' unit. They were replaced by nurses from the 2/6 and 2/7 AGH, specially selected for their ability to train orderlies who would be involved in initial patient care in forward areas. The 2/3 Casualty Clearing Station remained in Beirut, treating patients within their resources, most of these being malaria victims. In June 1942 it took over the old Italian Hospital at Tripoli and continued working there until early July.

Allied reverses in the Western Desert again prevented the return of the 9th Division to Australia. In the fluctuating fortunes of that war, the allies had again pushed Rommel back to Tripolitania, only to see his counter attack push them back again almost to the Egyptian frontier. Tobruk fell to the Germans this time. British and allied armies suffered heavy losses in men and armour before stabilizing the front line near Alamein. It was a desperate situation for the allies and by the end of June 1943 it seemed as if nothing could stop Rommel's victorious march to the Nile Valley and the Suez Canal. The 9th Division was transferred at speed from Syria to Amiriya, near Alexandria and from July to October 23rd when the famous battle of El Alamein, which virtually ended the North African war for the axis powers, began. During these months the allied forces, including the Australians, were engaged in constant battles with Rommel's forces — and casualties were frequently heavy. The medical support in the forward areas consisted mainly of Field Ambulances, supported by surgical teams from the 2/6 AGH and 2/7 AGH. Members of the AANS were not involved in these battles at the front. Evacuation was possible to the 2/6 AGH at Gaza Ridge which bore the brunt of the casualties. It was clear another Australian General Hospital was needed nearer the battle-zone, hence the move of the 2/7 AGH from Sidon to Buseilli in the delta area of Egypt. Within a month it held 600 patients and expanded to 1,000 beds, while the 2/6 AGH was ordered to expand to 1,500 beds. The 2/3 CCS (LtCol J.E. Gillespie) was also part of the medical team. It too was hastily withdrawn from Syria to Egypt and prepared for a mobile role. Once the axis forces were pushed back the 9th Division played little part in the final retreat to Tripoli. The 2/3 CCS was the only Australian medical unit to advance beyond Alamein to Matruh. Nurses were brought forward to this point, where they performed at a high level in treating casualties not far from the front line. However by the end of November the work of the 9 Division was completed and it moved back to Palestine. For 2/6 AGH and 2/7 AGH there was no rest, until all casualties had been returned to their units or evacuated to Australia. By the end of 1942, the war in the Pacific had reached a crisis point. Prime Minister Curtin, after long discussion with Churchill, Macarthur and world leaders, finally succeeded in having the 9 Division returned to Australia in February, 1943.

Ceylon

As a result of discussions by army medical authorities on the evacuation of casualties from the Middle East to Australia it was seen that the few hospital transports available were inadequate, so that major hospitals in Palestine and Egypt were becoming filled with patients awaiting evacuation. It was clear, that another 'half way house' to Australia was needed. Ceylon was finally decided upon as an ideal spot, removed from the battle-zone and within relatively easy commuting distance from Fremantle.

In an agreement between the Australian and Ceylonese Governments, the King George V TB Hospital at Welisara, some 12 miles from Colombo, was acquired as a staging hospital. Further buildings were erected to bring the capacity to 600 beds. It then fell to the 2/12 AGH (Col G.W. Macartney) to take over the task of setting up a staging hospital at Welisara. This period lasted from October 1941 until December 1942, by which time the Japanese war was engulfing the islands to the north of Australia and all Australian forces from the Middle East were recalled to Australia.(52)

At first the patients received were mainly those in transit from the Middle East to Australia, but as the Japanese war opened up they found other responsibilities. Evacuees from Singapore, casualties from Java and from naval engagements soon filled the wards. Fortunately in the confusion of the Stepsister Movement, the 16 and 17 Brigades of the 6 Division AIF, including the 2/1 and 2/2 Aust Fd Ambs and the 2/4 AGH, opened at St Peters in May. The AANS were in charge of Sister Ruth Baker who reported favourably on the work done by the AANS and by the AAMWS in this period. She commended the efficient work of Sister Vickers and Sister M. O'Loughlin and noted with great sorrow the death of Queensland Sister C.M. Thiedeke. There was great pleasure in the announcement of the marriage of Queenslander S/N M.E.F. Hitchings.(53)

The Stepsister Movement

The high level decision to divert the Australian troops in the Middle East (leaving the 9 Div to meet the German threat either in Syria or in Egypt) to the Far East to meet the Japanese threat was born in panic, developed in indecision and doomed to disaster. In this movement, members of the Army Medical Corps, including Sisters of the Australian Army Nursing Service were indeed fortunate in escaping with minimum casualties. While the entry of Japan into the war was not unexpected, the speed of the Japanese advance and their initial successes on land, sea and in the air necessitated hasty changes in world strategy against the three major Axis Powers, Germany, Italy and Japan.

An enormous armada of ships was required to assemble off Suez to transport more than 60,000 troops and their equipment somewhere to the Far East. It was planned to move the group in three large convoys and in so far as the Sisters were concerned they were involved in the movement of AGH's and CCS's. In the first convoy went the 2/5 AGH, the 2/11 AGH and the 2/2 CCS; in the second the 2/2 AGH, the 2/9 AGH and the 2/1 CCS, while in the last group went the 2/1 AGH and 2/4 AGH. There were of course many other medical units to which members of the AANS were not attached.

The first group moved out of Suez on the 30 January, 1942, the very day on which the Australians in Malaya had retreated across the Johore Causeway into Singapore Island. The second group left in the third week of February by which time Singapore had fallen, Darwin had been bombed and the war moved closer to Australia with the Japanese invasion of the Dutch East Indies and New Guinea. By the time the third group was scheduled to leave, the war in the Far East had moved so fast on all fronts from the Burma Road through to New Guinea, Guadalcanal and the islands of the Pacific, rapid and confusing changes of policy and overall strategy were a nightmare to those in charge of the vast armada of transports.

Of those medical units in the first group, the 2/2 CCS bore the brunt of the allied confusion in Sumatra and Java. Travelling in the *Orcades* the unit disembarked at Oosthaven in Sumatra, but then were told to re-embark and make for Batavia (Jakarta). Allied strategy at first aimed to hold Java against the Japanese and shipping built up at the port of Tanjong Priok. General Wavell realized the hopelessness of this plan and thought that Australian troops should be sent to Burma. Nevertheless, Australian troops disembarked in Java and the 2/2 CCS was sent to Bandung to set up a hospital there, with LtCol N. Eadie in command and the nurses in charge of Queenslander Sister Vi Paterson. As the forces were reorganized, LtCol E. Dunlop took over from LtCol Eadie. The military position deteriorated so rapidly that it was decided the nurses should be evacuated with all speed. They re-embarked on the *Orcades* on the 21 February together with six nurses who had escaped from Singapore in the *Wah Sui*, and after a perilous journey arrived safely at Fremantle.

The 2/5 AGH and the 2/11 AGH fared little better in these desperate weeks. They left Suez in the *Mauretania* in January and arrived in Bombay, apparently destined for the Far East. In fact, either by accident or design the baggage and equipment of 2/5 AGH went ahead on the *Orcades* and arrived with the 2/2 CCS and other units first of all at Oosthaven and then at Batavia. After trans-shipping at Bombay to the *City of Paris* and the *Esperance Bay*, the 2/5 AGH and the 2/11 AGH found themselves part of the rapid changes in strategy. With the deterioration of the war in the East they were diverted to Ceylon and soon afterwards sailed for Australia, where they arrived safely.

Other groups of nurses had startling experiences in these frantic moves to get Australian personnel safely home. When the 2/6 AGH took over 2/1 AGH at Gaza in March, the AANS from 2/1 AGH under Matron Fall embarked from Suez in the *Laconia*. Matron Sage was also on board. They arrived at Bombay without incident, where they split up into various ships of the convoy, the *Katoomba*, the *Duntroon* and the *Holbrook*. This convoy put in to Colombo to find the port still burning from the Japanese air-raid on 5 April. In view of the danger from possible Japanese naval units in the Indian Ocean, the convoy was diverted first of all to Mombasa, then to Durban before reaching Australia at the end of May, 1942. It was a perilous journey of some three months at sea, fraught with danger every day but thanks to the efficiency of the navy and of the mercantile marine no lives were lost.

While the 2/9 AGH and the nurses of the 2/2 AGH had a fairly uneventful movement from Port Tewfik to Adelaide during February–March 1942, the 2/4 AGH and members of the AANS with that unit had a

AANS — Departed Brisbane 31 Jan 1941.
L. to R: Sisters P. Mittelheuser, J. Tweddell, J.J. Blanch, C.S.M. Oxley, C.M. Delforce of 10th AGH with Sisters of 2/5 AGH I.F. Machon, D.M. Tomlins, E.M. O'Keefe and T.A. Skyring.

different experience. As they were in the last group to move, they were probably more fortunate, even though they were caught up in a welter of changing decisions. The 2/4 AGH received its first movement orders on 1 Feb, 1942 but these were cancelled, then reimposed, but it was not until 12 March that the members embarked on the *Westernland* at Port Tewfik for an unknown destination. The Sisters of the 2/4 AGH, prior to embarkation, found themselves in 'The Aviary', a wired-in area of about one acre resembling a large bird cage, with only one means of entry or exit. Early on the morning of 12 March they embussed in three-ton trucks, standing room only, down to the wharf for embarkation. This convoy arrived in Colombo Harbour on 25 March, obviously too late to join any possible reinforcements for Java. Shipping in Colombo was building up, as word was awaited on what to do and where to go next. It was with mixed feelings that the 2/4 AGH was ordered to disembark on 27 March, as it happened, only a few days before the Japanese air-raid on the Harbour.

While it was clear that the Australian Government wanted the 6th and 7th Divisions back as soon as possible, changing fortunes of war in the Far East constantly called in question the wisdom of this request. The 'Stepsister' Movement envisaged at least some of the troops going to Java in an attempt to stop the Japanese advance. Some of course did go, including a section of the 2/2 CCS and many were trapped by the Japanese in this futile exercise. There was still one school of thought that Australian troops ought to be sent to Rangoon to hold back the Japanese in Burma. The Burma Road to China was seen as an essential life line supporting China against the Japanese. The fall of Singapore, then Sumatra, then Java caused

Darwin 1940. 2/5 AGH; Sisters Short, E., Skyring, T.A., Tomlins, D.A., Machon, I.F., O'Keefe, E.M.

rapid changes of plans for the 'Stepsister' Movement, with convoys changing direction almost hourly. Once the decision was taken not to risk more Australian units in the defence of Java, the options narrowed down — Burma, Australia or back to Ceylon. In fact the British and American leaders supported by Churchill, succeeded temporarily in having the 7 Division Australian troops diverted to Burma, but Prime Minister Curtin would not agree and once again the convoys were re-directed. Some, as with the 2/9 AGH and 2/2 AGH nurses reached Australia directly, while others such as the 2/1 AGH and their nurses came home via South Africa. Still others, such as 2/4 AGH found themselves in Colombo. Overall, these few months in 1942 were 'touch and go' for Australian units of the 6th and 7th Divisions and it was more by good luck than careful planning that so many arrived back in Australia safely.

The tragic Malayan campaign

Of all the campaigns in which the AANS were involved in World War II, none was more tragic than the Malayan campaign. In all, twelve were drowned, twenty-two were shot and killed by the Japanese, thirty-two were taken prisoner and spent more than three years in captivity under conditions of great privation during which eight of their members died. Thirteen Queensland nurses were in this casualty list.

The story begins quite early in the war, before Japan attacked Pearl Harbour, when the Australian Government offered troops to assist in the defence of Malaya and the islands to the north. In January, 1941 part of the newly formed 8 Division was assigned to Malaya under MajGen H. Gordon Bennett. Accordingly on February 18, 1941 there arrived in Malaya by the *Queen Mary* a brigade group, including numerous medical units, for this story the important ones being the 2/4 CCS and the 2/10 Australian

AANS for Malaya leaving Sydney 19 February 1941. L. to R: Sisters Trotter, Grigg, Blanch, Adams, Tweddell, Blake, Doyle, Davis. (Photo Associated Newspapers).

General Hospital. Maj J.G. Glyn White was appointed DADMS and Col A.P. Derham ADMS. The nurses were in the charge of Matron O.D. Paschke., Matron of 2/10 AGH, and there were a number of Queenslanders in this group, including Sisters Trotter, F., Tweddell, J., Mittelheuser, P., Oxley, C. and Delforce, C.[54] The location for the 2/10 AGH was initially based not on strategic considerations, but on what suitable accommodation was available for a 400 bed (later 600) hospital in a malaria controlled area. The eventual site was at Malacca, in buildings which were part of the civil hospital, in many respects an excellent choice. There was one major drawback, which was to present serious problems later. Malacca was situated on the coast, within Malacca Strait, north of Sumatra and about 100 miles north west of Singapore. Unfortunately, the port was useless for evacuation, as large ships had to stand out to sea several miles. However road transport was reasonably good. The first few months were important in stepping up training and in preparing for any emergency. The normal sickness rate among the troops kept the nurses busy with cases of malaria, scrub typhus, dysentery, dermatitis, all of which required further training for members of the nursing staff.

The 2/4 CCS (LtCol T. Hamilton) with its quota of nurses was stationed first of all at Kajang, a few miles south of Kuala Lumpur. With the arrival of further reinforcements in August bringing the Australian force to full brigade strength, the 8th Division was given a clear responsibility for the defence of the southern part of the Malayan Peninsula, including in the area Johore and Malacca. Additional medical support units arrived and in September the 2/13 Australian General Hospital, but the establishment of the latter was a haphazard affair. However its senior officers were recruited from those already in Malaya, the CO Col D.C. Pigdon coming from 2/2 Con Depot and Matron Drummond from 2/4 CCS. There was no immediate site selected, so staff and nurses were dispersed to other establishments for 'experience', some to 2/10 AGH, some to Singapore General. By the end of November a site was selected across the causeway to

Australian nurses strolling through the ornate gardens surrounding Haw Par Villa Singapore 1941. L. front V. Bullwinkel, Rt front W. Oram.

Tampoi, seven miles from Johore Bahru, in an 'unfinished mental hospital'. Queenslanders with 2/13 AGH included Sisters Short, E., Smith, V., Muir, S., Hempsted, P.B., McElnea, V.I. and McDonald, G.M.

From the commencement of the war with Japan on the 7 December, 1941 to the capitulation of Singapore on 15 February all Australian medical units found themselves desperately busy as the casualties, both civil and military, increased. The pattern of evacuation involving Field Ambulances, MACs, MDSs has been well described by Col Walker, but this story concerns the fate of those members of the AANS attached to the 2/4 CCS, the 2/10 AGH and the 2/13 AGH. The three pronged attack by the Japanese down the Malay peninsula, coupled with the constant bombing of all centres through to Singapore soon created large numbers of casualties, with medical staff and nurses working round the clock in an effort to save lives. Soon it became apparent that the position of the 2/10 AGH at Malacca was untenable. To move a field hospital with 800 tons of equipment, 1,200 patients and several hundred medical, nursing and general staff was no easy task. A new site was found on Singapore Island, with the medical section at Oldham Hall and the surgical section at Manor House. Meanwhile the nurses were detached to the 2/13 AGh and to the

AANS with MajGen Gordon Bennett, Singapore 1941.
L. to R: Sisters Daley, Muldoon, MajGen Bennett, Sister Forsythe, Masseuse, and Sister McMahon. (Herald photo).

2/4 CCS. By the 15 January the hospital was operative again, but the situation was appalling as large convoys of wounded arrived from the north. The medical staff and nurses worked long hours under constant bombing, with additional problems as the water supply ceased.

The 2/13 AGH at Tampoi Hill across the Johore causeway fared little better. It took many of the patients from the 2/10 AGH while that hospital moved, but then came the time when it too was ordered back to Singapore Island. It was required to move back to St Patrick's School on the south side of the Island and accomplished this feat in thirty-eight hours! By January 26 it was ready to take patients and soon was overflowing, as there appeared to be no ships available for the evacuation of casualties. A convent school, a mile away, was taken over for additional casualties.

Meanwhile the 2/4 CCS at Kluang, with Sister K. Kinsella in charge, had been performing a key role in receiving casualties from all fronts and evacuating them back to the 2/13 AGH. It remained at Kluang until the last possible moment, January 19, before pulling out. Successive withdrawals were necessary until January 30, when the last of the British, Indian and Australian troops crossed over the causeway to Singapore. The first site for the CCS was so vulnerable it came under Japanese artillery fire.

Sister E.M. Wittmer, of the 2/13 AGH later gave a graphic account of these latter days in Singapore. She described how the 2/13 AGH arrived at its new site, St. Patrick's School, in great haste.[55] When they arrived there was nowhere at all for the nurses to sleep, shelter or eat. They slept on the lawns until the tents were erected. They were soon admitting casualties direct — many fractures, particularly legs and pelvis. As the Japanese planes came over day and night, bombing and strafing, conditions in the hospital became intolerable.

Sister Wittmer said the lack of hospital ships was a calamity. The whole hospital work at night had to be carried on in complete darkness. The only flare for a light was a small cupboard under the stairs where torches were used for filling syringes for injections.

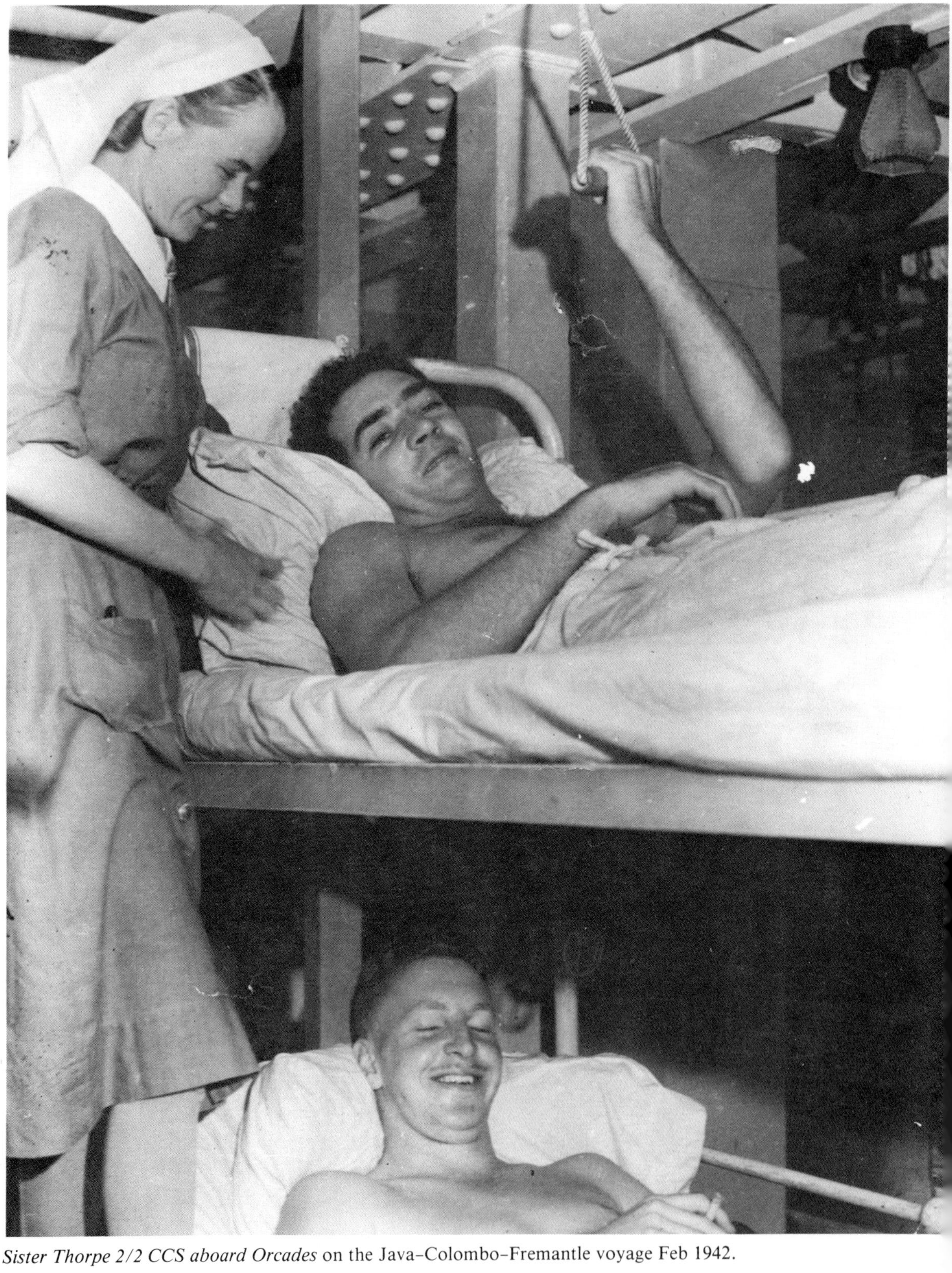

Sister Thorpe 2/2 CCS aboard Orcades on the Java–Colombo–Fremantle voyage Feb 1942.

Sister A. Anderson of the 2/13 AGH described the many serious problems faced by the medical and nursing staff at Tampoi.[56] They were always short of equipment, sterilizers had to be improvized and stock made from sheets. There was no water available in the wards until installed much later by British engineers. The Red Cross gave great help in supplying pyjamas, towels, sheets and pillowcases. There were many sick patients, malaria, typhus and dermatitis in the early stages. The scrub typhus victims were of the raving delirium type, uncontrolled by drugs and most of them died. Sister Anderson was in the resuscitation ward. They used mostly 'wet serum', but towards the end they had to use 'blood on the hoof'. There were plenty of giving and receiving sets. After December 7 conditions changed rapidly. As soon as it was known Japan was in the war patients made rapid recovery! Once the battle casualties began to roll in, the bed state built up rapidly. Sister Anderson commented, 'It was a matter for general criticism and wonderment that no hospital ships whatever were sent to evacuate the patients. There were men waiting for weeks in plaster who would have got back, if a hospital ship had come, but none did'.

Sister Irving, of 2/10 AGH, was transferred to the 2/4 CCS at Kluang while that hospital was withdrawn from Malacca to Singapore Island.[57] She described how the Sisters lived in a planter's bungalow amongst the rubber. They had air-raids every day, usually commencing at 4 am and at first they had to take shelter in the slit trenches. Big convoys were coming in all the time, bad casualties with arms and legs blown off. The wounds were treated with sulpha, and blood and serum were given. They had to evacuate patients as

AANS in Malacca 1941.
Rear: Sisters Tweddell, Delforce, Mittelheuser, Grigg, Adams.
Front: Sisters Ralston, Oxley, Matron Paschke, Trotter, Blanch, Calnan.

quickly as possible to the south, to make room for the next convoy of incoming casualties. Sister Irving paid great tribute to the medical officers LtCol Hamilton and Maj Hobbs and to her nursing colleagues of 2/13 AGH, 2/4 CCS and 2/10 AGH who worked day and night to help the wounded, while it seemed as if the world was falling apart around them. There was no time to think of personal danger. Seldom have nurses been asked to perform so close to the front line, yet their courage and devotion to duty never faltered.

These three medical units found themselves operating under appalling conditions. Casualties far exceeded available beds; water, gas and electricity were cut off. Still the medical care and the nursing went on, despite the darkness, despite the nearness of the enemy. There was a feeling of hopelessness about it all, as if waiting for the inevitable to happen.

By the second week in February it was clear that the end was not far off. The question of the evacuation of the nurses became one of great urgency and of great controversy. Eventually, the decision was made that they should go.

Once the decision to evacuate the nurses from Malaya had been taken, their chances of survival depended on the availability of shipping and the speed of the Japanese advance. Both these factors worked against the chances of the nurses getting out safely. On the 10 February, 6 nurses (including Sister Grigg, I.E. from Queensland) were sent off with 47 patients on the *Wah Sui*, a hastily prepared hospital ship, previously a convalescent ship for officers in Alexandra Hospital, Singapore. While waiting in Singapore harbour the *Wah Sui* was bombed but escaped damage. The nurses and the 350 people on board the crowded vessel had a terrifying journey to Batavia, but arrived safely on the 15th. There they found utmost confusion. Singapore fell on the 15th, by which time the allies were attempting a desperate defence of Java. The 2/2 CCS, including Queensland nurses had been sent to Bandung to work with No 1 Allied General Hospital, but as the position of the allied forces in Java was deteriorating rapidly their situation became precarious.

The Sisters from the *Wah Sui* assisted in the transfer of the wounded to the *HMIS Kapala* but were not allowed to accompany them, much to their distress. Instead once they landed at Batavia they set up a hospital at the Princess Juliana Convent. On the 18th they were sent to Bandung to help with the 2/2 CCS at the hospital but almost immediately all the Sisters were transferred back to Batavia to await evacuation. On the 21st they left on the *Orcades*, travelling to Colombo and then back to Adelaide.

While these members of the AANS survived the rapidly changing fortunes of war, their colleagues still in Singapore were not so fortunate. On the 11th February the names of half the nurses of the 2/10 and 2/13 AGH and 2/4 CCS were read out and these were told to be ready for immediate evacuation. There was great consternation at this announcement as the nurses believed their responsibilities lay with their patients — and they believed the morale of the troops would drop considerably once they realized the nurses were going. Still, they had to obey orders. At two hours notice, 59 nurses were embarked on the *Empire Star*, this group including Queensland Sisters Pugh, P., Selwood, M., Powell H., Adams, M. Daley, S. and Ralston, I.D.[(58)] By this time bombing and shelling was almost continuous and the whole place was black with smoke. Even getting the ten miles to the docks was a hazardous journey. They found conditions on the *Empire Star* appalling, with over 2,000 crammed into the small ship. The nurses were accommodated in the hold with hatches closed. The following day the *Empire Star* was bombed and machine gunned by hostile planes almost continuously. At one time 88 bombers attacked the ship inflicting damage and some casualties but fortunately the ship was able to keep going. Thirteen bombing casualties were buried at sea. Sisters came up from the hold to tend the wounded, two of them subsequently being decorated for their courage and devotion to duty under fire. The ship arrived at Batavia on the 14th February, unloading many passengers before proceeding to Fremantle where they arrived without further incident.

Sister Sheila Daley, 2/10 AGH, who was fortunate to return to Australia in the *Empire Star*, subsequently wrote a graphic account of her ordeal.

> We left Australia i.e. Fremantle in the 2nd week in October and took about 10 days to reach Singapore. As you know, I was in a British hospital there for a week and then got a lift up to Malacca where the 2/10th were established. The threat of war loomed close. The hospital personnel were confined to barracks, except on specified trips to pictures etc. The various admirers were all gone having been ordered to their battle stations. Even if it were a shock, there was some feeling of relaxation and relief when Japan attacked. The boys did not know how poorly they were

to be supported in the air and expected to thrash the Japs easily. The average soldier had no idea of the numerical odds. The papers had made them confident that there were around three million well equipped men at their stations in Malaya. The Japs advanced quickly and easily, soon taking aerodromes almost intact. Their methods you will have read about. In some cases, they filtered through dressed in Malay native dress on bicycles, one arrived with tommy gun, another with a case of ammunition and another with food. Having picked their positions, one would climb a tree and the food and ammunition handed up to him. The boys said no one man could have climbed the trees with the loads some were found to have taken up their trees. Then the others would filter back to their own lines for more. Dressed as Malays, whole bus loads of apparently innocent natives poured down behind the British lines. Then there would be an order to fall back instead of trying to get behind the Jap lines and cut off *their* communications.

The amount of scorched earth policy carried out was negligible-whole dumps of bombs, ammunition, petrol, uniforms etc fell into Jap hands. Soon the casualties were recognized as being caused by our own explosives. Even I was pessimistic before the Australians took part in the fighting. I wrote home and said "read 'what happened to France' again". It did. The fifth column was most active. Malays grew crops pointing at military objectives. They climbed rubber trees and painted their leaves with luminous paint, they placed banana leaves at a certain pattern near artillery positions, near anti-aircraft batteries, blocked roads, absconded from their work, signalled information to the enemy, enrolled themselves in the Malay volunteer defence corps and misled convoys, gave wrong orders and spread rumours.

The Australians went into action in the 2nd week in January. We had evacuated our patients to the 2/13th AGH at Jahore Dahru and got orders one day to be ready to leave in ¾ hour — we were, and drove down in a little over 6 hours in ambulances. The Japs broadcast that we had left and taken over a hospital ship which they regretted had been sunk in the straits. Work commenced to pour in, hours were longer, blackout almost complete. After some days, the Australians, who never fought as a body but as parts here and there, were orderded back. Many came to hospital exhausted, after days of retreat on foot with little food. The humidity was high but they stood up to it well. Every morning the Japs flew over to bomb the island. The opposition in the air grew less and less. Australian and New Zealand pilots were sent up in antiquated Buffalo fighters fit only for training purposes. Sometimes a couple of Hurricanes flew up and everyone felt cheered for about three minutes.

The sky was dotted with dots of smoke where anti-aircraft fire tried to keep the bombers off and break their formations; occasionally they did, but soon the Japs could fly leisurely above the ack-ack fire and drop their bombs accurately from many thousands of feet. Oil tanks at the naval dockyards burned for weeks. Crude oil burns slowly and with heavy clouds of smoke. We felt safe at the Hospital. The Japs appeared to respect it. All the time they came closer and made fresh landings. Our navy had withdrawn after the loss of the battleships without air support, it was obviously suicide to stay. But they failed to remove or destroy much of the equipment, much of it invaluable, stored at the docks and barracks. They may not have known the fall was coming so soon, at any rate, the day we left some of our boys were in town as we waited to make for the docks and they told us that for days they had been "put onto" carting waste from the naval dockyard area, while all that valuable stuff was left. If it was not destroyed, why not? Who was responsible for such foolish orders?

Still at Jahore Bahru on about the 27th Jan. The Japs sent over a message to get the hospital off the peninsular in 72 hours. We did. The 2/10th setting up in two buildings about the centre of the island and the 13th on the coast lower down. I had been working for about a week with what appeared to be a large boil or worse, near my left eye, but on arrival on the island, had it treated for a day, then opened. It was a filthy mess but soon looked like clearing up, so I took myself back to duty. At the end of the week, all the troops had crossed the causeway and some of it was blown up. No one seems to know why it wasn't all blown up. Then followed a few days lull in which the Japs massed to attack and their bombers came over all day. One bomb hit a building in the 2/13th and they sent their apologies over the air. At the end of the next week the section of the 2/10th in which I was working was shelled. Two men in the tents where I was, were hit and two men outside were killed. Everyone was badly shaken but carried on. There were heavy showers of rain every afternoon and we were working in slush.

On Tuesday, we were told to get ready to go. It came as a shock to me as I did not know for some hours that my name was on the list and I was not over anxious to go. The order was countermanded and we returned to duty till 10 pm. That night as we had our stuff moved to a bungalow alongside the hospital, we slept on the floor or anywhere there was space to lie down. The building rocked all night with the blast of the shells as they exploded somewhere overhead. At 5.30 am, we were ordered on duty without a wash etc. and to take our small cases. Breakfast was served somehow to hundreds of patients. I was just starting surgical dressings when I was ordered to get my case, with the rest of the others to go — half were left behind — and slept. Several ambulances arrived, we got in, some kit bags were thrown into a truck and we left. In Singapore, we sheltered in one of the hotels for a couple of hours while the Japs continually bombed the city and then marched out to the Cathedral grounds opposite. Our ambulances arrived, we got in and the convoy lined up. Then the 2/13th nurses or half their number, I ought to say, drove up in ambulances and they joined our convoy and we drove to the docks.

There was a raid in progress but after some hesitation, we went on board a ship which was tied up. The Captain and Air Force CO said "No" we could not go on board, but the officer sent down to put us on — said we were to go on, so go on we did. After about half an hour, we were taken down into the hold, and our kitbags and sleeping gear arrived and we sorted it out and picked our spots to sleep. Some of the girls just lay down and went to sleep. We were all dirty, untidy, sleepy and hungry and not caring greatly what happened.

There were air-raid alarms all the afternoon but we did not leave the harbour till 5.30 pm, and then anchored outside for the night as the minefields were unnavigable at night. We got some sleep that night in spite of the smell of the hold and the intense heat — we were wet with sweat all the time. Breakfast consisted of bully beef and rocklike biscuits, dixies of tea and water were handed down and we made the best of it. The Air Force were, though R.A.F., quite good to us and a corporal was appointed to look after us.

From about 9 am till 3 pm, the Japanese bombers were over-head. Our gun for dealing with submarines was soon out of action and our only defence was a number of machine guns rigged up on the open deck. They were manned mostly by volunteers from among some of the A.I.F. who had managed to get aboard. There were about 150 who had swooped down on to the wharf at the last minute and ran up the gangway, as they were preparing to take it up. The British tried to drive them off at the point of the tommy gun but they said they had faced worse, and were coming. As each man was wounded, another took his place on the decks at the guns and we had the satisfaction of hearing they got two Jap planes for certain. There were 92 bombers so they looked like doing what they wanted. We were in a convoy of four, two being naval vessels and one other like ourselves — a refugee ship. On board, were between 2,000 and 3,000. Most of these were R.A.F., among whom were some R.N.Z.A.F. and a few R.A.A.F. attached to the R.A.F. There seemed to be a good many British officers and some of their men; then there were a couple of hundred white women and their children, also 120 British nurses and Indian nurses, who shared the frozen meat hold with us, the ship being a cargo ship with accommodation for 24 passengers.

During the bombing, of which more was dive-bombing and machine gunning, 17 men were killed and 32 injured. When one large bomb hit the cabins near the engines room, 4 A.I.F. two of whom were brothers just disappeared. Either they vanished overboard or into invisible fragments. Many small bombs hit the ship as the Japs endeavoured to blow off the propellers but failed. There was a good deal of damage, and the Japs left us in clouds of flame and smoke thinking we were sinking. The next day we were left unmolested and as we retired were in sight of the port of Batavia. Next day, we drew up at the wharf among hundreds of ships and disembarked, leaving the remnants of the R.A.F. unloading. That night we spent in the hold of a Dutch ship. It was at least clean and wholesome and we lined up in queues for a plate of rice and stew and bread and butter.

Next day we were transferred after some trouble, back to the ship we had left the previous day. It was the only ship going to Australia and we were determined not to be left stranded in Java. Later we had leave and went ashore. Getting taxis we went into Batavia and had a good meal and stroll around. Next day, about 2.20 pm, we left and joined a small convoy. With it we went part of the way to Colombo and then turned south alone. The weather could not have been better. It rained for about 18 hours and clouds hung low over the water. Later it turned to beautiful sunshine and the trip was quite pleasant. I had another boil on my eye but it cleared up after a bit of treatment. We lay on the lounge and card room floors at night and took it in turns to wait on the other batch of girls for the next sitting and vice versa. The cabins were occupied by passengers who resented us using their bathrooms — we discovered that two only, of them had paid their passages!! I never used them but washed in the sink of the public retiring room.

It is hard to settle down knowing the other half of those two hospitals and the girls of the C.C.S. are dead or prisoners. It's a very thin line between life and death but most of us wish we had either stayed or else drowned together, and even that is not easy to take, though I did not at the time, expect to reach Australia and it did not worry me particularly and I'm not over happy to be here now leaving things and people the way we did.

The third group which left on the *Vyner Brooke* was not so fortunate. It was a small slow old ship dreadfully overloaded with more than 300 people mostly women and children and Australian nurses. On Saturday 14th it was discovered by the Japanese bombers and sunk in the Sunda Straits, near Banka Island, off the south-eastern coast of Sumatra. The nurses, some of whom had been wounded, assisted in the evacuation of the women and children before they left the doomed ship. It was a terrible sight as the survivors clung to whatever wreckage could be found, but the heavy currents soon dispersed them, some drowned, others drifted away to lonely deaths. They received no help or assistance from the Japanese. Two nurses reached the shore and were taken prisoner. Two life boats and others also reached the shore, including Matron Drummond and 21 nurses. When the Japanese arrived they bayonetted all the men and ordered the nurses to walk into the sea. They were then deliberately murdered by the Japanese, but the machine gun merely wounded Sister Vivian Bullwinkel and the story of her subsequent survival and experiences as a POW has passed into history as a glorious episode in the history of the AANS.[59]

Another group of nurses who landed further along the beach were taken to Muntok and imprisoned. One other life raft drifted away with Matron Paschke and five other nurses and was not seen again. Sisters Jeffrey and Harper were in the water 72 hours before they were able to swim ashore to become prisoners of war.

In cold statistics, the roll call for the AANS presented a grim picture:

AANS sailed on the Vyner Brooke	65
Murdered by the Japanese	21
Drowned	12
Became POW's	32

Subsequently eight died during the three years and seven months as prisoners of the Japanese. Of the 65 Sisters who embarked on the Vyner Brooke on 12 February, 1942 only 24 were left to be welcomed by Matron Sage in September 1945. Their experiences in captivity form another glorious chapter in the history of the AANS.

The Roll of Honour for Queensland nurses who lost their lives in this tragic episode or who endured the horrors of the Japanese prison camps reads as follows:[60]

Sr Calnan, E.	2/10 AGH	Drowned Sunda Straits 15.2.42	
Sr McDonald, G.M.	2/13 AGH	Drowned Sunda Straits 15.2.42	
Sr Hempsted, B.	2/13 AGH	Died while POW, Banka Is. 19.3.45	
Sr Mittelheuser, P.B.	2/10 AGH	Died while POW, Loeboech Linggau, Sumatra, 18.8.45	
Sr Blanch, J.	2/10 AGH	POW	
Sr Delforce, C.E.M.	2/10 AGH	POW	
Sr McElnea, V.I.	2/13 AGH	POW	
Sr Muir, S.J.	2/13 AGH	POW	Rescued
Sr Oxley, C.	2/10 AGH	POW	Sept
Sr Short, E.M.	2/13 AGH	POW	1945
Sr Smith, V.	2/13 AGH	POW	
Sr Trotter, F.E.	2/10 AGH	POW	
Sr Tweddell, J.	2/10 AGH	POW	

The question which has not been satisfactorily answered is why no hospital ship went into Singapore between 17 September 41 and the capitulation on 15 February 42, almost five months. Why did the urgent pleas of the medical authorities in Malaya go unheeded until it was too late?

The *Wanganella* might have been used more effectively in this emergency. It had sailed to Singapore arriving there on 17 September 1941 with the 2/13 AGH, returning with sick and other invalids, this being before the entry of Japan into the war on 7 December. Incredibly, this was the last visit of an Australian hospital ship to Singapore. In other words, once the action started and casualties mounted up in the hospitals, no hospital ship came to evacuate the wounded, despite repeated requests from the medical authorities in Malaya. Ironically, the next visit of the *Wanganella* to Singapore was in September, 1945 to rescue sick and wounded POW's. It is true that 118 were evacuated on the *Orion* on December 31, 1941 but at that stage the situation did not warrant the evacuation of nurses.

The *Manunda* sailed from Sydney to Darwin on 7 January, 42 arriving there on the 14 January, 1942 and for five weeks it was held there, virtually idle, while there was a demand for hospital ships in Singapore. It is true that the ABDA Command Headquarters in Batavia had signalled Australia on 12 February (the date of the sailing of the *Vyner Brooke* from Singapore) to hold the *Manunda* at Darwin until further notice, but by then it was almost too late, although the *Orcades* did not leave Batavia until 21 February. Unhappily, the *Manunda* was hit in the bombing raid on Darwin on 19 February. Had the *Manunda* been

AUSTRALIAN NURSES MASSACRED

COURIER MAIL
17 SEPT 1945

Mown Down In Cold Blood

8 A.M. Edition News

TWENTY-ONE Australian nursing sisters were murdered by Japs on Banka Island off Southern Sumatra shortly after the fall of Singapore.

They were lined up on the beach with their faces to the sea and mown down in cold blood.

The tragic story was released to-day with the dramatic rescue of the survivors from a foul prison camp in the heart of Sumatra.

The nurses were located by a war correspondent and an Air Force officer after an all night search, and were brought out by special R.A.A.F. aircraft, which flew into a dangerous airfield at Lahat, where the party had been assembled.

The Survivors

The Associated Press special representative at Singapore says that 24 Australian nurses, the only survivors of a party of 65 whom the Japs bayoneted and machine gunned on Banka Island beach on February 19, 1942, have been rescued from Sumatra.

They were found this morning about 100 miles west of Palembang and flown to Singapore this afternoon in a R.A.A.F. Dakota.

The massacred nurses were found lined up on the beach with their faces in the water.

Of the party of 65 nurses who left Singapore for Sumatra only 24 are alive, 12 are believed to have been drowned, and eight died in prison.

How the fate of the AANS massacred by the Japanese after the fall of Singapore in 1942 became known in Australia after the end of the war, September 1945. (Courier-Mail 17 Sept 1945).

THE MASSACRED

WHEN the nurses arrived at Singapore and were taken to a hospital full of A.I.F. men, who had been in prison camps in Malaya, there were heart-rending scenes.

As the nurses, some human skeletons and others like old women barely able to walk, shuffled up the stairs, there were cries from the men, "Give us guns. Let us at these b——s."

Some A.I.F. hospital patients obviously became hysterical and had to be controlled.

SOLE SURVIVOR

The party included Sister Vivien Bullwinkel, of Adelaide, who was the sole survivor of the nurses machine-gunned by the Japs.

Sister Bullwinkel was shot through the thigh and fell in the sea. She was washed out 30 yards to sea and left by the Japs for dead.

She staggered to the shore midst the bodies of her comrades and wandered in the jungle for two weeks before giving herself up because of hunger and exhaustion.

The whole story of the massacre has since been kept the closest secret, because of the fear that the Japs would murder her as an eye-witness of their shocking atrocity.

STOKER'S ESCAPE

At the other end was Stoker Ernest Lloyd, of H.M.S. Prince of Wales, who had joined Vyner Brooke. The Japs still had difficulty in getting compliance with their orders, so they motioned to the men to turn around and face the sea.

A Jap with a tommy gun came down the beach with the gun aimed. The men turned round, and the rating next to Lloyd said, "Here's where we get it in the back!"

Lloyd replied, "Well, I'm going to give it a go!" And he with the other rating rushed into the sea.

The gun then started firing, the Jap taking as the first target the other end of the line.

One at a time the victims were shot. By the time the steam bullets reached the last man left standing, Lloyd and his companion were 30 yards in the sea.

When his companion was killed the gun was turned on Lloyd. A bullet struck him on the shoulder, and a few seconds later he was hit on the head, knocking him unconscious.

When he came to Lloyd resumed swimming mid more bullets, but he got away. He landed farther down the coast, and after 10 days was captured and taken to Muntok labour station.

Before his recapture he returned to the scene of the tragedy and saw the bodies of his companions.

BODIES FOUND

He found the bodies of the nurses and others bearing bullet and bayonet wounds.

After killing Lloyd's companions the Japs returned, some wiping their bayonets. They then lined up the remainder, making them face the sea and motioned them walk towards the water.

A soldier with a tommy gun received his orders, and the shooting down of these nurses, civilians, and merchant seamen was carried out.

One sister tried to escape and failed.

After the shooting the Japs bayoneted those still alive.

The incredible escape of Lloyd had a parallel at this spot. Nurse Bullwinkel was shot through the body and fell into the water.

She recovered after the Japs had left and with the help of natives remained at liberty for several weeks before being gathered into the labour assembly station.

An American civilian, Mr. E. H. German, escaped the bullets, but was bayoneted through the chest. He survived, and about a week later was also shepherded into the prisoners' compound.

The Courier Mail 17 Sept 1945.

Two Queensland Sisters who died while POW's. (Photos Aust. War Graves Commission).

in Singapore early in February more wounded would have been saved. It too, was involved in 1945 in rescuing POW's from Singapore.

The third Australian hospital ship was the *Oranje*, a new vessel of some 20,000 tons offered to the Australian Government by the Netherlands East Indies Government in February 1941. The *Oranje* completed two trips to the Middle East by December, 1941. The rapid deterioration of events in Malaya prevented this hospital ship from being used before Java fell.

The fourth Australian hospital ship, the *Centaur*, was at that time more than twelve months away from commissioning.

Thus while the speed of events in January–February 1942 found the allies in disarray, still fighting rear-guard actions, still defending, still retreating, there appeared to be a lack of initiative, a lack of positive thinking which would have got a hospital ship — any ship — in to Singapore or Batavia to bring out the wounded and the AANS. This negative, defeatist attitude of the allies was to cost many lives during 1942 and 1943.

Once again, there was fierce debate on whether the nurses should stay with their patients, even if this meant capture by the enemy, death or an even worse fate, or whether they should be evacuated. For the nurses' part their views were clear, they had a duty to remain with their patients and they were prepared to risk their lives in the fulfilment of their duty. So much was implied in the oath of their calling. The ADMS, 8 Division, Col A.P. Derham, was greatly concerned about the welfare and safety of the nurses. As early as 18 January, when the 2/10 and 2/13 AGH and the 2/4 CCS were still in Malaya, he recognized the danger from the onward march of the Japanese.

The facts regarding the confusion surrounding the evacuation of the AANS from Singapore were later recorded by Col A.P. Derham, ADMS, 8 Aust Div in a letter to Col Sage, Matron-in-Chief, AANS, 9 March 1946.[61] He stated that between 20 January and 25 January, 1942 he recommended officially to

Miss Evelyn Conyers, Matron-in-Chief during World War I, places a wreath at the Edith Cavell Memorial, Melbourne, in memory of those Sisters who had given their lives in World War II. April 1942.

MajGen Gordon Bennett that AANS personnel should be evacuated from Singapore by the first available hospital ship. He repeated this again on 25 January and 30 January but on each occasion the request was refused by the GOC on the grounds that it would have a bad effect on civilian morale in Singapore. Col Derham made a third appeal on 8 February 1942, again refused. He then conferred with LtCol J. Glyn White, DADMS, Admin HQ, AIF, Malaya asking him to get as many nurses away with the next batch of casualties on the pretext that they were travelling on duty. 'This he agreed to do. I told him I would take full responsibility if such action were disapproved by higher authority,' wrote Col Derham. Thus it happened that on 10 February, 6 AANS embarked on the *Wah Sui*. Col Derham then informed the GOC and asked what was the policy for the evacuation of the remainder of the AANS. The GOC agreed they should be evacuated as soon as practicable. By this time chaos and panic reigned. Key officers were not available. On 11 February, 1942, 59 AANS embarked on the *Empire Star* at short notice and on 12 February, 65 members of the AANS embarked on the ill-fated *Vyner Brooke*. Col Derham concluded his report to Matron Sage by stating that when they learnt of the fate of the nurses on the *Vyner Brooke*, while they were in captivity in Changi, they made several appeals to the Japanese to have the remaining captive nurses repatriated. These appeals went unheeded by the Japanese.

Gen Bennett, as is well known subsequently escaped and returned to Australia in circumstances which are still argued over to this day. Those who were fortunate enough to return safely to Australia found a great deal of public disquiet about the disastrous campaign in Malaya. One whole division had been wiped out, either killed or captured. There was hardly a home in Australia not affected in some way by this

Sister J.E. Grigg back from Singapore lays a wreath in the crypt at Anzac Square.

tragedy. The public looked for scapegoats — Gen Bennett, the British, the AIF commanders, the Government. Those who escaped and returned to Australia came in for much vilification, particularly by successful AIF troops returning from the Middle East. Those nurses who returned from Malaya, Singapore and Java did not escape this criticism. Their ultimate humiliation came when some of the nurses received white feathers!

One of the finest tributes to the courage and devotion of the AANS in Malaya was paid by Col A.P. Derham in a letter addressed to the Victorian Branch of the British Medical Association:

> The work and conduct of the members of the Australian Army Nursing Service in Malaya were so magnificent at all times that I can hardly trust myself to speak of them. Their supreme courage and devotion were never more inspiring than when our hospitals came under heavy fire. It was then that they set an example of calmness and courage to many a shaken soldier which brought tears of pride and admiration to the eyes of their male colleagues, including myself. The fate of some of them, who were evacuated too late, is the blackest page in the black tragedy of Malaya. Their memory will always be the most sacred trust of those for whom they so gladly risked their lives.[62]

This most tragic episode in the history of the Australian Army Nursing Service should never have happened. It resulted from sheer bungling and inefficiency by those in command in Malaya whose inconsistent directions led to such confusion that there was no hope of a safe evacuation for the remaining Sisters in Singapore. It was but sheer good fortune that those who escaped on the *Wah Sui* and *Empire Star* lived to serve again with the AANS. This in no way exonerates the Japanese for their total disregard of the rules of war and for their crimes against humanity, for which some of them paid the supreme penalty after the war.

References

1. Australian Archives, Melbourne, MP 527/26/1401.
2. Ibid.
3. GOGGINS, M.E. Member of old Warwick family. Name appears on Warwick War Memorial. See Fn 55 Chap 3.
   McLEAN, C.E. Probably Sister McLean who served at Rabaul in W.W.I.
   SMITH, A. See Fn 7, Chap 3.
4. BASSETT, M.E.V. See Fn 78, Chap 3.
   CAIRNCROSS, S.L. See Fn 82, Chap 3.
   WILSON, E., and
   WILSON, MAJORIE and
   WILSON, MADELINE. See Fn 48, Chap 3.
5. LIMPUS, E. See Fn 44, Chap 3.
   HARTE, M. — Probably an incorrect spelling for S/N May Hart. See Fn 9, Chap 3.
   McLENNAN, J. See Fn 46 Chap 3.
   SKYRING, M. See Fn 48, Chap 3.
   WILSON, MAY. See Fn 48, Chap 3.
   WILSON, MYRTLE. See Fn 48, Chap 3.
   BLACK, C.R. See Fn 43, Chap 3.
   WILSON, LILIAN. See Fn. 48, Chap 3.
6. Information from a local resident (Maryborough).
7. SQUIRE, DAISY W. b 1892 Springsure Qld; trd Chns Hosp Bris; AANS WWI, service Egypt, Salonica, France; After war pte nursing & staff Bris Gen Hosp; also trd as masseuse, Syd.
   (i) On the War Memorial in Goondiwindi appear the names of Sisters Allen, E; Allen, F; and Drake, P.C.
   (ii) *The Church Chronicle*, 1 August 1917.
   (iii) *Lest We Forget*. A Study of War Memorials in Qld, Bris. 1983. This contains no ref to AANS War Memorials.
8. War Memorials, AWM File 527/12/167 of 14 June, 1923.
9. HANRAHAN, ETHEL FRANCES, OBE, MID (Twice), b. Brisbane, 6 May, 1909, Trd Bris Hosp 1930–34, Staff Rosemount Repat Hosp 1935–39. Enlisted AANS, 15 Dec 1939 and sailed to Eng with 2/3 Fd Amb and 2/3 AGH in May 1940. Nov 1940 to ME for service with 2/2 AGH at El Kantara, (MID twice). Mar 1942, Matron 117 AGH Toowoomba and Women's Hosp Redbank, (Qld).
   April 1944, Matron 116 AGH Charters Towers and Cairns.
   June 1945, Matron 107 AGH Darwin and Principal Matron Northern Territory. Promoted LtCol, Matron 112 AGH Greenslopes and Principal Matron, Northern Command.
   1947 discharged from AANS, appointed Senior Matron, Repat Gen Hosp Heidelberg. 1963 OBE. d. 17 Aug 1981.
   DEAN, EDNA M. Educ & trd Townsville; AANS 1939–46; Service 2/2 AGH, Eng, ME and Aust; 1946–1948 Repat Hosp Bris.
   DOIG, EDNA NELL. RRC, FNM, FCNA. b. Bris 21 June, 1915, daughter J. Doig, ed All Hallow's Convent; Trd Bris Gen Hosp, Women's Hosp Melb and Berry St Home, Melb. AANS WWII, 1939–49. 2/2 AGH, Eng, ME, Sing, Japan. Served BCOF Japan, Kure, Deputy Matron 1951 Aust Reg Army. Princ Matron Northern Commd Qld 1955, Southern Cmd 1960, Eastern Cmd 1961.
   Matron-in-Chief & Dir R. Aust Army, N.S. 1961–70. Colonel. Trustee numerous ex-service orgs; incl Trustee Shrine of Remembrance since 1970–.
   HARLAND, VERA MAY. b. 19.6.15. AANS 1939, service 2/2 AGH, ME and Aust.
   HARVEY, FLORENCE, E. (Joan). Trd Bris Gen Hosp 1933–1937; AANS 1939–1946; Service 2/2 AGH Eng, ME, and Aust.
   HELY-WILSON, UNA b. 12 Nov 1910. Trd Bris Gen Hosp 1932–36, Midwifery 1936, Child Welfare 1937. Enlisted AANS 15 Dec 1939. Service in Britain 2/3 AGH, ME 2/2 AGH, Australia at Watten and Rocky Creek (Qld). Dis 6 Mar 1946.
   HOEY, IRIS. b. Bris 2 Sept 1913. Trd Bris Gen 1932–36. Staff Bris Gen 1937–39. AANS 8 Dec 1939–4 April 1945. Service in Eng, ME and Aust, 2/2 AGH. Repat Gen Hosp Greenslopes 1946–75. Sen Sister & Asst Matron.
   KEYS, L. b. Bris 11.4.13; trd Bris Gen Hosp 1932–1936; AANS 1939–1942; served 2/2 AGH Eng, ME and Aust.
   MARKS, I.A. Trd Chns Hosp Sydney; AANS 1939–1945; Served 2/2 AGH, Eng, ME and Aust.
   REID, MURIEL VIVIENNE, b. 25.1.04; trd Bris Gen Hosp; staff various Qld country hosps; AANS 1939–46; 2/2 AGH, ME, UK and Aust; RGH Greenslopes.

PETERSEN, F.M. AANS 1939–1946; 2/2 AGH Eng, ME and Aust; post-war Matron RGH Concord and Nursing Adviser, Comm Gov, Canberra.

COLEMAN, DOROTHY, M., Trd Bris Gen Hosp and on staff there prior to enlistment; AANS 1939–1945; 2/2 AGH Eng, ME and Aust.

DICKSON, JEAN CATHERINE. ARRC, b. 10 Oct 1913 Toowoomba, Qld; Trd Gympie Gen Hosp 1933–37. 1937–38 Lady Musgrave Hosp, Maryborough Qld. 1938 Private nursing Brisbane, Feb 1939 Repat Depart, AANS 15 Dec 1939 - 11 Feb 1946, service ME, Aust & Eng with 2/2 AGH. 1946–1974 Repat Depart. CMF 1951–56 (Major).

KNUDSEN, V.M. AANS 1939–1940.

STEWART, HEATHER. Trd Bris Gen Hosp 1933–1937; AANS 1939–1945; service 2/2 AGH Eng, ME and Aust.

THOMPSON, IVY. 1934–38 Trd Maroochy Dist Hosp. 1938–39 Theatre Sister Maryborough Hosp Qld. AANS 1939–1945. Service UK and ME, 2/3 AGH and 2/2 AGH, Aust 2/4 AGH Redbank, 77 ACH, 7 ACH & 112 AGH Greenslopes. After war, nursing staff Vic Hosps, tutor at Nambour Gen Hosp and Matron CWA Hostel.

MILLER, PHYLLIS BERNICE. b. 6 Nov 1912. Trd Bris Gen Hosp 1933–37; 1938 Midwifery Women's Hosp. Melb; 1939–46 AANS. Service 2/2 AGH, Britain, ME, El Kantara, Queensland, Watten, Rocky Ck, Nth Qld; 1946 RGH Greenslopes; 1961–72 Matron Repat Hosp Kenmore.

11. *Courier Mail*, Jan 18, 1940.

12. War Diary 2/1 AGH, AWM 11/2/1.

13. McCONNEL, AMY: from letters to her mother in Brisbane.

14. AWM File 509/2/1. See also DRL 2518, Third Series, Personal records of Matron Muriel Doherty, RRC, AWM File 419/27/5.

15. SINCLAIR, LUCILLE. AANS 1940–1946; service 2/2 AGH Eng, ME and Aust; post-war Comm Bank, Brisbane, d 1979.

BLAIN, E.M. (Sally), Trd Bris Gen Hosp 1934–1938; AANS 1940–1943; service 2/2 AGH, ME and Aust.

GLASGOW, MARY (Meg), MID, RRC, AANS 1940–1946, service 2/2 AGH Eng, ME and Aust.

DAVIS, RUBY. Trd Rockhampton Base Hosp, staff Rockn, Tully and Innisfail Hosps prior to enlistment; AANS 1939–1947; service 2/2 AGH.

For another version of AANS life with the 2/2 AGH in the Middle East, especially Gaza, Nazareth and El Kantara, see Burchill, Elizabeth, *The Paths I've Trod*, Spectrum, Melbourne, 1981.

16. ROE, DOROTHY GRACE. Trd St Martin's Hosp Bris 1927–31; Midwifery Lady Chelmsford Hosp Bundaberg; Private nursing. AANS 1940–1947; service with 2/5 and 2/6 AGH in ME, Greece, Eritrea, NG, Labuan. Later service at Heidelberg, Greenslopes and Yeronga Hospitals.

BROOMFIELD, DORIS. FCNA. b. 8.11.1905. Trd Mt Morgan Hosp; Midwifery at Lady Chelmsford, Bundaberg; Child Welfare Bris; AANS 1940–46, service troopship Queen Elizabeth, 2/7 AGH, Rehovot, Sisters' Staging Camp, Gaza; 2/1 AGH in W Aust, 116 AGH Charters Towers and 112 AGH Greenslopes. After war Nursing Admin, Coll Nurs Aust; Matron Maryborough and P.A. Brisbane.

BURNETT, DOROTHEA ANNE, MID. b. 19 Oct 1914; Trd Bris Gen 1933–37; Staff Bris Gen 1937–40; AANS 1940–47, ME, Greece, Crete, Eritrea, NG, Morotai, Australia, mostly 2/5 AGH.

CASEY, ELLEN. Trd Bris Gen Hosp and on staff. AANS 1940–1946; Service 2/5 AGH, ME, Greece, Crete, Aust & SWPA; post-war Royal Bris Hosp.

EXTON, ADA, Margaret (Peg). Trd Hillcrest Hosp Rockn, Midwifery Bris Women's Hosp; AANS 1940–1945; service 2/5 AGH, ME, Greece, Crete, Aust & SWPA; Repat Depart 1946–1976.

HOOKE, HELEN, MID. Trd Brs. Gen Hosp; AANS 1940. Service Darwin, Greece, Crete and SWPA.

HOOKE, JOAN. Trd Rockhampton Gen Hosp; AANS 1940. 2/5 AGH. Service Greece, Crete and SWPA.

PATRICK, EVA. Trd Bris Gen Hosp. Private nursing. AANS 1940–1946; 2/5 AGH, ME, Greece, Crete, Aust & SWPA.

TAYLOR, MARY. Trd St Martin's Hosp Bris. Staff Atherton Dist Hosp.

ZIELKE, J.B. Trd Bris Gen Hosp staff Sandown Priv Hosp, Southport.

17. War Diary, 2/1 CCS, AWN 11/6/1.

18. War Diary, 2/2 CCS, AWM 11/6/2.

PATERSON, VIDA MARGARET. b. 13 Jan 1913; Trd Bris Gen Hosp. 2.2.36. Staff Bris Gen Hosp 6 yrs. Midwifery Q. Alex Hosp Hobart 1936. Royal Soc for Welf for Mothers & Babes Sydney; Staff P.A. 4 yrs; AANS 1940–46. Service 2/2 CCS, ME, Tobruk, SWPA.

THORPE, M.F. b. 30 April 1914. Trd Bris Gen Hosp, AANS 1940–45, service 2/2 CCS, ME, Libya, Tobruk, Egypt, Australia, Java, NG.

WALLACE, MARY, MID. b. 19 Mar 1913, trd Gen Nurs St Martin's 1932–36. Midwifery, Lady Chelmsford Hosp, Child Welf Bris 37, Bundaberg 1938. AANS 1940–46. Service 2/2 CCS, 2/7 AGH.

MARSHALL, MARGARET. AANS 1940–1946; service 2/2 CCS, ME, Tobruk, Java, Aust & SWPA.

FINLAY, MARY. Trd Bris Gen Hosp, AANS 1940–1946. 2/2 CCS ME, Tobruk, Java, Aust & SWPA.

PYM, PHYLLIS. b. 10 Mar 1914, trd Bundaberg Gen Hosp 1932–36; 1936–37 Lady Chelmsford Hosp, Bundaberg. Moree Gen Hosp 1938–40. AANS, May 1940 to Sept 30 1945. Egypt, Libya, Palestine, PNG.

GOLDEN, RHODA. Trd Bris Gen Hosp; AANS 1940–45. 2/2 CCS ME, Tobruk, Java, Aust.

WILSON, HEATHER. Trd Bris Gen Hosp. AANS 1940–1945; service 2/2 CCS ME, Tobruk, Java, Aust.

McDONALD, MARJORIE MARY. b. Dec 1911, trd Bris Gen Hosp. Queen Vic Hosp Melb, (obstetrics). Staff Bris Gen Hosp pre-war. AANS 1942–46. Service 6 CCS, 2/2 CCS, 2/4 CCS.

BEST, KATHLEEN ANNIE LOUISE, RRC. b. 28 Aug 1910. Trd Western Suburbs Hosp Sydney & Crown St., AANS 1940–42; Matron 2/5 AGH, ME, Greece; 1942–43 Controller AAMWS; 1943–44 Asst Adj Gen Women's Services; 1944–49 Asst Dir Dep Post-War Reconstn; 1951 Dir, Women's Royal Aust Army Corps.

19. For an account of the Sisters' work at Tobruk, see Goodman, Rupert, *A Hospital at War*, pp 62–65.

20. Ibid, p 64.

21. Ibid, pp 64–65.

22. AWM File 253/4/1, also 11/1/92.

23. Ibid.

24. AWM File 481/12/3.

BOWE, ETHEL JESSIE ('BOWIE'), RRC, FNM, FCNA. b. 27 May 1906, Maldon (Vic); Trd Melb Hosp 1927 and on staff 1931–35; Sister Tutor Perth 1936–39; AANS 1940, Senior Sister 2/2 AGH; Service in Eng & ME; Matron 2/2 AGH Sept 1940; 1943 Deputy Matron-in-Chief LHQ; Princ Matron ALHQ (Morotai); 1946–7 Matron 115 Heidelberg; 1948–50 Chief Nurse 1 RO; 1950–51 Matron Austin Hosp; 1951 Deputy Matron-in-Chief ARA; 1952 Matron-in-Chief, Royal Aust Army Nurse Corps; Hon Col.

REID, THELMA, Trd Bris Gen Hosp and on staff there. AANS 1940. (Sister to Reid M.V.).

25. 2/3 AGH War Diary, op cit.

26. Ibid.

27. Ibid.

28. MURRAY, HANNAH, b. 23 Nov 1899. Trd Charters Towers & Toowoomba Gen Hosp; Matron Mitchell, Taroom & Collarenabri Hosps. Member QAIMNS 1939–1946. Service Dunkirk, Tobruk, ME. d. Toowoomba 6 July 1976.

ELLIOTT, FRANCES SIDNEY, MID. b. 18 Mar 1905, trd Bris Gen Hosp 1928–32m midwifery St George's Melb; private nursing, staff Nambour Hosp, QAIMNS 1939–46. Service France, West Africa, Egypt, Eng.

WHITTEM, LETITIA. b. 23 July 1908. Trd Roma; QAIMNS 1940–46. Service in Egypt, Britain, Germany. Nursing exper. various hosps in WA.

WILLIAMS, MARIE. Trd Rockhampton.

TWINE, MAVIS. b. Brisbane & trd at Bris Gen Hosp. In Britain at outbreak WWII and served 7 yrs with QAIMNS. Took part in evacuation at Dunkirk with the British Army. Later served in ME and on Malta.

29. Correspondence from Sr Murray to her relatives and friends in Queensland.

30. Op cit, Fn 28.

31. Op cit, Fn 28.

32. Correspondence from Sr Twine.

33. Op cit, Fn 28.

34. War Diary 2/5 AGH. AWM File 11/2/5.

35. War Diary 2/5 AGH. AWM File 11/2/6.

36. War Diary 2/3 CCS. AWM File 11/6/4.

37. Diary, Sister Gordon, Muriel Maud, AWM. b. 22 Jan 1912.

Trd Bris Gen Hosp & Bris Women's Hosp 1934; AANS 1940–43. Service 2/6 AGH, ME, Greece, Alamein, Aust; Prom Capt; 112 AGH Greenslopes.

SOORLEY, JOAN. b. 31/3/1910. AANS 1940. Service 2/6 AGH, ME, Greece, Crete, Aust. 113 AGH, Concord.

CARMICHAEL, EVELYN. b. 26 May 1912. Trd Bris Gen Hosp 1934-38, Women's Hosp 1939. Staff Bris Gen Hosp & Women's Hosp. AANS 1940-1946, Service 2/6 AGH, M.E., Egypt, Greece, Palestine, Australia.

38. ABBOTT, JOAN STEVENSON, RRC. FNM. b. Bris 1899. Trd Bris Gen 1920. Gold Medallist. Midwifery Lady Bowen Hosp; 1929 1st Tutor Sister BGH & Infant Welfare Sister. AANS 1940-45. Matron 2/6 AGH, ME, Greece, Aust; 1943 Princ Matron AANS, Qld. After war LtCol & Princ Matron CMF Northern Command. Awarded FN Scholarship Royal College Nursing, London. Staff Bris Hosp Nurs Sch. Later Staff Nurse Comm Savings Bank, Brisbane. Appd Princ Matron CMF, 1962 Hon Col, Pres ATNF & RANF. Awarded Florence Nightingale Medal 1957.

39. The evacuation from Greece:— 2/5 War Diary, op cit. 2/6 AGH War Diary, op cit. 5th Aust Gen Hosp, Greece, AWM, 403/7/7. Appreciation of the Nursing Staff, Greece 1941, AWM 534/3/7. Narrative of the Campaign in Greece, 1941, AWM 534/5/14.

For another vivid account of the evacuation from Greece see Jones, T.M. and Idriess, I.L. *The Silent* Service, A & R Sydney 1944, Chap XIX.

40. BURNETT, DOROTHEA, See Fn 16.

41. Diary of Sr Burnett.

42. MACHON, IVY FRANCES. Trd Bris Gen Hosp 1935-39. Service Gregory Dist Hosp (Qld). AANS 1940-47 and Interim Army 1947-48; Service Darwin, 2/5 AGH, ME, Greece, Aust & SWPA.

43. Report by Matron Best on the evacuation of the AANS from Greece and Crete 1941. AWM 422/7/8.

44. Ibid.

45. Report by LtCol Coppleson to DMS, AIF. AWM 411/1/44.

46. CAVENAGH-MAINWARING, PATRICA. b. 14 April, 1910 at Herberton (NQ) and lived at Chillagoe. Trd RPA Sydney. Service AANS 1941-45. 2/5 AGH, Greece, Aust, NG & SWPA.

HAWKEN, PHYLLIS &

WILSON, DOROTHY. Two of the first masseuses appointed. Service with 2/5 AGH in Greece, Crete, ME and Eritrea.

47. 2/5 AGH Unit War Diary, op cit.

TOMLINS, D.M. Trd Mater Hosp and served with 2/5 AGH.

48. Queenslanders listed in these units included:

ADAMS, ELLEN MARY. b. 27 May 1910. Trd Wilga Pte Hos, Toowoomba, 1930; obst Mother's Hosp T'ba, Child Welfare trg Hobart. Pte nursing NSW and at 'Glandore', (Gympie). AANS 1941-45. Service 2/9 AGH Nazareth, 2/5 AGH Armidale, Bootless Bay NG, and at Warwick, Ekibin & Greenslopes.

ALLEN, EDITH MURIEL, MID, BEM. b. 20 May 1907. Trd Bris Gen Hosp, general, midwifery & child welfare. AANS 1940-46, service 2/1 AGH Gaza, Merridan (WA), Port Moresby, NG & SWPA. MID for services in NG. After war with Comm Dep Health, Bris. Awarded British Empire Medal on retirement.

ANTHONY, HILDA JANE. b. 25 Nov 1908. Trd Royal Hosp Women Syd 1932-33, child welfare Bris 1934-5. Staff Maryborough Gen Hosp Qld 1935-40. AANS 1941-44. Service 2/2 AGH, ME, Aust, Redbank at Camp Hosp, 117 AGH Toowoomba, 2 Aust Wm's Hosp.

BALDOCK, AGATHA MARION. b. 29 Sept 1913; Trd Mater & Bris Wm's Hosp 1932 and on staff there until enlistment. AANS 1941-46. Service 2/1 AGH, ME, GAZA Rehovot, Merridan (WA), Port Moresby, SWPA. Prom Capt.

BEATTIE, THELMA JANET. b. 19/5/1908. AANS 1940. Service ME, 2/6 AGH, Greece, Crete, Aust, 2 AWH. Prom Capt.

CRUDGINGTON, ALMA. b. 18 May 1912. Trd Rockhampton Gen Hosp 1930-34. Pte nursing Yeppoon & Beaudesert; AANS 1940-43, 2/9 AGH, ME, Palestine, Egypt, Aust, NG.

DENMAN, JEAN. b. 30 June 1916. AANS 1941. Service 2/9 AGH, ME, NG, Aust, 2 AWH.

DUFFIELD, EDITH MERLE. MID. b. 29/1/05. Trd Mater Hosp Bris, & Lady Bowen 1936, private & general nursing Glen Rowan & Fermoy. AANS 1940-46. Capt service with 2/1 AGH. ME, Gaza, Merridan (WA), NG, Bougainville & SWPA. After war private nursing in South Africa & Britain.

FREEMAN, MENA. b. 6 Dec 1915. AANS 1940. Service 2/6 AGH, ME, Greece, Crete, 110 AGH Aust.

HALL, ELEANOR, b. Toowoomba; trd RPA Sydney 1935-39; AANS 2/6 AGH, ME, Greece, Crete & Aust.

NETTERFIELD, M.M. Trd St Martin's Hosp Bris; AANS 1940-1945; 2/11 AGH, service in ME and NG, 47 ACH Koitaki; After war opened Moorlands for Legacy.

HOOPER, KATHLEEN E.M. b. 26.7.12. Trd Maroochy Dist Hosp. Nambour 1931-35, Lady Chelmsford Hosp Hosp Bundaberg, staff of Maternal & Child Welfare on enlistment. AANS 1940-46. Service with 2/1 AGH, Gaza, Merridan (WA), Port Moresby, Bougainville.

LEAR, MARTHA ALICA. Trg Ingham Dist Hosp 1925–29. Royal Hosp for Woman Sydney 1930. Balby Clinic Bris 1936. AANS 1940–45. Service 2/5 AGH, ME, Greece, Armidale, Aust, NG & SWPA. 30 yrs service in Maternal Child Welfare until retirement in 1971.

LEVARING, JOCELYN, A. AANS 1940–1945. Service 2/6 AGH ME, Greece, Crete, Aust & SWPA.

LONERGAN, ELLEN LORRAINE. b. 16.9.15. Trd Bris Gen Hosp & on staff there and at Liverpool NSW. AANS 1939–44. Service 2/5 AGH, ME, Greece, Crete, Aust, NG & SWPA. Prom Capt.

MILNE, JOYCE MADELINE. b. 29.7.16. Trd Bris Gen Hosp 1937–40 and staff nurse there on enlistment. AANS 1941–44. Service 2/5 AGH.

PALMER, ALICA ELIZABETH. b. 2.3.1910. Trd Bris Gen Hosp. AANS 1940–46, Capt. Service ME, Palestine, Greece, Crete, Qld, Morotai, Labuan. After war School Health, Dist nursing & asst neuro-surgery.

RIDDELL, GWENDOLINE ELAINE. b. 22.1.1910. Trd Bris Gen 1930–34 & Bris Women's Hosp. AANS 1940–46. 2/6 & 2/12 AGH, Syria, Greece, West Desert, Palestine (Greek Commem Medallion). Sen Sister Comm Serum Lab 1946–69.

CHAMBERS, ELLEN EMILY (Nell). Trd Bris Gen Hosp 1926–30; Mat & Child Welfare, Lady Bowen Hosp; Psychiatric Nursing, Toowoomba Gen Hosp; service with Northern Territory Nursing Service; AANS 1940–1942 2/9 AGH; service ME and Aust.

VIGAR, ELLA JEANIE. b. 10 Aug 1905. Trd Warwick Gen Hosp, midwifery Toowoomba Mothers, owner of 'Mylo' Pte Hos; AANS 1941–45. Service ME, 2/5 AGH, Nazareth, Palestine, Aust & NG & SWPA.

49. WALKER, A.S. Medical Services of the RAN and RAAF. AWM 1961. p 441.

CARMODY, MARGARET CATHERINE. ARRC. b. 6 Mar 1914. Trd Mater Hosp Bris, midwifery at Lady Bowen Hosp. AANS 1940–46. Service in ME, Aust, NG, and SWPA, with 2/2 AGH Greenslopes. After service with Repat returned to Aust Reg Army 1951. Appoints included Matron 1 & 2 Camp Hosps, Asst Dir AANS Eastern Command, Matron 2 Gen Hosp, Asst Dir AANS Southern Comd, retd 1969. d Bris 27 Aug 1975.

50. Queensland members of the AANS promoted to the rank of Captain at this time included Sisters Blain, Burnett, Carmichael, Casey E.L., Coleman, Couche, Davis R., Dean, Dickson, Doig, Exton, Finlay, Freeman, Glasgow, Golden, Gordon, Hall Harland, Harvey, Hely-Wilson, Hoey, Hooke H., Hooke J., Levaring, Marshall, Miller, Newitt, Palmer, Patrick, Pym, Reid M., Riddell, Sinclair, Soorley, Steen, Stewart, Taylor, Thompson, Wallace, Wilson H., Zielke.

51. 2/3 CCS Unit War Diary AWM 11/6/4.

52. 2/13 AGH Unit War Diary, Ceylon. AWM 403/7/21. 481/12 and SP 515.

53. HITCHINGS, M.E.F. Trd Bris Gen Hosp 1934–1938. AANS 1940–1945.

WARFIELD, NORMA. Trd Dalby Gen Hosp 1933–37. Bush nursing Staff Dalby Hosp. Midwifery at Bris Women's Hosp. AANS 1941–48, 2/12 AGH Ceylon, Aust, NG, SWPA, Balikpapan, Concord & Greenslopes. After war Marooma Pte Hosp, Toowoomba & Oakey.

JACKSON, DOROTHY AGNES. b. 25 Sept 1907. AANS 1941. Service 2/12 AGH Ceylon, Aust, 112 AGH.

THIEDEKE, C.M. dau Carl & Anna Thiedeke, Wellington Point, Bris. Trd Bris Gen Hosp 1935–39. AANS 1940, service 2/12 AGH, Ceylon. Died of illness there on 27 Sept 1942. Sister Thiedeke is buried in the Colombo Kanatte General Cemetery, Ceylon.

Voluntary Aid Detachments (VAD's) and Australian Army Medical Women's Service (AAMWS).

VAD's had been involved in the First World War. The first VAD to go from Queensland was Lydia Grant, who died of illness at Manchester, 1 April 1917. Another Queensland VAD who achieved distinction was Annie Darvall, a B.A. from the University of Queensland (1914). She served with the BEF and was mentioned in despatches, (1918). Between the wars VAD's were involved in training through the Red Cross Society and the St John Ambulance Brigade. Two VAD's were sent with the 2/2 AGH to the Middle East in 1940 and in October 1941 a contingent of VAD's was sent to the 2/12 AGH in Ceylon. In November 1941 further drafts were sent to the 2/1 AGH at Gaza and to the 2/7 AGH at Rehovot. In 1942 VAD's who by then were serving in many Australian Army medical units became members of the Australian Army Medical Women's Service. See Goodman, R.D. *A Hospital at War*, pp 138–144.

54. Queensland members of 2/10 AGH:—

CALNAN, Ellenor, drowned when Vyner Brooke bombed and sunk by Japanese 14 Feb 1942. Name appears on The Singapore memorial.

TROTTER, FLORENCE, b. 4 Oct 1915, trd Bris Gen Hosp 1935–39 and on staff of Gen Hosp until 1940. AANS 1941–46, service 2/10 AGH Malaya. Survived bombing of Vyner Brooke & taken POW. Released 16 Sept 1945. After war established employees medical section at Allah & Stark's Brisbane.

TWEDDELL, J. b. 3 July 1916. Trd Bris Gen Hosp 1935–39, and on staff of Gen Hosp until 1940. Survived bombing of *Vyner Brooke* and taken POW. Released 16 Sept 1945. After war completed radiography cert. and employed with Qld Radium Inst.

OXLEY, C.S.M. MID. b. 7 June 1912, trd Glen Innes Dist Hosp 1930–34 and Women's Hosp, Syd 1935–36; after pte nursing in London & Qld. Enlisted AANS. Service 2/10 AGH, Malaya. Survived bombing of *Vyner Brooke* and taken POW. Released Sept 1945. 1946 Staff Repat Hosp Brisbane. 1947–49 Matron Selangor Hosp Nambour. 1949–52 Industrial nursing, Brisbane.

BLANCH, JESSIE J. Trd Bris Gen Hosp 1932–36; Staff Bris Gen Hosp 1936–40; AANS 1941–46, 2/10 AGH Malaya, survived bombing of *Vyner Brooke*, Pow 3½ yrs. Rescued Sept 1945, pte nursing after war.

DALEY, Sheila. Trd Bris Gen Hosp 1935–39; 2/10 AGH Malaya, returned to Aust on *Empire Star*.

MITTELHEUSER, PEARL BEATRICE. dau John & Margaret Mittelheuser of Bundaberg, trd Bris Gen Hosp & on staff of that hosp until enlistment. AANS Malaya 1941. Survived bombing of *Vyner Brooke* and taken POW. Died of illness in Sumatra prison camp 18 August 1945. Buried in Jakarta War Cemetery. Her name also appears on the Bundaberg War Memorial. See also *White Coolies* (p 142) and *While History Passed* for accounts of her death.

DELFORCE C. Trd Stanthorpe Base Hosp; midwifery Bris Women's, pte nursing prior to enlistment, AANS, 2/10 AGH Malaya, taken POW 1942, released 1945.

Queensland Members of 2/13 AGH

McDONALD, GLADYS, MYRTLE, dau of John & Charlotte McDonald of Bris. Drowned when the *Vyner Brooke* was bombed and sunk by Japanese off the coast of Sumatra, 14 Feb 1942. Sister McDonald's name appears on the Singapore Memorial.

SHORT, EILEEN, trd Kingaroy Hosp; Matron Isisford Hosp; AANS 2/13 AGH Malaya; survived bombing of *Vyner Brooke* and taken POW; released 1945.

SMITH V.E. MID. Trd Cairns Dist Hosp 1935, general nursing in Qld until enlistment. AANS 1941–46, service 2/13 AGH Malaya; survived bombing of *Vyner Brooke*, completed Midwifery and Child Welfare Certificates, School Health Service, Tutor Sister Bris Gen hosp and Psychiatric nursing.

MUIR, SYLVIA. b. 24 Aug 1915. Trd Bris Gen Hosp 1935–39. Midwifery at Townsville Gen Hosp & district nursing until enlistment. AANS 1941–46, service 2/13 AGH in Malaya. Survived bombing of *Vyner Brooke* and taken POW. Released 1945.

HEMPSTED, PAULINE BLANCHE. dau Percy and Bertha Hempsted of East Graceville, Bris; trd Bris Gen Hosp 1934–38. AANS 1940. Service 2/13 AGH Malaya. Survived bombing of *Vyner Brooke* & taken POW. Died of illness at prison camp, Banka Is. 19 Mar 1945. Sister Hempsted is buried in Jakarta War Cemetery, Indonesia.

McELNEA, V.I. Prior to enlistment Matron, Dist Nursing Assoc, Milton. AANS 1941–1945, 2/13 AGH. Taken POW 1942, released 1945.

55. Sister Wittmer's account. Sister Wittmer, of 2/13 (formerly with 2/10 AGH) went to Jahore Bahru in Nov 1941. Her account is one of a number of interviews with AANS (former POW's) with AWM.

56. Sister Anderson's account. Sister Anderson, from Melbourne, served with 2/13 AGH at Jahore Bahru in Nov 1941. Her account of her subsequent capture is with AWM. See also statement by Sr N. James, AWM.

57. Sister Irving's account. Sister Irving, from Wagga, left Malacca on 1 Jan 1942 and went to 2/4 CCS at Kluang. Her account is from subsequent report with AWM.

58. RALSTON, I.D. Matron Lady Musgrove Hosp Maryb; AANS Singapore, 116 AGH, Charters Towers. After war opened nursing home at Coorparoo.

SELWOOD, MARGARET. Trd St Helen's Hosp Bris 1939; Midwifery Women's Hosp 1940; Theatre Sister at General 1940–41; AANS 1941–44. Service 2/13 AGH Singapore; 2/10 AGH at Johore Bahru; on return to Aust Feb 1942 service at Northam, WA, 6 CCS Ipswich, 106 CCS Townsville and at Ramu NG.

POWELL, JULIA. b. 13 Feb 1909, Staff Barcaldine 1933, Lady Goodwin Maternity Hosp 1934; AANS 1941, 2/13 AGH Malaya, 8 ACH Aust. Prom Capt.

ADAMS, MONICA. b. 9 April 1915. AANS 1941, 2/10 AGH Malaya, 103 AGH and 2/1 AHS.

PUGH, P. b. 1916; Trd Bris Gen Hosp 1934–1938; midwifery Bris Women's Hosp; service Aboriginal settlement prior to enlistment. AANS 1941–1945, service 2/13 AGH Malaya, retd 1942; Northam WA, 6 CCS Ipswich, 2/2 AGH Rocky Creek & 2/14 AGN Singapore.

59. BULLWINKEL, SISTER VIVIAN, MBE, ARRC, ED, FNM, FCNS b. 18 Dec 1915. Trd Broken Hill Dist Hosp 1934–39. Staff pte hosp Hamilton and Jessie McPherson Hosps Vic; AANS 1941 — 2/13 AGH Singapore & Malaya, sole survivor Banka

Is massacre, POW 1942–45; after war staff Repat Hosp Heidelberg, member numerous professional & community organizations.

60. The AANS in Singapore. AWM 509/1/1.

61. Correspondence, Col Derham to Matron Sage 9 Mar 1946.

62. DERHAM, A.P. 'Singapore and After'. *The Medical Journal of Australia*. Sept 21, 1946, pp 400–401.

Chapter Five

Continuing Traditions (ii):

World War II (1942–1945)

Australia, New Guinea and the South-West Pacific

It was a dramatic moment, like a tableau, nobody moving or speaking for a minute. Suddenly we all started talking and laughing together with Matron (Sage) trying to talk to each one of us at once. 'Where are all the others?', she asked. Poor Sister James, we let her do the talking. 'This is all', she said simply. 'There are twenty-four of us'. Sixty-five had left Singapore in 1942.

Sister Betty Jeffrey, White Coolies.

Queensland in the War Years

When members of the 2nd AIF arrived back in Australia in 1942 they found remarkable changes in the Australian attitude to the war. When they left for overseas in 1940 the war to many at home was far away in Europe, it was still a 'phoney' war. Indeed there had been much debate on the question whether Australian troops should be sent to fight on some far-off battlefield. The returning troops found that the war had arrived on Australia's doorstep and that there was fear, even panic, that Australia was in imminent danger of invasion. The fall of Singapore on 15 February and the subsequent loss of the Netherlands East Indies (Indonesia) opened up Australia as the next target for the Japanese. The bombing of Darwin on 19 February highlighted Australia's defence weaknesses. Australia's best divisions were still overseas, its home defence was in tatters, its air force ill-equipped and its navy, on its own, no match for the Japanese. The further bombing of West Australian towns, Wyndham, Broome, Derby, Port Hedland and Exmouth Bay and the Japanese landings on the northern coast of New Guinea kept Australia guessing. It was not so much a question of when the Japanese would invade but where. One school of thought argued that West Australia, including Perth and Fremantle was the most likely, giving Japan control of the Indian Ocean. Another possibility was Darwin and the Northern Territory. But in the end the defence authorities saw Queensland, with its long coastline, its many unprotected beaches and numerous harbours as the most likely invasion target. The defence of Eastern Australia, to protect the major industries located in NSW and Victoria took priority. The plan to concentrate troops in the south-east of the continent led to a claim that it was intended to defend Australia south of Brisbane should the Japanese invade. The 'Brisbane Line' controversy remained for many years, although there is no evidence to support it.

Whether the fear of invasion was whipped up by Prime Minister Curtin to attract more American aid or to get all Australians behind the war effort or to support the many restrictions on trade and industry and on individual freedoms is a matter of debate. In so far as the Queensland public was concerned the fear was very real indeed. People fled from the coastal towns to safer areas inland. Schools were closed and the pupils moved to safer areas. Slit trenches were dug, air raid shelters appeared on North Quay, in Albert Park and around the crowded cities and suburbs. Air raid wardens were appointed in selected areas and the public had to become used to air raid sirens. Windows were boarded up or taped and householders told what to do in an emergency. Restrictions were placed on travel interstate without a permit — 'border-hopping' was illegal. Rationing was introduced in essential items of food and clothing. Food hoarding for emergencies was popular and black marketing of goods in short supply became widespread. Petrol was a scarce commodity and returning troops were intrigued to find cars fitted with 'gas-burners'. Men and women not in the armed services were called up for essential work in munitions factories or in special 'reserved' occupations.(1)

1942 remained a critical year for Australia. The fear of Japan remained a constant factor in the minds of all Australians, as reverse after reverse continued. The Japanese attack across New Guinea brought them within sight of Port Moresby. However, the Japanese naval force believed to be headed for Port Moresby was destroyed in the Coral Sea battle (5–8 May). This euphoria was shortlived as the midget submarine attack on Sydney Harbour (31 May), the shelling of Sydney and Newcastle suburbs, the submarine attacks on shipping off the eastern Australian coast and the bombing of Townsville in July were ever present reminders of the proximity of the Japanese and of the vulnerability of Australian defences.(2)

Queensland in 1942 became the front line in the desperate attempt to halt the Japanese. Barbed wire defences were hastily prepared on many of the beaches. The 6th and 7th divisions were rushed to Queensland. The arrival of the American troops was a great morale booster, as they flocked into camps from Wacol through to the Atherton Tableland, where airstrips were laid down at remarkable speed. The success of the Americans at the Midway battle (4–6 June) in crippling Japan's aircraft carrier task force was a major factor in giving more time to the defence of Australia. However it was many months before Australia was secure from invasion. The defeat of the Japanese at Milne Bay (25 Aug–6 Sept) and the halting of the Japanese army in New Guinea when it was within sight of Port Moresby (Sept) turned the tide. Nevertheless it was touch and go in Guadalcanal where Japanese and American forces slogged it out, with heavy casualties on both sides, (Aug 42–Feb 43) before the Americans were successful. Thereafter Queensland became the springboard for the allied counter attack on Japanese strongholds to the north of Australia.

A dramatic change in the number and location of troops in Queensland had occurred during 1942. In January there were only 35,000 troops (AIF and AMF) in Queensland, compared with 105,000 in NSW and 82,000 in Victoria. By the second half of the year the 6th and 7th divisions had returned, with many of the troops stationed in Queensland. The compulsory call-up added many more thousands to the militia. Women's organizations had been included in the services — AWAS, VAD's, AAMWS, as well as the WAAF and WRAN services. But the biggest impact was the introduction of American forces, especially army and air force personnel, into Australia and their movement into Queensland. In the second half of 1942 Queensland was virtually an armed camp, from Warwick and Toowoomba on the Darling Downs through to Thursday Island in the north. As camps developed, so the medical services had to be provided. Queensland Line of Communication Area or QL of C as it became known proved to be a massive army organization controlling the troops and their movements in Queensland. The monthly report on the AANS, for example showed that in December 1942 some 600 members of the AANS were stationed in Queensland.

The extent of the build up of troops in Queensland during 1942 and consequently of medical units including AANS may be seen from the following table. Not all of the nurses were from Queensland of course, but the majority were.

Table 3
Monthly Report AANS HQ QL of C December 1942

Hospital or Unit	Number of Nurses
HQ QL OF C	2
112 AGH Greenslopes	64
116 AGH Charters Towers	60
117 AGH Toowoomba	50
2/2 AGH Watten	121
2/4 AGH Redbank	65
2/11 AGH Warwick	52
2/14 AGH	26
2 ACH Chermside	9
3 ACH Enoggera	20
4 ACH Exhibition	12
5 ACH Atherton	20
7 ACH Redbank	14
8 ACH Warwick	9
10 ACH Coorparoo	10
46 ACH	11
47 ACH Goondiwindi	16
56 ACH Cowan	4
77 ACH Tenterfield	14
2 Aust Women's Hosp Redbank	5
2/1 Aust Hosp Ship Manunda	15
1 Aust Amb Train	2
5 Aust Amb Train	2
1 MD Con Details	1
AAMC Trg School Tenterfield	1
X List	1
	600

The needs of the services for trained nurses had to be balanced against the needs of civilian hospitals. There had been several top level discussions on this issue. The plans of the Central Medical Co-ordination Committee with respect to rationalizing the demands of civil and service authorities for medical personnel, including nurses, brought about a top level conference of DDMS in April 1941. While the conference discussed a wide range of issues concerning the proper allocation of medical resources, the problems of the nursing profession were not overlooked. The Matron-in-Chief of the AANS, Miss Sinclair-Wood, was called to give evidence. At that time 636 nurses had been sent overseas and she estimated a need for a further 300 a year. Her biggest problem was getting theatre Sisters for home hospitals, such as Heidelberg. The problem of supplying civilian hospitals was discussed. Another difficulty was that with over 1,300 nurses on the waiting list for the AIF, how should they be selected. There was some dissatisfaction that many nurses who had enlisted in 1939 were called up and posted to Camp Hospitals, while later enlistments had been sent overseas after a few weeks. The Matron-in-Chief was asked to look at this — and also at the suggestion that Sisters should have a Mess separate from that of the officers.[(3)]

When the AANS returned from the Middle East they found changes in the administration of the nursing service in the Queensland Line of Communication (QL of C). Miss M.K. Caldwell who had been appointed to the AANS reserve on 6 August 1935 (AAO 266/35) succeeded Miss E. Paten as Principal Matron QL of C on 1 Feb 1941 (AAL 26/41), a position she held until her retirement to civilian nursing on 17 April 1943. On that date Miss J. Abbott succeeded Miss Caldwell as Principal Matron. Miss Abbott was a well-known Queenslander who had served in the Middle East as Matron of 2/6 AGH and whose work had earned her a RRC.

The question of commissioned rank for the AANS came up for consideration again in July 1942.[4] In addition to the case in England, Matron-in-Chief,Miss Sinclair-Wood drew the attention of the DGMS to the fact that a member of the AANS had recently been suspended for reprimanding a nursing orderly who had been neglecting his duties. In addition a decision had already been taken to grant commissioned rank to officers in the AWAS (Australian Women's Army Service) and to female physiotherapists in the Australian Army Medical Corps. Legally, it was still held that members of the AANS were not part of the Australian Military Forces.

Nevertheless, there was still a great deal of uncertainty about the degree of control or command which women officers of the AANS or AAMWS were authorized to exercise over members of their respective services. The Matron-in-Chief at LHQ, the Principal Matron at HQ L of C Area, the Matron or Senior Sister in a medical unit did not, under GRO's, exercise any powers of command and were not authorized to award punishment for offences committed by officers of the AANS or by members of any other service. At all times they were advisers to the senior officer commanding the appropriate army unit to which they were attached. The senior officer had no authority to delegate such a function to the senior member of the AANS concerned.

Of course, the official recognition of members of the AANS as officers of the military forces brought with it formal responsibilities in matters of discipline. In writing to the DDMS, New Guinea force in May, 1943 the DGMS put the position quite clearly:[5]

> Members of the AANS are now officers of the Military Forces and the procedure applicable to male officers applies to them. All communications of an official nature to higher authority must now be submitted through the Commanding Officer of the unit in which they are serving. This does not prevent the Matron of a hospital communicating with the Principal Matron of an L of C Area on matters of technical nursing services nature not involving questions of policy or routine administration affecting other branches or services.

While this created few problems for members of the AANS, Matrons had to tread a thin red line in dealing with higher authority. It speaks well for the discipline of the unit as a whole that there were few personality clashes between Matrons and Commanding Officers of army hospitals.

One of the problems confronting the Matron-in-Chief on the return of AANS and troops to Australia in 1942 was the implementation of the marriage regulation. This was a desperate period of the war and many of the AANS wanted to marry and some in fact did so, even if that meant automatic resignation from the army. The Matron-in-Chief was very sympathetic to the plight of the nurses, believing they should not postpone marriage until the end of the war, which at that time was thought to be some years away. At the same time she did not wish to lose experienced and competent Sisters. Eventually AMR and O No 1085 was amended making resignation no longer mandatory on marriage, giving them the opportunity of making a request to remain in the service. Of course, in army parlance information about marriage had to be submitted on Casualty Returns for amendment of next-of-kin and allotments in the records. A subsequent list showed there were 38 married Sisters with Q, QX or QFX numbers still in the service.[6] The marriage regulation remained a difficult problem. Married members tended to be kept in base areas, but their presence in units was always clouded in uncertainty. There was always present the possibility of pregnancy, while nurses expected to be granted leave when their husbands (if in the services) were on leave and looked for discharge if their husbands were discharged.

While desperate fighting was going on in New Guinea and in areas to the north of Australia, army medical authorities were taking stock of the hospital facilities available in Queensland and New Guinea. In August 1942 a report to the DGMS, based on an estimate of hospital beds for 8 per cent of field formations and 4 per cent for troops in base areas disclosed 6,920 beds would be needed for the 100,000 troops involved. The beds available from the medical units neatly fitted this estimate! The only recommendations in this report were to set up the unit at Glennie (Toowoomba) as an orthopaedic hospital, to build a 60 bed women's hospital at Redbank and to provide additional facilities at all convalescent depots. These estimates were to be substantially increased in New Guinea and Queensland during the next twelve months. By April 1943 there were available almost 12,000 beds in army general hospitals in New Guinea and Queensland, without taking account of those in casualty clearing stations and camp hospitals. Nor did this figure include those available for RAAF or RAN casualties.(7)

The question of promotion became an important issue for all members of the AANS in 1943, as many had had several years service. The recommended criteria were efficiency and qualifications, period of overseas service in a theatre of war and finally — seniority. Among the Queenslanders listed for promotion in August 1943 were the following:

To be *Principal Matron* (LtCol)
QFX6297 Joan Stephenson Abbott HQ QL of C

To be *Matrons* (Majors)
QFX6121 Dorothy Grace Roe, 2/6 AGH
QFX6103 Isabella Annette Marks, 7 Aust Camp Hosp

Earlier in the same year the following Queenslanders had been promoted:

To be *Majors*
QFX44617 Anna Ruth Bonner 3 ACH
QFX140423 Grace Irene Sheahan 112 AGH
QFX6110 Florence Mary Petersen, Nol Ortho U
QFX6107 Ethel Frances Hanrahan 117 AGH
QF70215 Ida Elizabeth Horton Pearce 116 AGH
QFX50137 Kathleen Cahill 5 ACH(8)

Camp Hospitals(9)

The large number of service units scattered throughout Queensland during the war years in camps of varying size necessitated the provision of medical services. In the beginning it was possible to improvise by using buildings in Showgrounds or in school or church halls, but there was no uniform standard or design. Treatment varied from the bare minimal, a little better than an RAP to operating facilities which brought them close to a small base hospital. Mostly, camp hospitals catered for short term medical cases and were staffed on the size of 30, 60 or 120 beds. Depending on the size of the Camp Hospital, staff consisted of medical officers (sometimes visiting), orderlies and one or two nursing staff. There was little glamour in working in these hospitals, mostly routine work, with staff isolated and 'on their own' in terms of facilities and equipment. They were the forgotten people of the army medical services.

Camp Hospitals flourished in Queensland during the war years, the best known being those at Chermside,(10) Cairns,(11) Warwick,(12) Goondiwindi,(13) Cowan Cowan,(14) Canungra, Mt Isa and Toorbul Point.(15)

Sister J.B. Casey(16) was posted to the 7 ACH at the Brisbane Exhibition Grounds, which covered all troops in transit and those from the Brisbane area. Routine injections and vaccinations had to be given, provision had to be made for immediate treatment of medical cases and for the occasional routine surgery, such as appendicectomy. Sister Casey was appointed Theatre Sister which was established in Kodak House in the Exhibition Grounds. Fortunately the Brisbane General Hospital nearby provided sterilization facilities for gowns, linen, dressings. A copper was used for 'boiling up' the large bowls but finding suitable 'lifters' or tongs to lift out the bowls was a problem. Eventually, Sister Casey came across, in a second-hand shop in George Street, a pair of Marcel Curling Tongs, 1920 vintage! At this stage, at the end of 1942, this camp hospital was in charge of Matron K. Cahill, with a staff of 2 Sisters, several VAD's and a number of orderlies. The CO, Capt J.K. Mowat was in effect part-time as he was still in private practice.

Sister Gladys King(17) was on duty when the 'Brisbane riot' took place, with many Australian servicemen requiring medical attention. Sisters in other metropolitan hospitals, especially those with facio-maxillary facilities, also reported numerous casualties with fractured mandibles arising out of the Brisbane riot. It was believed that facio-maxillary units in American army hospitals were equally busy.

AANS Redbank Camp Hospital 1941. Visit by the Governor of Queensland, Sir Leslie Wilson Pte Thomas (Patient), Sister M. Carseldine, Sir Leslie Wilson.

In addition to 2/4 AGH, Redbank Camp also boasted the 7 Australian Camp Hospital and 2 Women's Hospital which later moved to Yeronga. With the large influx of women into all branches of the armed services the question of providing separate hospital facilities for women was hotly debated. Some saw this as unnecessary duplication of specialized facilities while others recognized the problems of admitting women to hospitals where the vast majority of patients would be men. The No 1 Women's Hospital was established at Claremont (WA) and No 3 at Concord. When the larger base hospitals were developed at Greenslopes, Concord, Heidelberg, Daws Road Hollywood and Campbelltown the practice developed of reserving a ward or wards for female patients. The 2 Women's Hospital at Redbank had the distinction of having as its Matron, Sister E.F. Hanrahan, distinguished Queenslander, with a long record of army nursing experience behind her. At the 7 Aust Camp Hosp at Redbank Major G.B.V. Murphy successfully carried out insulin coma treatment in psychiatric cases.

Col Walker referred to the ever-present danger of an outbreak of cerebro-spinal meningitis in army camps such as had occurred during the 1914–18 war. In the early stages of the war the construction of adequate camp facilities barely kept pace with enlistments, a period of some concern to medical and army administrators until proper hygiene standards were accepted, better nutrition was observed and the sickness rate kept to a minimum. There were spasmodic cases of cerebro-spinal meningitis in crowded cases such as Redbank but these were quickly isolated. Sister McCallum reported on the nursing care at Enoggera Camp Hospital when despite the primitive conditions, not one case — meningitis or other — was lost.

5 Camp Hospital, Sheridan St, Cairns, formerly Edge Hill School.

Ipswich figured prominently in the provisional Order of Battle in Feb 1942, probably based on the assumption of an enemy landing on the Queensland coast. No 6 CCS (Haughton Valley), Ipswich was brought to full war establishment to operate in conjunction with camp hospitals and general hospitals being established on the Darling Downs. Sister Cahill was one of those sent to Ipswich in June 1942 when it was proposed to establish a general hospital there, but for some reason this idea was dropped and subsequently a larger hospital was built at Redbank to be taken over by 2/4 AGH when it returned from Ceylon later in the year.

Nevertheless it was at Ipswich that No 5 ACH was eventually established before it was sent north to Rocky Creek, a busy little unit of 200 beds. At this time this area of the Atherton Tableland was a hive of military activitiy, with preparations being made also for the 2/2 AGH and the 2/6 AGH. No 5 ACH then moved on by road convoy to Cairns to take over the Edge Hill State School, where later on malaria experiments were carried out.[18] Late in 1944, No 5 ACH moved on again to Charters Towers, as a small 90 bed hospital. An event to be remembered there was a shocking accident when a large truck containing Air Force personnel, men and women, returning from an Air Force function collided head-on with an army truck. There were many casualties, a situation which stretched the resources of the small camp hospital to the utmost. Sister Cahill[19] recalled the tragedy:

> Suddenly trained nurses came from everywhere to help us — housewives, Sisters on holidays (one was even AWL from her unit — never did find out what happened to her). Our own MO's and a team of surgeons from Townsville used the operating theatre non-stop for about 36 hours. It was a remarkably efficient response to an emergency situation.

One of the most unusual stories to come out of the war concerned Sister Edna Gilbert and her work on Thursday Island. When the war broke out she was on the army reserve list but she was not called up for

Inspection of Camp Hospital, Enoggera by the Governor-General, Lord Gowrie, 1941.
L. to R: Lord Gowrie, Sisters McCabe, Milne, McCallum, McCready, Campbell and Eagles.

service until October 1940, when she was advised to report to Miss Paten, then Principal Matron AANS, Brisbane. At that time Sister Gilbert was on the staff of Infant, Maternal and Child Health Section of the Department of Public Health, stationed in Toowoomba. However, at her medical examination, she was declared medically unfit for army service.

In April 1941 Sister Gilbert was appointed Matron of the Hospital at Thursday Island, some 30 miles to the north-west of Cape York. The island is about 1½ miles long, ¾ mile wide and surrounded by seven small islands, one of which Horn Island was to be of military importance. The population at this time was 300–400, mostly Europeans, Malays and Chinese plus about 400 Japanese engaged in pearl fishing. The Australian army had sent up a small garrison of 600 personnel and the Navy had a small complement there also.

All went well until 7 December 1942, when Japan entered the war. The Navy brought in the pearling luggers, the Japs were taken south for internment and all women evacuated, except for essential personnel. Sister Gilbert continued to run the civilian hospital assisted by Sister D.E. Francis caring for many evacuees from New Guinea. However the hospital was taken over by the army in March 1942, although the civilian medical and nursing staff remained, pending service staff appointments. Before this happened, the Americans arrived and set up their hospital. Then the air-raids started, with the following air-raid alarm instructions alongside a decrepit bath tub:

> Should a Jap plane come in sight, Beat the bath tub with all your might, Blow the whistle, blow like hell. If you've got one, ring a bell. Grab your rifle, gas mask and smokes, Don't waste time cracking silly jokes, Run for the jungle, helter skelter. If you have not time, make for the shelter. Stay under cover, until you hear Blasts on the whistle, which means, "All Clear".

Ambulance Train, Ipswich, 1942.

In June 1942, Sister Gilbert was again examined for army service and declared fit! She flew to Brisbane to meet Miss Caldwell, Principal Matron. After becoming a member of the AANS she was sent back to Thursday Island hospital — on army rates of pay![20]

In November 1942, 1 Camp Hospital arrived at Thursday Island en route to Horn Island. It was quite a Queensland affair, with Sister Thorburn in charge of the nursing staff, assisted by Sisters Heers and Althaus. Others served there during the war years.[21] Sister Kettle and Sister Francis continued to serve on Thursday Island during the dangerous years of 1943 and 1944 when they were both transferred to the 2/2 AGH at Rocky Creek.[22] The Camp Hospital on Thursday Island ceased to function on 25 April 1946 and was handed back to civilian authorities. Sr Gilbert, having been discharged from the army, was there to receive it back!

South of the border at Tenterfield was 77 Aust Camp Hospital, unusual in that at one stage all the Sisters were Queenslanders, Sisters Barnard, L.B., Beardmore, F.E., Daly, E.A., Fletcher, D.E., Lipstone, R.A., Moss, M., Page, F.M., Prideaux N.A., Reid, R.A., Searle, M.A., Sier, E.J. and Stevens, E.D.H.[23] The hospital was located in the centre of the town in Stannum, a gracious old home owned by the Misses Reid, which was taken over by the army for essential purposes. The spacious grounds were soon filled with army tents, each one housing several patients, most of them with minor complaints. Those who were there recollect the extreme cold under which they worked and lived. Sister Fletcher wrote:

It was never really warm, but at night it turned really cold. The boys would then lace up the tents so tightly from the inside we could barely find a hole to shine the torch in and ask if everyone was all right. Night duty was an ordeal, with thick heavy frost which often remained until midday. The office and surgical ward were on the ground floor of the old house which was warmed by a large wood fire in the kitchen.

In this isolated unit, seemingly far removed from the actual fighting it was still a case of survival, for everyone seemed neglected and forgotten. It was a case of improvisation, initiative and self-help for which Australians are renowned. The ambulance was used to go out to a dairy to collect the daily milk. The 'locals' were always very helpful in assisting the hospital. Sister Fletcher recalled how one elderly couple offered them the apple crop, as they couldn't pick it or market it themselves. They refused any payment but the Sisters made it up by giving them sweets from the canteen, sweets having been off the shelves of the local shops 'for the duration'.

Base Hospital — Greenslopes

Very early in the war years there was considerable debate on the best means of providing medical facilities for service personnel, sick or injured in Australia, likely to require long and specialized hospital treatment. The civil hospitals were considered but the demands of the civilian population were considerable. Some assistance was possible from existing Repatriation Hospitals in each State but these were outside the direct control of the army authorities. As the war escalated, the number of militia and AIF in Australia increased and the normal rate of sickness made greater demands on service medical institutions. The despatch of the 6 Division overseas, with the possibility of returning casualties made a decision on this problem a matter of urgent priority. Not until July, 1940 was a decision taken by Cabinet to build an army hospital in each capital city for the use of service personnel during the war years and which could in the post-war years be handed over to the Repatriation Commission. So the plan for the establishment of a base hospital in Brisbane eventuated.

At first the Brisbane hospital was planned for 200 beds, capable of expansion to 800 beds. Initially, there was some confusion over the best site but eventually Greenslopes was selected.[24] This site of 28 acres had been set aside for War Service Homes, but it was close to the city and to transport, convenient to city trams and bus. The sloping site on the hill backing on to Mt Stephens caused some delay, but eventually a plan was envisaged with the wards on different levels. There was still some reluctance to go ahead with such a major undertaking, possibly because of the danger of enemy attacks within five miles of the city and possibly the 'Brisbane Line' philosophy might have had an influence. It was November 1940 before Thiess Bros moved in to excavate the site, to build the administrative block and the tiered wards.

Meanwhile the authority to establish 112 AGH (later 112 AMH Greenslopes) was given and staff selected. Col D.G. Croll and Capts E. Stanford and J. Green with 21 OR's were appointed in March and took up residence under a grandstand at the Exhibition. It then moved to Enoggera, then back to the Ernest Baynes Grandstand at the Exhibition to take over the Camp Hospital there. Its next move was to the old military hospital at Kangaroo Point (so well-known to World War I Sisters), at that time known as 'Yungabar' and used for immigration purposes. In January, 1942 the advance party moved in to Greenslopes and patients were transferred from Kangaroo Point on 2 Feb with Col A.W.L. Row as CO. Sister J. Weaving was in charge of the first group of AANS which included Sisters A. Thorburn, F.M. Allison, K.I. Hansen, M. Hawkins, F. Stobert, R. Seamark, A. Creagh, F. Alcorn and M. Cumming. Sister M.E. Alcorn was in charge of the second block which opened at the end of the month, followed by the third block the following month with Sister B. Philp. Sister G.I. Sheahan was appointed the first Matron of 112 AGH in March and held that position throughout the war years. She was assisted by Sisters H.V. Compton, J.A. Cameron, V.M. Paterson and B.I. Hayes. At this time some beds were reserved for RAAF patients who were nursed by Sisters F.M. Amps, E. Maher and E.F. Neilson.[25]

It was a busy time during 1942 and 1943 to establish such a major base hospital. AANS personnel found that working in a military hospital far from the front line, in the midst of a city teaming with troops from many allied forces was no picnic. When they had the occasional day off they went for a quiet cruise up the Brisbane River to Lone Pine or for a picnic and swim at the popular Oasis at Sunnybank. Sister Audrey Gray recounted how on one occasion she shyly asked one man in swimming togs if he would kindly take a photo of the nurses. This he agreed to do. Some time later when she saw him in uniform she realized then it was none other than General MacArthur! He was very nice about it and introduced the nurses to his wife and son.[26]

As the war in the north developed, 112 AGH was an important link in the line of evacuation of patients stretching from New Guinea back to the mainland. As the air evacuation scheme developed, many cases requiring specialized treatment were brought direct from Port Moresby. Hospital ships and sea transports likewise brought back many patients. With so many cases of gun shot wounds of the mandible and face being admitted a facio-maxillary and plastic surgery unit was established. Sister G. Hair was the first Sister in charge of this unit and when the Blood Bank was set up, Sister M. Henning was in charge.[27]

Matron Sheahan was very proud of the high standard of nursing achieved by members of the AANS in these pioneering months of 1942 and 1943. In her Report at that time she said:—

> The history of the Australian Army Nursing Service in aspiring to and achieving a high standard of efficiency has been well maintained. Sisters have given generously of their skilled nursing care to these patients. They have worked zealously and loyally and with cheerful co-operation — and all they desired and expected was worth while work. Their discipline has been good and they truly have upheld the traditions of the Australian Army Nursing Service of the First World War.[28]

A fine tribute indeed!

The Union Jack Club

When members of the AANS were on leave in Brisbane they found numerous opportunities for relaxation in homes and hostels provided by welfare organizations, the Red Cross and similar associations. Of all the facilities provided, none was more popular than those provided by The Union Jack Club. This had a most interesting history. In 1940, a group of nurses many of whom were returned Sisters from the First World War and who had been doing welfare work for soldiers saw the need for the establishment of a hostel for men of the services on leave in Brisbane. They got in touch with nurses in the public and private hospitals throughout Queensland and received great support for the proposal. So The Union Jack Club was established in premises in Wickham Terrace, formerly the Alexandra Hospital. It provided bed and board, reading rooms and entertainment for some 27,000 guests during its first year of operation. When it became obvious that the Club needed a larger building it acquired a large stone building on Wickham Terrace at a cost of £5,200 ($10,400), most of which was paid for by the Queensland Turf Club. One of the founders of this club was Sr Winifred Payne.

In writing to the Queensland Turf Club at the end of its first year in this building, the following reference was made to nurses.

> As no provision is made for returned Sisters it is the intention of the Committee to preserve the stone building as a Nurses' Club for members of the Australian Army Nursing Service and Air Force Nurses when they return from active service.

The Club then secured premises at New Farm which was called The Union Jack Army Nurses' Club. Over the next few years many hundreds of nurses enjoyed the facilities for recreation offered by this Club.[29]

One of the hospitals which the Australian army inherited from the Americans was at Ekibin, in the block bounded by Toohey's Road, Sexton Street and Mains Road. It was described at the time (1945) as being about one mile south-west of 112 (Brisbane) Military Hospital 'from which it is separated by a deep gully. It is surrounded by forest trees, giving a country effect and has a pleasing north-easterly aspect'. The site had been occupied by the American 155 Station Hospital, was completely sewered, was linked to the Brisbane water supply and the hospital had its own steam heating system. It had a comprehensive system of wards and specialized facilities, capable, in the view of DDMS, Q L of C of holding a 600 bed AGH plus a 200 bed orthopaedic division. The Director of Hygiene added a rider that the mosquito problem in the area was not serious and was controllable, although he noted that 'anopheles mosquitoes are scattered thinly in the area'. The local dairy would also have to have its hygiene standard improved![30]

AIF and AMF Hospitals

The critical years in Queensland following the entry of Japan into the war called for planned control over medical resources and equipment, both for civilian and military purposes. In turn army medical authorities had to plan for adequate medical facilities wherever troops were gathered. The escalation of the war in 1942 and 1943, the increased number of troops in Queensland and the evacuation of casualties from New Guinea brought about the establishment of major army hospitals throughout Queensland. These included

2/2 AGH	Watten	1,200 beds
2/4 AGH	Redbank	600 beds
2/6 AGH	Rocky Creek	600 beds
2/11 AGH	Warwick	600 beds
2/14 AGH	Charters Towers,	Townsville 600 beds
116 AGH	Charters Towers	800 beds
117 AGH	Toowoomba	1,200 beds
1 Orthopaedic Hosp Toowoomba		
2 Women's Hosp Redbank		

The return of AIF general hospitals from overseas supplemented those AMF hospitals already established. Most of these finished up in Queensland and Northern Territory before going to New Guinea. One major hospital (2/1 AGH) finished up in WA.

After their return from the Middle East in May 1942, and their well-earned leave, members of 2/1 AGH, including the AANS, were posted to a little known place, Merredin, in Western Australia. In fact, the hospital was divided into two sections, with another 250 bed hospital established at Guildford Grammar School (previously No 5 U.S. Stationary Hospital). There this valuable and experienced hospital remained until July 1943. No one was quite sure why such a unit was selected for such an obscure part of Australia, when all the action appeared to be in Queensland, Northern Territory and New Guinea. One theory was that a Japanese landing was expected in Western Australia, another than many naval battles might be fought in the Indian Ocean, while yet another was that there would still be many casualties to come 'home' to Perth from the Middle East. Whatever the reason, it is true that there were many troop formations in the West and an adequate army medical service was necessary. With 2 CCS's, 1 base hospital, 3 general hospitals, 1 women's hospital and 7 camp hospitals it appeared that the situation was adequately covered. As it happened, the worst fears never eventuated and many thousands of troops as well as AANS and medical services soon became bored with the lack of action.[(31)]

Queensland members of the AANS with 2/1 AGH at this time included Allen, E.M., Baldock, A.M., Byers, J., Duffield, E., Foley, B., Goodman, H.M., Hooper, J., Geraghty, M., McConnell, A.B., and Suttie, C.M.[(32)] For them the time was not wasted. It was a relaxing period after the strain of Gaza, a time for further training and tutoring a new staff. New buildings were planned and erected; better accommodation was needed for the 60 AANS than that at Parbury House. Matron Sinclair-Wood, Matron-in-Chief visited the hospital in August 1942 to boost the morale of the AANS and to resolve urgent problems. Life at Merredin was described in great detail by Sister Amy McConnell who wrote regularly to her mother, her family and her friends in Brisbane. She told her mother about the countryside and the beautiful wildflowers, about Queensland friends she had met and the sad news of those mentioned as casualties. She wrote of the simple pleasures enjoyed on her days off — a visit to an ABC Symphony Concert, her impressions of the latest films, of little humourous anecdotes which might amuse her mother. Occasionally she wrote about her work in the wards, of the many malaria patients, of the men desperately ill with scrub typhus.

- I wish we were doing something useful. This is nothing but a Con Depot and in the eastern States all the hospitals seem to be frantically busy. If we were busy we would not grumble so much.
- All the nurses have gardens round their tents which make it more homely and attractive with ferns and aspidistras. Unfortunatley we have to carry water in canvas buckets a fair distance and then water the plants from punctured jam tins — effective but very tedious.

- Yesterday the Principal Matron of Western Command, the VAD Commandant and our own Matron paid an official visit so everything was spit and polish. We even had a home grown salad, and flowers on the mess table so everyone was duly impressed. Usually on such occasions the weather is perfect but this time Providence was on our side as the day was a positive scorcher and the PM was wearing a woollen coat and skirt! Then a terrific storm burst and the visitors were marooned in the sick nurses' tent. Nearly all the tents leaked at the brigadings and water rushed in underfoot. They had to stay for dinner in the mess which was under water too. We were a pretty sight as most of us had been out in the storm deepening the trenches round our tents!
- We had a 'house-warming' party for our new mess hut. The building is neither lined nor ceiled so our footsteps echo hollowly. It will be good though to walk on boards instead of through mud in wet weather as was the case in our old mess. We all felt sentimental in abandoning it!
- We've heard a lot about the riots between the AIF and the Militia in Queensland. One story says a CO issued an order than on no account were his men to call the Militia "Chocos" (Chocolate soldiers). The next day his men were calling them "Koalas" — because they were protected by the government, could not be exported, were not to be shot at and they lived on leave(s)!
- We are very busy at the moment as we have a number of casualties from the islands suffering from a disease called scrub typhus. It is hard nursing and we do what we can to alleviate the suffering, but mostly it's hopeless.
- The patients here object to wearing hospital blues and to add insult to injury, red ties, especially as some of the people hereabouts mistake them for POW's and want to know why there are no guards with them!
- I enjoyed the gossip in your letter, so did some of the other Queensland girls. I am nursing now with Sr Goodman. Remember her up at the Ridge? Keep sending papers when you can as news from the eastern states is so rare here. If it weren't for the mail and the papers we'd be cut off completely from the outside world. There is a wireless set but it's pretty hopeless most times.
- The event of the week was the installation of electricity in our bathrooms and other buildings. Mr Curtin, our PM, will not allow it in our tents and tented wards so we continue to use hurricane lamps. Now that darkness comes early it is very gloomy in the wards, as only four lanterns are permitted for each 24 beds.

2/2 AGH Watten[33]

The 2/2 AGH had some remarkable experiences during their service in Queensland. After their perilous return trip from the Middle East, as part of the Stepsister Movement in March 1942, they were asked to set up a field hospital at Watten, in Queensland. Where was Watten? Never heard of it, said the usually well-informed Queensland Sisters. It proved to be a small village of 30 people, some 30 kms south-west of Hughenden. Even Hughenden was but a vague name, somewhere 'out west'. In fact it was 1,867 kms from Brisbane and nearly 400 kms from Townsville! No one was quite sure of the logistics of this site and rumours abounded. It was on a 'track' some 700 kms SW of Cairns. Perhaps if the Japs invaded there would be a rearguard action towards the centre of Australia. Perhaps it was only a staging area, to 'toughen them up'. If so, it was not for jungle warfare as the almost treeless plains rolled westwards.

Just when a few huts had been built and tented wards were erected, an extraordinary event occurred. In an area where rainfall was patchy and seldom more than 450 mms a year, the place was hit by a devastating cyclone and by flood rains. Tents were demolished and equipment ruined. The hospital had to be disbanded and the 2/2 AGH was moved to a new site at Rocky Creek on the Atherton Tableland. This was between Atherton and Mareeba, less than 200 kms from Cairns and as one Sister said — 'nearer the front line'.

One of the Sisters who made no claim to immortality as a poet expressed the thoughts of the unit in verse, entitled 'How We Weathered It in Watten'.

Squelching round in mufti, though mufti is taboo,
But after all, if the heavens fall,
What can the Matron do.
Hauling on a tent flap with the ridge pole on your head,
While your cabin trunk in ooze has sunk.
And the river rose round your bed.
You had nothing to eat and your face was ruined.
You couldn't reach the "what not" for the river got their first.
You can boast your Battle for Britain
And your service in Moresby, too.

I don't care for I did my share
When I weathered it in Watten in 1942.

Redbank, 2/4 AGH, 2 AWH, 7 ACH.[34]

Another experienced medical unit to serve in Queensland during the critical years 1942–1944 was the 2/4 AGH. It had behind it experience in the Western Desert, in Barce and Tobruk, in the Middle East at Jerusalem and more recently in Ceylon. Its new assignment was to establish a 600 bed hospital at Redbank, between Brisbane and Ispwich. While it was not an ideal spot for a hospital, being bitterly cold in winter, very hot in summer and subject to the wild westerly winds which blew for their accustomed three days. When the 2/4 AGH arrived in October, 1942 they found the hospital still under construction. Even before the hospitals moved in, the cockroaches were there before them, large ugly dark brown specimens with voracious appetites. No amount of spraying ever got rid of them. Another pest which upset many of the nurses were the large green frogs and the nasty cane-toads. The hospital had the difficult task of setting up in the midst of a very wet, wet season, while still accommodated in tents. Staff had to dash through the rain to toilets and showers, where, as one Sister described it, repulsive looking toads 'leered from the rafters and perved in the latrines'.

The 2/4 AGH at Redbank was almost at the end of the evacuation line from New Guinea. Patients came by land, sea and air transport from the north, by motor ambulance convoys, by ambulance trains, by hospital ships and by air rescue teams. At Redbank they were classified in accordance with their need for immediate treatment or for movement to convalescent depots or for evacuation to southern base hospitals. This pattern of admissions and evacuations was followed with monotonous regularity over the next two years, when some 20,000 patients passed through the hospital. Nevertheless, there was no lack of devotion and dedication by the members of the AANS, assisted by AAMWS and nursing orderlies. Six surgical wards contained many amputees and bed-ridden patients who required much personal care and intensive nursing. At one stage there were 60 patients in this block, only one of whom was not bed-ridden and few could feed themselves. With his arm held high in an 'aeroplane splint' and his plaster oozing stinking flesh, this one ambulant patient was a happy volunteer helping with the washing up and in feeding his mates.

In these early days, before hot water was connected to the hospital, the coppers outside the wards were essential items. Those on one side were kept for boiling bedpans and other such hospital equipment. Coppers on the other side provided the boiling water for tea. On one occasion a Sister found a soldier from a nearby camp quietly getting hot water from one of the coppers used for the bed pans. She shouted to warn him but he quickly made off with his precious pot of boiling water. She often wondered whether he enjoyed his mug of tea!

While this famous unit was initially established in Victoria in 1940, mostly with Victorian, South Australian and West Australian members of the AANS, many of these had been promoted by 1942 and had moved on to other units. Reinforcements were readily available from Queensland. Sisters Ramsay, A., Redman, E., Tinney, H., Smith, A., Carson, A., Doherty, M. and Glen, I.M.[35] Redbank, for a number of years a major recruiting campsite and the home of GDD (General Duties Detail) was also the site for 2 Women's Hospital and 7 ACH.[36]

Atherton Tableland: 2/6 AGH, 2/2 AGH

A most secret study of the Atherton Tableland area as a suitable locality for a base hospitalization scheme for the New Guinea force was carried out late in 1942 by a team of top army medical administrators. They reported favourably on a plan, recommending one 1,200 bed hospital, one 600 bed hospital, one convalescent depot for 1,000 personnel. In addition there was a site for US Army medical facilities. These medical units could service at least two divisions, while a third division could be located in the Townsville — Charters Towers area. The Atherton Tableland became well-known to thousands of Australian and allied servicemen during the war years.[37]

2/6 AGH Rocky Creek[38]

It was April 1943 when the 2/6 AGH arrived in Rocky Creek after their prolonged stay in the Middle East. They had remained in Gaza after the 6th and 7th Divisions had returned to Australia under the

Stepsister Movement. The 2/6 AGH had a major role in receiving patients from Alamein. The Atherton Tableland in North Queensland, 2,000 feet above sea level was totally different from Gaza. The Tableland, with its mild climate, free from mosquitoes, was an ideal camping area for Australian troops preparing for the counter attack against the Japanese to the north of Australia. The 2/2 AGH and the 2/6 AGH were necessary medical units both for the needs of the many thousands of troops on the Tableland and as part of the line of evacuation from New Guinea. The 2/6 AGH was sorry to lose Matron Abbott who had been promoted to the position of Principal Matron, Q L of C. Joan Abbott had had an outstanding record as a nurse, nurse educator and nurse administrator in Queensland prior to enlistment. As Matron of 2/6 AGH she had been an inspiration to all in Greece and in the Middle East. As it happened another Queenslander, Sister D.G. Roe from 2/5 AGH succeeded her as Matron. 1943 was a very busy year for AIF hospitals on the mainland and within weeks, the hospital had extended to 900 beds. Strangely, the 2/6 AGH also had to suffer the violence of a tropical storm. In September, 1943 a storm demolished part of the hospital but fortunately there were no casualties. The Sisters, accustomed by this to untoward happenings of violence in one form or another went about their business with their usual calm. Some idea of the difficulties under which they worked may be gained from the fact that it was not until October 1943 that electricity from the Barron Falls Hydro-Electric Corporation was switched on to the hospital site. This significant event coincided with the arrival of the first battle casualties from the 2/9 AGH, by then established at Port Moresby. The bed state soon climbed to over 1,300. One dramatic event which occurred locally and which caused the Sisters much hard work was the unfortunate accident when a Japanese mortar bomb, being demonstrated, exploded, injuring 44 Australian servicemen. The 2/6 AGH was to spend eighteen months at Rocky Creek, until it closed on 27 October 1944, during which more than 28,000 patients passed through the hospital. That represented many, many hours of faithful nursing by many Queensland nurses who served at Rocky Creek during this period.

2/14 AGH, 116 AGH(39)

Medical units seemed to be constantly fighting the elements, whether the heat and sand of the desert, the ice and snow of Greece and Jerusalem, the jungles of Ceylon, Malaysia and New Guinea or the vagaries of the weather in Northern Australia. As these medical units were invariably housed in tents, extreme conditions posed severe problems for staff and patients alike. In North Queensland, where so many troops were stationed during the war years, cyclones wrought havoc. One of the worst hit Townsville in 1944, where the 2/14 AGH suffered severely. This hospital had been erected at Cape Pallarenda, a few miles out of Townsville, across the road from the sea. It was an idyllic situation but conditions were primitive. Townsville was in the grip of a severe drought, resulting in a water shortage, made worse by the demands of thousands of troops in the area. The Sisters were rationed to two gallons per day for personal use. This was carried to their tents in canvas bags by the orderlies, but they had the time consuming task of filling them from a tap which trickled water for half an hour a day. Sometimes Sisters returned to their tents only to find wandering stock had quenched their thirst from the precious water.

Then one hot February morning the cyclone struck. Sister Henry,[40] who had trained at Charters Towers and who had seen storms come and go, related the events of this day.

> I was on duty in a malaria ward when three stretcher cases were brought in. The first patient was being lifted on to the bed when a sudden gust of wind rattled the tent. A minute later, a second heavier gust tore the tent from its poles. Staff and patients were enveloped in a billowing and flapping mass of canvas. While we were trying to extricate ourselves from the tangle, a third gust struck the tent, bringing heavy rain which stung with its force. Part of the tent blew away leaving sixty very ill patients out in the blinding rain. The orderlies battled hard to get all patients to a hut about fifty yards away. Sister Norman and I emptied the icebox and filled it with valuable medicines, while into the breadbox went the charts and medical records. These had to be tightly secured as the wind was so strong that bedpans, bottles and loose furniture were hurled across the road, across the sand into the sea.

Meanwhile urgent help arrived and ambulances took the patients into the Townsville Hospital. A passenger train about to leave for Brisbane was taken over, the passengers emptied out and patients put aboard for a quick trip south. Such were the exigencies of war! The wind and the rain stopped at daylight to reveal an odd collection of towels, clothing and underclothing hanging in the treetops beyond the Sisters' tents. The new hut built for the Sisters' dining room had almost disappeared in a lake of rainwater when the roof collapsed.

Sister Henry and Sister Martin were despatched to Charters Towers with the remaining patients, where they were accommodated with 116 AGH. When they returned to 2/14 AGH they found all the damage had been repaired and that the hospital was open for business once more.

Charters Towers

Charters Towers, 135 kms SW of Townsville was selected as a suitable site for a general hospital. It was far enough removed from possible bombing raids on Townsville, its climate was much cooler than that on the tropical coast. Moreover it had numerous large buildings, including boarding schools which could be used for medical purposes. So 116 AGH was required to set up a hospital of 800 beds there, moving up from Townsville.

Sister T.C. Troedson, who trained at the Brisbane General and served with the 116 AGH and later with the 2/6 AGH at Labuan has made some pertinent comments on the early days at Charters Towers.

> When we arrived in Charters Towers the hospital was not ready for patients and I found that doctors, orderlies and clerical staff were busy with spades, axes and rakes clearing the racecourse and erecting tents. Eventually we took over All Souls' Church of England School and the Mt Carmel Catholic School, both of which had been boarding schools. During the Japanese attack on Darwin we were given the "Red Alert" and the whole hospital was forced into the slit trenches. No other information was given to us as to what was going on. Soon afterwards 50 nurses arrived at 116 who had been evacuated from the Hospital at Katherine (NT). They had been sailing between Townsville and Bowen for days as apparently the Army could not make up its mind what to do with them.

At this time everywhere was haste and confusion. Ordinance could not keep up with the demand. 116 AGH found it difficult to get adequate supplies of blankets, linen, even pyjamas. In the beginning they had only 12 pairs of pyjamas so these were divided up, some given tops, some bottoms! Milk was in short supply and the hospital pathologist was shocked by the number and variety of bacteria in the local milk. As goats ran wild in Charters Towers there was a demand for goat's milk, but of course goats had first of all to be caught and then milked!

Sister M. Kay,[41] who trained at Stanthorpe and later served with the 2/12 AGH in Balikpapan, was another who served with the 116 at Charters Towers. She found the operating theatre under the school at Mt Carmel and tented wards on the racecourse. The horse stalls were used as dressing rooms for treating patients. It was certainly very austere living for Sisters and patients alike, with primitive facilities such as primus stoves in daily use. These Sisters soon mastered the art of improvisation and maintained a reasonably high standard of nursing, despite the conditions.

Sister Troedson, made some further observations on these two hospitals:

> All Souls had a junior dormitory with a cement bath, NO HOT WATER, in fact no HOT water in any bathrooms . . . the Matron's and 2 IC's bath water was heated in a washing copper each morning by an general duties orderly. Along the shower arm used to sit 26 green frogs who had to be chased from their perch before any one could have a bath . . . All through my army career Bird Baths were the order. A basin full of warm water, body soaped all over, then the basin poured all over. At 2/6 AGH even on the coldest of days we learnt to take cold showers and at night we could put on the bed 8 blankets. Without a doubt this was one of the greatest hardships.
>
> The Army built wooden wards at All Souls and we built an operating theatre underneath the school at Mt Carmel. It was such a shame to shift the brothers from Mt Carmel as their rooms had never been cleared of their books etc. and believe me it was a major undertaking to get this part of Mt Carmel ready for patients. The Officers' ward and the Sisters' sick bay were stationed in Mt Carmel.
>
> To return to the 2/6 AGH whose matron at that time was "Judy" Abbott and who had been my tutor Sister at RBH was marvellous, as this was a very well organized AGH. Miss Abbott left to become Head of the Service and then Dorothy Roe, whom I had known when specializing at St Martin's Hospital took over and we had a very happy unit.

AANS — 116 AGH Charters Towers 1942.
Rear: Sisters Spearritt, Asmus, Brownsdon, Cottell, Lyons, Scott.
Centre: Sisters Burns, Brosnan, Larsen, Hanlon, Brook, Oliphant, Briant, Goos, Webb, Black, Scott, Stuhmcke.
Front: Sisters Troedson, Gracey, Kenny, Whelan, Matron Pearce, Col Wilson, Sisters Ralston, Adams, Clements, Elms.

> We knew we were going overseas, and one day we were loaded into the ambulance train and set out for Cairns. I think it took us 14 hours to get down the range and into the Convalescent Home in Cairns. After a night's sleep we were loaded on the Ambulance train and back to Rocky Creek, where we stayed for another TEN MONTHS, hospital all packed and ready to go. The civilian hospitals did not want us as we could be called on any moment to get on to the Hospital Ship. We did NOTHING for ten months.

During this time 116 AGH had to nurse American airmen, injured or wounded in the many sorties from adjacent airfields. Many Australian patients were flown down from New Guinea, with limbs in plaster and suffering from malaria and other tropical diseases. In these grim months the Sisters worked long hours with devotion and care, but they never lost their sense of humour — nor did the typical Australian patient, no matter how severe his injuries. Sister Kay relates one humourous incident of the time. One day a patient was waiting at the end of a queue for a minor operation when he felt the urge to urinate. Going outside he found a jam tin near the coppers where the orderly had been tending the fire. There was apparently some inflammable liquid still in the tin into which he urinated and finally dropped his cigarette butt. Suddenly the tin became a can of flames. The poor man had to be treated for third degree burns to the most private part of his anatomy. The unfortunate patient exclaimed later 'I never knew that urine was explosive!'

In April 1944 Sister E.F. Hanrahan became Matron of 116 AGH, a further step in her outstanding nursing career. Many other Queensland Sisters had long experience with 116 AGH at Charters Towers.[42]

The last hospital to leave the Tableland was 47 Camp Hospital which rendered long and efficient service until its closure in 1945. Its Matron, Sister G.M. Taylor was a Tasmanian, formerly attached to the 2/2

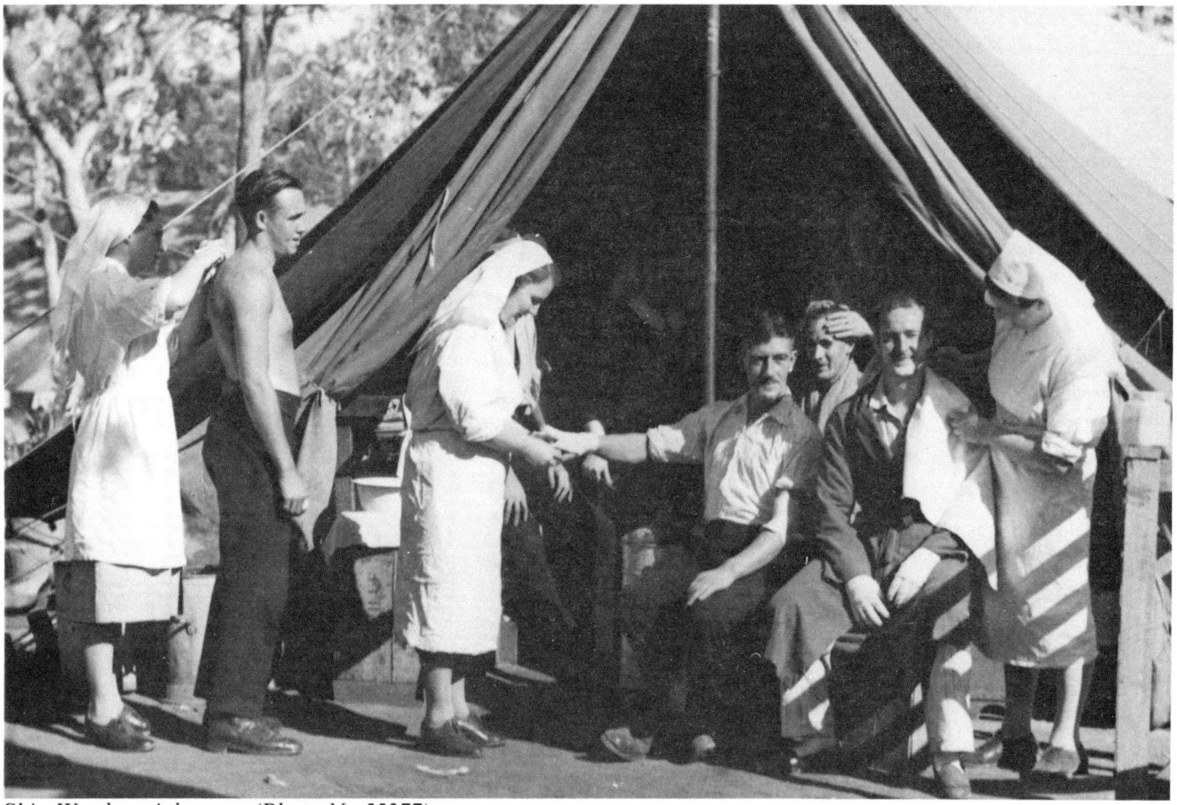

Skin Ward — Atherton. (Photo No 55277).

AGH in the Middle East. She later became Principal Matron in Tasmania. Many Queensland Sisters who served at 47 ACH had great respect for her efficient service.

2/11 AGH and 2/12 AGH, Warwick[43]

The Darling Downs area in south-eastern Queensland was an ideal spot for medical treatment. It was much cooler than in Brisbane, although much colder in winter. The two major towns, Toowoomba and Warwick had all the necessary facilities and besides, there were numerous private boarding schools and colleges which were taken over by the army. In addition, the cynics observed that it was free from any enemy bombing and behind the much publicised 'Brisbane Line'.

After its fine period of service at Alexandria and its return to Australia as part of the Stepsister Movement, the 2/11 AGH found itself at the end of 1942 at Warwick, committed to a 600 bed hospital. Scots College, a Presbyterian boys' boarding college, was taken over and with the addition of tents a start was made.

When the 2/11 AGH moved to New Guinea in September, 1943 to take part in the Huon Peninsula Campaign, its place in Warwick was taken by 2/12 AGH. This hospital had seen long service in Colombo and needed time for re-grouping before embarking on any further service in the SWPA. The war against the Japanese demanded a new approach, not only by combatant units, but by medical personnel also. They were put through commando-type training 'to toughen them up'; they were taught survival skills in water, they were taught survival skills in the jungle; they were taught self-defence and how to handle a pistol! Sister Kay recalled her experiences in Warwick:

> I was transferred to the 2/12 AGH at Warwick which was located at Scots' College. We lived in Nissen huts amongst the gum trees nearby and when it rained heavily water would drip from the rafters onto our heads whilst we occupied the stretchers. We used ground sheets to cover the bed clothes but could never avoid the persistent drips coming onto

our pillows or heads. Late afternoon one day a hurricane hit the camp and some of the tent wards being used as Orderlies' Quarters were inundated with water and some personal belongings were swept away. Next day some male members of the staff were climbing trees to retrieve some of their clothing. A tree branch came down on our hut and pierced the roof but no one sustained any injury. We spent some time prior to going to Balikpapan preparing for the adventure. We spent some time on the Parade Ground doing Physical Training under the guidance of the Physiotherapist. We had to become qualified in swimming. This entailed going to the local baths and training each day. Non-swimmers were all taught and had to qualify. We were taken on route marches. One enthusiastic Sgt Major took us on a five mile route march over rough bush roads which resulted in most of us being treated for blistered feet as our footwear was too light. That put an end to route marches. We were given pistol practice and all duly qualified as competent in that field. We were issued with Smith & Wesson pistols and were instructed how to clean and care for them. A common sight was to see the Sisters sitting on the steps of their huts with a piece of wire, lint and oil, cleaning the barrel of their pistols. We were taken to a quarry a few miles away from the camp for target practice where we would buckle on thê holster and under instruction would go through the motions of aiming at the target with dummy cartridges. Later we were divided into groups and had mock attacks. This all created a great deal of amusement and we became known as the "Pistol Packin Mummas!" Once we mastered the art the pistols were returned to the Q Store and we never had occasion to need them to be re-issued. The purpose of the exercise was for self-defence.

117 AGH Downlands and Glennie[44]

Another group of some thirty Queensland Sisters found themselves in unexpected surroundings, as the war situation in the Pacific worsened early in 1942.[45] In his history of Downlands College, Father J.F. Mooney relates how the Rector of the College, Father J. Doyle learned that the Army was about to take over his College as well as the Church of England school Glennie, as a military hospital. Thus a Catholic boys' boarding school and a Church of England girls' boarding school became the nucleus of the 117 AGH in April, 1942. At that time it was the most northerly of the Australian General Hospitals. The CO, Col R. Fowler, a Melbourne specialist faced the difficult task of establishing a field hospital under war-time conditions. The experiment of separating medical and surgical divisions in widely scattered buildings was not successful. Soon afterwards Sister E.F. Hanrahan arrived as Matron, accompanied by a large group of nurses. Many of these had been hastily called up, as near panic developed in political and army circles. The nurses quickly adapted themselves to the demands of nursing tropical cases of malaria, dengue and dermatitis. Before long many of them had transferred to the AIF and been given their QFX number. Over the next twelve months, as the fighting in New Guinea escalated, many of the sick and wounded found their way down the line of evacuation to the safety, security and coolness of Downlands and Glennie, or as they knew it, 117 AGH. They came by plane to the mainland or by sea on one of the Hospital Ships, probably to the 2/4 AGH at Redbank or to the 2/1 CCS at Ipswich. The journey to the Darling Downs was completed either by Ambulance or by Ambulance Train to Toowoomba or Harlaxton. Because the location of the hospital was ideal in many respects, extensions were planned, increasing the capacity to 1,200 beds. Not until the end of 1943, with the tide of war turning in the South-West Pacific and with many more hospitals established in Queensland and in New Guinea, was it considered safe to abandon the concept of a hospital at Toowoomba. 117 AGH was disbanded in November, 1943.

Sister Cahill was one of those sent to Glennie early in 1942 to assist in the development of 117 AGH. Referring to this period, she wrote

Not ideally suitable for a hospital, it was nevertheless adaptable. At first we were accommodated in tents — there were no floorboards available at that time. Cold showers and miniature hurricane lamps made life very difficult.

Ambulance Trains

The line of evacuation from New Guinea to Queensland during 1941 and 1942 varied in its method of transport, sea or air, but the major principle involved was to get the patients to the nearest point on the mainland as soon as possible. Thus the ship or aircraft could turn around quickly, ready for the next batch of casualties. So coastal towns such as Cairns and Townsville were the first staging posts. Thereafter the Queensland Railways had the responsibility, in conjunction with the army of servicing the Ambulance Train on its journey south. Sister Nancy McCallum, formerly of the Brisbane General Hospital, but then a member of the AANS attached to No 1 Ambulance Train, travelling between Cairns, Townsville, Brisbane, Toowoomba and Warwick has vivid memories of her experiences. In those days such a journey was an epic indeed.

> We were very busy at this time as we had casualties coming in from New Guinea, the Milne Bay Show and the Coral Sea battle. At first we brought Australians & Americans together but as another war was being fought on the train we subsequently brought a train load of American and a train load of Australian wounded — separately! The casualty sheets were interesting in that the majority of Australians were battle casualties or very ill with malaria, dysentery, tropical ulcers, scrub typhus, etc, while a few were diagnosed "neurasthenia". With the Americans the number suffering from neurasthenia was far greater.

Sister McCallum paid tribute to the magnificient work done by the Queensland Railways in getting the Ambulance Trains through when there was excessive traffic on the line. She recounts the problems which the wet season brought.

> Of course the Burdekin Bridge was under many times. On one occasion we were held at Ayr — eventually a pilot train was sent over the Burdekin and we were allowed through with the water still lapping the rails. We were thankful to get across, especially when we heard the following goods train caused the bridge to collapse. Another time we had to go out through Winton and Longreach to get through and we found a Ball was in progress at Longreach. It was in the small hours of the morning but the Red Cross invited everyone to join in. It was amazing how everyone suddenly felt better — even the stretcher cases!

The Ambulance Train was a mixture of glamour and austerity, of convenience and inconvenience. The Sisters were accommodated in a VIP compartment used formerly by the Prince of Wales (later King Edward VIII). There was a red carpet, embroidered with the Prince of Wales feathers on the floor, with its own toilet, hand wash basin and cold shower. Sister Pollock, G.I., who also saw a period of service on the Ambulance Train described the primitive working conditions in other parts of the train.

> The cars were mostly old second class sleeping cars with six berths in each compartment and a very narrow passage in between. The moment we reached the cooler climates the walking cases in the top berths were stricken with rigors and raging temperatures, due to malaria. Sponging, changing and making them comfortable in a rocking train was no end of a job. As there were only two Sisters on for day and one for night duty on the train, plus a number of orderlies, the work was consistently demanding.

In those anxious months there was always the fear of sabotage, not a difficult operation on those lonely stretches of uninhabited country. Sister Pollock recalled one event which might have ended in tragedy.

> On one occasion, we were suddenly awakened in the dead of night and almost thrown out of our bunks. There was a series of bumps — then complete darkness. By the time I got to the window to see what was happening, I saw the train crew racing up, waving their lanterns. It seems all thirteen carriages had gone over several pieces of iron which had been placed on the line, but fortunately none had been derailed. The train was able to proceed slowly to Rockhampton where it was carefully inspected before being given the all-clear to proceed to Brisbane and Warwick.

Sister Breen R.M. was another Queenslander who served on No 1 Ambulance Train. One assignment which she recalled was the transport of patients from Charters Towers to Townsville and then on to the *Centaur*. She and Sister Pollock went aboard to meet their AANS colleagues and to see that their patients were made comfortable for the trip to Brisbane. It was on the return trip near Brisbane that the *Centaur* was sunk with tragic loss of life.

Another Queenslander, Sister Lucas, M.K. in between service on the *Manunda* also experienced life on an ambulance train. No 3 AAT serviced the route Brisbane, Sydney, Goulburn, Albury where the patients were transferred to a Victorian Ambulance Train. (All through the war years the break of gauge at Albury necessitated a change of train.) The train was made up of nine "wards", one to each carriage with three tiers of beds on either side. With 40 beds to each ward the train moved a large number of patients on each trip. An operating theatre and sterilizing room were also attached. With only one doctor and 3 Sisters (one for night duty), plus two male orderlies for each ward, the staff were kept very busy. Temperatures varied

HOSPITAL SHIP SUNK

299 LIVES LOST NEAR BRISBANE

Torpedoed Without Warning: Went Down in 3 Minutes

The Australian hospital ship, Centaur, was torpedoed and sunk by an enemy submarine without warning 40 miles east of Brisbane on May 14, at 4.10 am.

The vessel capsized and sank within three minutes after being hit. Of the 363 members of the crew and medical staff and nurses on board, only 64 were rescued. The remainder were lost.

Hospital Ship Sunk

2/12 AGH — March through Warwick 1944. (Photo No 84676).

from almost freezing to an outside temperature of 115 degrees F at Wagga Wagga. (There was no air conditioning in those days). Other Queenslanders who served with Ambulance Train formations included Sisters Brown, R., and Graham, G. (No 1 AAT) and Sisters Thorburn, A. and Weaving, J.[47]

Hospital Ships and Sea Ambulance Transports

During 1943 there was opportunity to review the use of hospital ships in the evacuation of patients from New Guinea. With a lull in fighting and increased use being made of aircraft, especially in short hops to Townsville, three hospital ships were considered sufficient, 2/1 AHS *Manunda* (capacity 322), 2/2 AHS *Wanganella* (430) and 2/3 AHS *Centaur* (284). These operated between Darwin, Merauke, Port Moresby, Milne Bay, Oro Bay and Brisbane. There was also a small vessel, 2/4 AHS *Stradbroke II* (27) used to pick up sick and wounded in outlying islands and return them to a major base. With the majority of AIF returned from the Middle East the hospital ships could be used more efficiently with a quicker turnaround from New Guinea.

In view of subsequent events it is relevant to trace the history of this ill-fated vessel, the *Centaur*. On 18 April 1943 the OC of the 2/3 Aust Hosp Ship *Centaur* Lt Col C Hanson reported to the DGMS in the first six weeks after the ship's commissioning. The ship left Melbourne for Sydney on 12 March, where certain minor deficiencies were rectified, the medical staff, including nurses, embarked and the ship sailed on to Brisbane and Townsville. This was virtually a trial run for the embarkation of patients and their subsequent disembarkation at Newstead Wharf, Brisbane. The OC was quite happy with the arrangements and on 13 April it was in Port Moresby, returning to Brisbane on the 18th. However, there was one disquieting note. He pointed out that when the ship was full, some 120 personnel would have no other life saving apparatus than rafts to which they could cling. The whole of his staff and 50 patients would have no boat accommodation. The STO in Sydney offered to make larger rafts available for the next voyage. As well as patients, the ship carried medical and dental personnel. It also had provision for 100 tons of cargo in the lower holds and 60 tons of light cargo between decks.

Precise details concerning the sinking of the *Centaur* were conveyed by the Secretary, Department of the Navy, to the committee of inquiry into Japanese atrocities headed by the Chief Justice of Queensland Sir William Webb. The *Centaur* sailed from Sydney at 1044 local time on 12 May 1943 en route to Cairns and Port Moresby. At 0410 hours on 14 May, when it was about 23 miles abeam of Point Lookout and 40 miles east of Brisbane, it was attacked, without warning, by a Japanese submarine. It was hit by one torpedo and sank in 3 minutes. Of the 332 persons aboard there were only 64 survivors, including one woman, Sister E. Savage, AANS. Eleven members of the AANS, a number of medical staff and members of the 2/12 Aust Fd Amb lost their lives. The weather was fine and the visibility good, the vessel was brightly illuminated and properly marked. Notice of the use of the hospital ship had been conveyed to the Japanese on 5 Feb 1943. The survivors, who saw the enemy sub surface shortly after the attack, were not picked up until 1430 hrs on the following day by a passing allied vessel. One Queensland Sister, Joyce Wyllie was lost on the Centaur.(48)

In the subsequent inquiry into the sinking of the *Centaur* the ARCS representative is alleged to have said on two occasions:

(a) that he knew positively that the Centaur was carrying personnel and equipment in contravention of the provisions of the Red Cross Convention and that these facts were known to, but suppressed by the Australian authorities.
(b) that this knowledge came to the Japanese and they were undoubtedly in their rights in sinking her.
(c) that "the Japanese were alright" and would not sink hospital ships without just cause.

There was no evidence to support this claim. It would appear extremely unlikely that at that stage of the war the situation was so desperate that a hospital ship would have been used for purposes outside that

A group of allied nursing sisters. L. to R: American Navy Nurse, American Army/Air Force Nurse, AANS Sister Gray, Australian Navy Sister, Australian Air Force Sister. (112 AGH Greenslopes 1944).

Sister Wyllie's name is commemorated on the Sydney Memorial, a register of those who lost their lives in the South-West Pacific Area and who have no known graves.

agreed upon in the Geneva Convention. In any case, as Japan was not a signatory to the Convention it could hardly have had any rights in the matter.[49]

The hospital ship *Manunda* was well-known to members of the AANS. In the early stages of the war it made several trips to the Middle East to bring back sick and wounded, occasionally running into danger, especially in the Suez Canal when enemy mines were dropped. After refitting at the end of 1941, the *Manunda* was sent to Darwin early in 1942. As explained earlier there was urgent need for a hospital ship in Singapore but authorities believed it was too risky to send the *Manunda*. So it happened that the *Manunda* was in Darwin Harbour on 19 February 1942, when the Japanese bombers struck. Among those killed in the raid was Sister de Mestre, M.A. of the AANS, while Sister Blow, C.S. was seriously injured. After rapid repairs the *Manunda* was able to reach Fremantle safely and subsequently to Melbourne for further repairs and refitting. Før the remainder of the war the *Manunda* was on the run to New Guinea, Port Moresby, Milne Bay and the South-West Pacific Area, occasionally running in to extraordinary situations but thereafter escaping damage or casualties.

Sister Lucas was one of a number of Queensland AANS members who saw service on the *Manunda*. She recalled the commander of the ship LtCol J.B. McElhone and Matron Schumack who received the RRC following the Darwin bombing. During 1943–44 she made 10 or 11 round trips to Port Moresby, with the four days at sea each way always potentially dangerous. Frequent life-boat drills were essential. The return trips were always hectic, with the *Manunda* loaded with sick and wounded. As the front line moved northwards so the *Manunda* followed up with visits to Buna, Lae and Madang to collect the wounded, sometimes waiting for smaller ships to bring patients from Bougainville.[50]

Queensland Sisters also served on two other hospital ships, well-known to allied troops during the Second World War — the *Wanganella* and the *Oranje*. The part played by the *Wanganella* in Singapore (15–17 Sept 1941) has already been mentioned, following which trips were made to Suez, Colombo and Darwin. The *Wanganella* was frequently seen in Port Moresby and Milne Bay and as the war front moved northwards, in other New Guinea ports, Morotai, the Solomons and Borneo. Sisters spoke highly of the discipline and service on this hospital ship, commanded by LtCol R.L. Lenihan and later LtCol F. Brown Craig. One experience they all remembered was the explosion on the waterfront at Bombay in April 1944 in which many hundreds of people were killed or injured. The whole resources of the Hospital Ship were diverted for use as a civilian emergency hospital. Staff worked at top level for almost 24 hours before the victims of this catastrophe were all accommodated. The *Oranje's* major contribution was in bringing wounded back from the Middle East, but once the Japanese war commenced the Australian personnel were withdrawn and this Hospital Ship ceased to be part of Australia's medical system.[51]

Sister Ismay Fowler had a different war experience in the little known 'black' ships, nursing sick and wounded.[52] She was officially a member of No 1 Sea Ambulance Transport on board SS *Taroona* during 1944. This ship transported troops to the war zone and on the return journey brought back sick and wounded and other troops, some of whom required medical attention, but not serious enough to be sent back by hospital ship. Sometimes these ships sailed in convoy, sometimes they made the dash to the islands on their own. The *Taroona* made a number of trips to Milne Bay which always made a deep impression on Sister Fowler.

> On this voyage we had a number of American Army soldiers and seven American Nurses on board. The Ship's Captain related the incident of his entry on this voyage into Milne Bay. So much had changed since his previous visits. The American Armada had arrived: huge, big, little and tiny ships were everywhere. Across the Bay came a flashing signal to the Captain of *Taroona* to the effect:— "Identify yourself and state your business" Captain's reply was immediate. He had a contingent of American troops on board. He was relieved. The *Taroona* went to its usual anchorage at Lyle Wharf deep into the Bay. It was a beautiful spot with a tropical island on the bayside: on the shoreside was a huge tree draped with orchids in bloom. Close by was a smaller tree hanging over the water with a tree house which had been used as a signal station. This area was flat terrain bearing the scars of terrific battle — no end of coconut palm stumps, their tops having been shot off or others with bedraggled fragments. It made one shudder to realize and visualise that site. Yes, a battle which had cost many lives.
>
> I visisted the cemetery at Milne Bay. I was shocked to see the number of graves. On leaving the cemetery a spot cleared from dense jungle and surrounded by rubber trees, we drove away along a track with shadows falling from the bower of foliage. Later the light of a beautiful sunset came into view. With reverence I remembered those who paid the supreme sacrifice. They will be remembered.

On another occasion Sister Fowler was on board the *Taroona* when a cylcone struck Townsville, just as it was due to embark troops. She described how the Navy quickly got the bamboo fenders down between the side of the ship and the wharf to prevent damage to the *Taroona*. The *Canberra* was also moored close by, having disembarked 2/9 AGH AANS on their return from New Guinea. Very quickly these valuable ships cast off their moorings and headed out to sea to ride out the cyclone. This was the same cyclone that did so much damage to 2/14 AGH at Cape Pallarendra.

Some unusual experiences

Queensland Sisters also served in a variety of medical groups as the army medical service set up new specialized units. Often in ones and twos they followed their professional interests in dental units, blood banks, facio-maxillary and plastic surgery units, convalescent depots, dressing stations and with small surgical and medical terms. Sister R. Seamark, after serving in the facio-maxillary and plastic surgery ward at Greenslopes was transferred to 102 CCS and sent per hospital ship, *Manunda*, to Thursday Island. From there she was sent on to a lonely outpost at Red Island Point at Jacky Jacky, Cape York. With others in the team she lived in tents until a hut was built. Her patients came across Torres Strait from Dutch New Guinea, sometimes to a small airstrip nearby, known as Higginsfield. There were also patients from the troops in the area, one from the explosion of a primus stove and another from wounds received when grasped by a crocodile! As she said, life was never dull at Jacky Jacky![53]

The AANS also became involved in a number of 'top-secret' activities, 'hush-hush' jobs which at the time were regarded as very important by the army's strategic planners. Security was all-important in those days, as no one could be certain that there might not be 'fifth-columnists' lurking in the background ready to divulge vital information to the enemy. The ever-present fear that gas might be used prompted experiments on the best way to treat expected casualties. In Routine Orders of 2/2 AGH on 30 October 1944, appeared a notice calling for volunteers for chemical warfare experimental research. The experiments involved exposure under supervision, to blister gas. The notice confidently expected that there would not be a high percentage requiring hospital attention! There was a small element of risk with burns but no risk of permanent injury. They were committed to absolute secrecy on all matters associated with the experiment. The troops were given special concessions — extra leave and one shilling (ten cents) extra pay per day. So 1 Aust Fd Exper Stn, Innisfail was set up.[54]

Thus an unusual nursing experience befell Sisters B. Parker and Bewglass who were sent to the General Hospital at Innisfail on this special secret mission,[55] On the top floor of the civilian hospital, screened and isolated from the rest of the patients they were required to nurse those soldiers who had taken part in these gas experiments. The Sisters were under strict orders not to disclose any information which came to their knowledge on the nature of the illnesses of the troops or any condition requiring treatment or nursing.

A Sister who was there reported as follows:—

> We were told nothing officially. We treated patients who were admitted with various skin lesions. Something different was applied to each lesion to see which was the most effective. Photographs were taken before and after treatment. Medical Officers came from South Africa, Scotland and England as well as from our armed services. There was also one Aboriginal patient. After this we were returned to our parent unit, 116 AGH and immediately reposted. It was an extraordinary episode as on arrival at Innisfail no one handed over to us and on our departure we handed over to nobody.

During the war there were on-going experiments by the LHQ Medical Research Unit to evaluate antimalarial drugs. To this end, volunteers were called for among Australian troops to participate in such experiments under the supervision of the above unit which was attached to 116 AGH. The men were required for up to six months and were exposed to malaria infections of malignant tertian or benign tertian drugs and a proportion of them developed malaria. The volunteers were assured of first-class medical and nursing care and promised 21 days special leave. Nurses who were aware of the experiments were required to keep the details secret. Sister Beryl Burbidge was brought back from New Guinea to Cairns to take charge of the nursing staff in the Medical Research unit doing 'hush-hush' work in malaria control. Experts from many allied countries were there to watch the outcome of these experiments.[56]

Another unusual task befell members of the AANS in Queensland during the war years. In April 1944, civilian nurses at Goodna Mental Hospital, not far from Redbank Camp went on strike as a protest against alleged inadequate staffing at that institution. Members of the AANS were asked to volunteer to relieve the situation. The response was overwhelming and on 17 April 1944, 30 members of the AANS marched into Goodna Mental Hospital. The staff shortage was eased soon afterwards and the nurses returned to their units on 13 May. There was continual pressure from civilian authorities towards the end of the war for the release of more army nurses to relieve the civilian shortage. In Jan 1945, AANS members from 2/4 AGH and 2/11 AGH were 'loaned' to Manpower authorities for varying periods to help out in public and private hospitals where need was urgent.[57]

Members of 2/5 AGH returning from Morotai in 1945 had a narrow escape from death or serious injury when the aircraft burst into flames soon after taking off from Darwin. The pilot managed to land safely, while the crew smashed the perspex machine-gun turrets at the rear to enable the nurses to get clear. Seconds later the plane blew up.

Sisters Mary Wallace and Dorothea Burnett remembered an unusual experience which befell them during their stay in Brisbane. They were suddenly called to a top secret job at Stuartholme Convent, at one time an American hospital. It transpired that during August–September, 1944 a Japanese ship transporting allied POW's from Singapore to Japan was sunk, presumably by an allied naval vessel. Some time later about 80 survivors were picked up by another naval craft and brought back to Brisbane. Intelligence believed important strategic information might be gained from these men and in any case it was essential that the Japanese should not know there were any survivors. So they were whisked away to a 'top security' camp at the Convent. Seven members of the AANS and senior medical staff were called to the DDMS office with instructions that for three weeks they would be working under strict security regulations and their whereabouts were not to be divulged. Sister Wallace related the circumstances:—

> We were taken to Stuartholme Convent where there were temporary army buildings. The Convent was ocupied by two Nuns and two elderly gardeners who lived in a small cottage. The situation appeared ideal for security — the roads were hidden by bushy slopes, the gates and entrances were covered by army guards, the provisions came by special transport — and nothing left the grounds. The Sisters had to set up a hospital for 81 patients and we could ask for anything to make the hospital comfortable. A mobile pathology lab was set up and amenities of all kinds provided. When the men arrived they were obviously debilitated but their morale was high. Mr Justice Webb and Intelligence Staff spent each day interviewing the men.
>
> When the mission was completed it was discovered that those men who had homes in Brisbane had found the escape goat tracks out of the Convent and spent each night at home, returning before dawn! So much for security!

Sister Casey was another who was involved in a 'hush-hush' assignment. In 1940, before Japan came into the war, she was asked to be prepared to leave Australia at a moment's notice. 'You leave on the Northern Mail tonight at 8 pm and you are to be met at Bowen. Pick up your berth and train ticket at MacDonald-Hamilton Shipping Co.' After a hurried departure and a tedious trip she arrived at Bowen at 2 am and no to meet her! So she booked into a hotel and next morning wandered down to the only ship tied up at the wharf — the SS Zealandia. The Captain was very apologetic, not having been informed that she was on the train! When the ship put out to sea the Captain informed them they were going to Manila to bring back evacuees from Hong Kong. When they arrived at Manila a medical officer from the Red Cross arrived giving details of the patients and their medical history.

> I found the medical team was responsible for 100 adults and 192 children — 13 pregnant ladies, one elderly lady haemorrhaging TB, several babies with feeding problems and several passengers with dengue and malaria. What an assignment!

Sister Casey had to deal with other crises on the voyage — an outbreak of measles, crew reactions from small-pox injections, another desperately ill with pneumonia.

However, the voyage ended safely, the ship calling at Brisbane, Sydney and Melbourne where RAAF personnel for Darwin were embarked. Sister Casey left the ship in Sydney, with the observation:

> On looking back over this hurried trip I realize how lucky we were that there were no serious accidents — and no emergencies, submarines or German surface raiders. We put up with black out conditions, wireless news we could receive but transmission was forbidden. Still at that age I was young and full of confidence.

No doubt there were many such 'hush-hush' assignments performed by the AANS during the war, many forgotten, some even unrecorded in the need for secrecy in war-time. However there is no evidence to support the claims by one writer that there was a commando type AANS group ready to take on any dangerous assignment, that members of the AANS parachuted behind the lines in Greece or that some were taken prisoner at Aitape and sent to Sandakan. This is pure fiction.[58]

Darwin and Northern Territory

The Northern Territory was always regarded as vulnerable to attack by an enemy power, but successive governments did little to bolster its defences in the pre-war years. Indeed to those in the populated areas of Sydney and Melbourne it was regarded as an area unsuitable for 'whites', with extreme viewpoints expressed that it would be best to hand it over to cheap Asian development. One of the best known nurses in the Territory during the late 1930s was a Queenslander, Miss E. McQuade White.[59] After training at the Brisbane General and serving on the staff of that hospital, she joined the Northern Territory Medical Services as it was then known. Darwin Hospital in 1937 was very primitive but it provided Sister McQuade White with wide experience in civilian nursing, especially with Aborigines. She learned to cope with the exigencies of the 'wet' seasons, with cyclones, with emergencies of all kinds. She learnt to know the Territory well, as duty often took her to Katherine, Pine Creek, Tennant Creek and Alice Springs. She travelled in all kinds of aeroplanes, bounced along in old utility trucks and worked with the Flying Doctor. All these experiences in civilian nursing in the Territory are faithfully recorded in her booklet, *Reminiscences of an Army Nurse*.(60)

At the outbreak of the war against Germany, militia units were sent to Darwin and the Northern Territory Force was gradually expanded. Raw, unseasoned troops soon fell victims to the many tropical diseases of the Territory, and they were not helped by poor hygiene, lack of personal discipline and inadequate medical facilities. In September, 1939 Miss McQuade White received immediate notice of her appointment as a Staff Nurse in the Australian Army Nursing Service but curiously another arm of the Commonwealth would not release her from her position with the Northern Territory Medical Services, owing to a shortage of civilian nurses in that field.

Meanwhile the Army was forced to move rapidly on establishing hospital facilities in Darwin. It was however not until the middle of 1940 that an old hospital at Bagot Compound, previously used for Aborigines was acquired. For some months a small detachment of nurses from the 2/5 AGH was posted there but they were soon re-posted to their parent unit in the Middle East. Bagot Compound was inferior by any standards, with accommodation for 150, the infectious patients in tents and the remainder in hutted wards. From December 1940 until April 1941 it was known as 119 AGH.

On 12 April 1941, the 119 AGH with Sister E. McQuade White as Matron sailed from Sydney for Darwin, a journey of some ten days. Unfortunately a strike on the Darwin waterfront delayed unloading of essential equipment for five weeks. It was found that Bagot Compound, between the civil and RAAF aerodromes was inadequate for nursing accommodation, so they had to be accommodated at the Quarantine Station, with a six mile drive each way each day. It was on one of these journeys that Sister M. High was seriously injured in a truck smash, losing her right arm. The Bagot site presented many problems. The septic system was faulty, drains had to be dug, laundry facilities were primitive and the 'wet' season added to their problems. The despatch of four Sisters and eight men to provide an emergency service at Katherine added to the initial difficulties. Moreover the hospital was committed to receiving RAAF and RAN patients, as well as civilians.

A better site for an army hospital was selected at Berrimah, nine miles south of Darwin and early in 1942, Major Cole, Matron McQuade White and 6 Sisters moved in, with an initial intake of 300 patients. The entry of Japan into the war on 7 December, 1941, together with air-raid alerts and warning sirens prompted all to dig slit trenches. With the rapid advance of the Japanese and the fall of Singapore, all civilians in the Territory were evacuated. The 119 AGH found it was operating in three sections, one at Kahlin Hospital in Darwin, one at Bagot and another at Berrimah. This was the situation on that eventful day, 19 February 1942, when the Australian mainland was bombed, for the first time.

Matron McQuade White's *Reminiscences* provide a graphic account of the events of that day.

> At 10.05 am an air-raid warning was heard just as enemy bombers and fighters came over the town of Darwin. For a few minutes it was not realized by staff and patients that the armada of planes overhead was the enemy approaching. The sound of bomb explosions and ack-ack fire soon disillusioned everyone. Patients were placed under beds and those who could, made their way to the long grass. Others scrambled to the few slit trenches which had recently been dug.[61]

It was not long before the casualties began to arrive, mainly from the RAAF drome which was a prime target. Every bed was in use and even the verandahs at Berrimah were soon filled, as were the wards at Kahlin and Bagot. Sisters worked without a break for thirty-six hours, while physiotherapists and ancilliary staff helped in the care of the wounded. Many cases came in from the bombing of ships in the harbour.

The bombing of the Hospital Ship, *Manunda* was a terrible shock to all members of the AANS. Twelve members of the staff were killed, one officer, one member of the AANS (Sister de Mestre, M.A.) and ten members of the crew. Sister L. Blow was also seriously wounded. Many patients were killed or wounded, suffering particularly from burns. The 119 AGH at Kahlin, Bagot and Berrimah was used initially to accommodate the casualties but as the *Manunda* was quickly made seaworthy, these were evacuated to the Hospital Ship which sailed for Fremantle. As Matron McQuade White wrote:

> Loading of the wounded was most difficult and was carried out with great dexterity on the part of the officers and members of the crew of the "*Manunda*". The wharf was still burning on the afternoon of the 20th February, and as I waited with 4 Sisters from the 119 AGH who had volunteered to help the staff of the *Manunda* nurses, the whole scene was one of devastation; bodies were washed up on the beaches. Men were collecting the dead and placing them on barges for burial at sea.[(62)]

However, the ordeal was far from over. The following day, Berrimah Hospital was again attacked from the air and machine-gunned at low level. One patient on a stretcher under the bed, was fatally wounded. All the available evidence suggests that the conduct of the medical and nursing personnel under fire was beyond reproach but the same could not be said for other persons in the area, as mild panic developed and many fled south to escape the conflict. Indeed the Kahlin Hospital was not only damaged by enemy action but looted by Australians! Berrimah Hospital was obviously vulnerable and had to be evacuated. It was clear that the site at Adelaide River, 70 miles south of Darwin had to be developed as quickly as possible, for in the event of a major battle developing for Darwin, following an anticipated airborne invasion by the Japanese, a Field Hospital was a necessity.

On the evening of 7 March, 22 Sisters and 50 patients were transferred by ambulance to the Adelaide River site for the 119 AGH. With Col E. Culpin and later Col J.R. Donaldson as CO, LtCol C.A.M. Renou in charge of the surgical section and J.H. Halliday the medical, 119 AGH soon established itself as a first class medical unit. It was not easy for a hospital to operate on three sites, Kahlin, Bagot and Noonamah. The move to Adelaide River was not without its problems, however, for the first site on the north side of Adelaide River was found to be unsuitable, so the Hospital packed up and moved to the south side! Matron McQuade White described some of the difficulties confronting the Sisters:

> The first site at Adelaide River was low-lying with thick bush; sanitary and bathing conditions were shocking or perhaps I should say, non-existent. Much of our equipment had been lost, not due to enemy action, but to looting. Messing arrangements were very primitive. Laundry was a greater problem than ever before. A kerosene tin was a great prize for any one lucky enough to obtain one. During this period the weather was very bad and many tempestuous nights of wind and rain were experienced. The Sisters worked on undismayed, and endured hardships in the roughest and murky sludge.[(63)]

By the end of 1943 life in the Northern Territory had settled down, with the army somewhat like an army of occupation, waiting for something to happen. Fears of a Japanese invasion were past and bombing raids had been forgotten. Large numbers of troops still occupied major areas, partly as a preparation for movement to the islands, partly as a warning to any potential aggressor that Australia had the means to protect itself. The health of the troops became an important issue, with diarrhoea and dysentery ever present. Malaria in some areas was always a possibility, accidents all too common and prickly heat an occupational hazard to many. For others boredom and a desire to be 'with the action' created mental attitudes and problems of morale which could destroy the will to serve and fight.

The report by the Principal Matron, Northern Territory Forces, LtCol P.J. Haines, in December, 1943 is an interesting overview of how the AANS were coping in this situation.[(64)] 101 AGH (Katherine) was staffed mostly by South Australian members of the AANS. It was clean and well kept, with neat and attractive gardens developing. Most Sisters were in huts and it was hoped to have complete hutted accommodation before the next big 'wet'. A tennis court was laid out — and a quiet room provided where Sisters could write or read or sew. At 109 AGH (Alice Springs) most of the patients were civilians and the staff showed great skill and tact in handling them. Women's and children's wards were kept very busy, with

119 AGH Adelaide River (N.T.) 1943. Matron E. McQuade White (R.)

an obstetrical ward functioning well and the infant welfare service in great demand. The townspeople were very grateful for this unexpected first class medical service. 129 AGH (Darwin) was kept busy with medical cases at a time when many staff were on leave. A sick Sisters' ward was opened. Accommodation was a problem and various plans were suggested, including a takeover of a small cottage occupied by the Navy. At 45 Camp Hospital (Larrimah) the daily average of patients was so small it was difficult to find enough work to keep the staff busy. There was time for painting and cleaning up the hospital, giving it a bright and fresh appearance. On the other hand 55 Camp Hospital (Tennant Creek) was kept fairly busy. Kitchen and diets were satisfactory, the patients content and the staff happy. It appeared to be running very smoothly. Queensland Sisters, Elizabeth Lyon, Mary Luddy, Phyllis Smith, Isobel Hanlon, Mary Purkess and Bridget O'Leary were stationed at 74 Camp Hospital, waiting for emergencies which never arose.[65] If AANS members had any complaints they related to uniforms and equipment. No grey raincoats were available, a necessity in the wet season, shoes were not AANS shoes and often lacked arched support. Large size gumboots were not available and often the clothing was unsuitable for the Northern Territory climate.

Stirring the washing in the old coppers, somewhere in the Northern Territory c.1943.

When Sisters were appointed to operational areas in Central Australia, Northern Territory and in other parts of northern Australia and Torres Strait they had to undertake the care of civilian and native populations as well as military personnel. They found themselves dealing with infant welfare, obstetrical cases as well as the usual water-borne, fly-borne and mosquito-borne diseases.

The AANS in New Guinea

With the outbreak of war in the Pacific, New Guinea became an important line of defence for Australia. Port Moresby (Papua), Rabaul (New Britain) and Kavieng (New Ireland) were key centres which had been reinforced mostly with militia troops even before December, 1941. Medical units, including some nurses, were rushed to New Guinea to service the medical needs of the increasing number of troops in this area. Further reinforcements including nurses were sent to Rabaul, almost to the same place where Australian nurses had been in 1914. A 60 bed hospital was set up, staffed by personnel from the 2/10 Field Ambulance and including 6 members of the AANS. However, the Japanese continued their onward march through Timor, Ambon, Kavieng and on to the Rabaul garrison by January 1942. Everywhere the story was the same — brutal murders, bayonetting and shooting of defenceless prisoners. The massacre at Rabaul has been described elsewhere.[66] Many of the medical staff, even wearing Red Cross brassards and having surrendered, were savagely cut down. The 6 nurses of the AANS under Sister K. Parker embarked on the *Naruto Maru* and taken to Japan as POW's, where they endured the perils and tribulations of captivity until rescued by the Americans at Yokohama in August 1945.

Sister Anderson was one of the Sisters taken at Rabaul and subsequently imprisoned in Japan. In her report she described the hardships she and others endured — digging air-raid shelters, carrying bundles of cut wood, shovelling paths through the snow for the guards in knee boots to walk through, while they were barefooted, glueing envelopes at 1 sen per 1,000, inadequate sleeping quarters, flimsy clothing, thin diet of bread and watery soup. Bathing and toilet facilities were primitive, the latter being pit toilets, 'emptied by the local gardener whenever his vegetables need nourishment'. While the story of their colleagues in Malaya has been well told, little has been written about the privations of the 6 AANS, POW's in Japan.[67]

The Japanese launched the first air-raid on Port Moresby on 3 February 1942 and it seemed in those dark weeks as if a major disaster might befall New Guinea. Mindful of the reported atrocities committed by the Japanese, the authorities decided to evacuate all army nurses from New Guinea until the military position was clearer. These were Sisters Gillanders, Chadwell, Graham, Cashin, Weaving and Whelan. The nurses left Murray Barracks under protest as they realized the heavy load which would be thrown on the orderlies and on the medical staff. Once again, it was argued that nurses should not be in forward positions where they were in danger of captivity or injury. Sister G. Pollock, who had trained at Maryborough, was one of those nurses sent to Port Moresby soon after Japan entered the war. On evacuation, she was airlifted to Cairns and then sent off on the long train journey to Brisbane. On arrival in Brisbane she was stricken with malaria and had great difficulty being admitted anywhere for treatment. Sister Pollock paid great tribute to Sister M.D. Gillanders who until the Sisters were returned to New Guinea made countless sterilized swabs and dressings and sent them back to the orderlies at Murray Barracks.[68]

46 Camp Hospital

When the war against the Japanese erupted it was clear to the Australian authorities that urgent measures would have to be taken to protect Australia against invasion. In this context the islands to the north of Australia were critical barriers in defence planning. While the Dutch East Indies, as they were then known (now Indonesia), were regarded as the first line of defence, Australia realized that Papua and New Guinea,

Original members of AANS in Port Moresby, 1941.
Rear: Sisters Gillanders, Chadwick, Graham.
Front: Sisters Cashin, Weaving, Whelan.

over which it had control, would need to be reinforced as Australia's front line. Port Moresby and Rabaul were seen to be centres of strategic importance. Rabaul had in fact been reinforced in March 1941, while Moresby was in the military sense part of 8 Military District. At this early stage civil and military administrations operated throughout New Guinea. Medical Officers of the 9 Fortress Company and the 49 Battalion were conscious of the dangers of malaria, scrub typhus, hookworm, dysentery and tropical ulcers. However, in the fairly quiet days of peace, preventive measures in both the civil and native populations kept the area reasonably healthy.

The DGMS, MajGen Downes, visited the Moresby and Rabaul areas frequently to upgrade army hospital and medical services, in conjunction with improvement of facilities by the civilian authorities. The RAAF also provided medical facilities for its personnel. As more reinforcements arrived during the latter months of 1941, the incidence of malaria increased. It was becoming apparent that there would need to be a unified approach between the civil and military authorities to control this disease. After the attack on Pearl Harbour by the Japanese on December 7, 1941 events moved rapidly in Moresby. The 8 Military District was expanded to include the 30 Brigade, the 39, 49 and 53 Battalions, plus supporting medical units. The fall of Singapore on 15 February, the loss of 8 Division, the disastrous Japanese raid on Darwin on 19 February, air-raids over Port Moresby and the loss of Rabaul were all pointers to the need to halt the Japanese in New Guinea. At this point the civil administration was suspended in Papua and New Guinea, which meant that among other things, all medical matters came under military control. The main army hospital at Murray Barracks was transferred 21 miles up the Laloki River to become the 46 Camp Hospital.(69) This was supplemented by the Red Cross Convalescent Depot and the development of facilities by the 3 Field Ambulance. With the invasion of northern New Guinea at Lae and Salamaua, and the consequent threats to Moresby it was decided to withdraw the 20 nurses from Murray Barracks and 46 Camp Hospital to Australia. Authorities were still conscious of the tragic loss of nurses in Singapore and Malaya. Some idea of the fears of the authorities are contained in a letter from Sister G. Muller who related her experiences at this time. Sister B. Alfredson wrote in similar vein.

> Sister Tinney and I were sent up to Port Moresby on a small ship called the *Malata*. There were about 60 troops on board but no medical officer. We were to relieve six Sisters who had been there some time, at Murray Barracks. Some time later 14 more Sisters arrived and we were billeted in houses at Ella Beach. We were there for the first bombing raid on Moresby, 3 Feb, 1942. When things looked dangerous we were evacuated back to Greenslopes for a time. Then three of us, Matron Bonner, Sister Campbell and myself were sent back, but again were evacuated at the time of the Coral Sea Battle. Then 8 of us were sent back to join the 105 CCS with LtCol Swinburne.(70)

Events during 1942 appeared to be moving towards some dramatic climax. As it happened, most events were in favour of the allies — in May, the Japanese invasion force aimed at Port Moresby was defeated in the battle of the Coral Sea, in September the attempted Japanese landing at Milne Bay was repulsed and the Japanese attack over the Owen Stanleys was halted almost within sight of Port Moresby at Ioribaiwa. These were desperate days in New Guinea, the story of the Kokoda Trail having now passed into the annals of Australian history.

2/9 AGH Port Moresby(71)

As casualties mounted in New Guinea it was obvious that medical facilities in the Moresby area would have to be reinforced and upgraded. A general hospital was urgently needed, so the 2/9 AGH was brought up from South Australia in August 1942. The site selected at 'Seventeen-Mile', presented many problems for the erection of a tented hospital. The undulating site, surrounded by eroded hills and gullies was a natural reservoir when the heavy rains came. It was anticipated that the hospital would be connected to a piped water supply, but this did not happen for many weeks. Despite all these difficulties the hospital was admitting patients within ten days, at the rate of 46 patients per day. Within three weeks 605 beds were equipped and 560 occupied. This was a remarkable achievement by the engineers, by the men of the unit and by the physicians and surgeons. Surgical work was performed under great difficulties as it was several months before an adequate theatre was erected. The physicians also faced major problems in handling

AANS Brisbane Dec 24, 1941. Sent to Port Moresby. Evacuated to Australia after fall of Singapore and Rabaul.
Back Row: A. Woolley, E. McCready, G. Pollock, G. Kirwan.
Middle Row: Matron A. Bonner (centre) A. Carey, A. Bleney.
Front Row: B. Alfredson, M. Henry, R. Campbell, M. Haworth, M. Luke, D. Harrison.

malaria and dysentery cases, even among the hospital staff. Sanitation was unsatisfactory, flies were a constant menance and much had to be done to combat the mosquito problem. By the end of September, 1942, the establishment of the hospital was increased to 800 beds and later 1,200.

The most extraordinary fact surrounding this rapid development was that it was done without any members of the AANS! It was still thought that the military situation was too unstable to risk bringing up the nurses, so from August to 29 October the nursing work had to be carried out by orderlies. While these carried out their duties with goodwill but little training, it threw added responsibility on the medical staff to carry out procedures normally performed by nurses. Undoubtedly the presence of nursing staff would have meant a more efficient hospital and probably would have hastened patient recovery. The 68 nurses who arrived on 29 October found that life at 2/9 AGH was far from easy.[72] These nurses, under Matron N.M. Marshall were posted to 2/9 AGH at 'Seventeen Mile' while 4 nurses under Sister M.L. Dunman went to 5 CCS (later 105 CCS).

With the increase of other medical units, the 46 Camp Hospital, the original base hospital of Moresby, returned early in 1943 and took over the role of a convalescent hospital. It set up such camps at Rouna Falls and at Koitaki, both in established buildings. Unfortunately, as the military situation deteriorated and the Japanese approached the area the hospital had to be withdrawn closer to the Moresby area, at Three Mile (Kila). The 113 Convalescent Depot, one of the earliest established in New Guinea, became a major base in the Sogeri Valley, but it too was evacuated when the Japanese threatened to overrun the area. As the situation improved the Camp Hospital was re-located in the hills at a spot known as King's Hollow. There they received not only army casualties but men and women from the RAAF and even civilians. Matron M.M. Mullane and a small group of nurses worked hard and long under very primitive

2/9 AGH Seventeen Mile, Port Moresby 1943. (Photo No 53199).

conditions but their work was lightened somewhat by the service of native boys, the only hospital in the area at that time to have such assistance. Their mess was also improved by the addition of furniture from wrecked civilian homes and by an iced water fountain last seen in a well-known Moresby hotel. Early in 1943 the nursing staff was supplemented by a group of AAMWS, the first to see service in New Guinea.

The move of AANS staff to tropical areas necessitated a major change in their traditional uniform to cope with the problems of malaria. It consisted of a grey cotton safari jacket, drill slacks, brown boots and gaiters, a khaki slouch hat with the AANS band instead of a puggaree and a grey cotton beret for use in the wards. This jungle uniform was a far cry from the ankle length dresses of AANS in the First World War!

It was no picnic for the nurses in the grim months of 1943. It required not only strength of body but firmness of will to carry on, as the war swirled about them in New Guinea. Sister M.E. Carseldine of 2/9 AGH, described in vivid terms her impressions of nursing in a surgical ward at that time.

> There were 44 beds in the ward, in these were the seriously wounded, while on ambulance litters under the beds were the not so ill. There was one orderly and myself and thankfully some walking wounded who helped. The theatres worked all night and on moonlight nights the wards would be in darkness. Once when I got to the ward entrance the stench would hit me, my stomach would do a flip and I would think I cannot go in there tonight. Then I would say please God give me strength for if I do not go in there who will. Once inside your own feelings were forgotten and if you had time after midnight you would eat supper in the midst of it all. Supper for the 28 nights consisted of tinned tomatoes on toast (burnt bread made by putting bread on a fly swat over a primus), always made by a walking wounded.[73]

The rain and the mud in and about the 2/9 AGH at 'Seventeen Mile' were constant discomforts for the Sisters. They ploughed their way in heavy gumboots to the wards, often carrying oxygen cylinders or other equipment. They squelched their way to the mess or to their quarters where picturesque grass huts seldom stopped the rain from blowing in on every side. Rats, scorpions, insects and vermin of all kinds were frequent visitors. After dark, mosquitoes swarmed in, necessitating rigid personnel discipline in dress and in taking the requisite dose of atebrin. At this time there was a heavy influx of casualties from the Kokoda Trail, with the medical and nursing staff at 105 CCS and 2/9 AGH working round the clock to treat the wounded. The CO of 105 CCS, LtCol T.G. Swinburne reported that many lives were saved due to the excellent work of the AANS staff.

Despite all these difficulties the Sisters always tried to make their quarters homely, just as their mothers and aunts had done in Heliopolis a generation before. Gardening was a popular pastime, as everything seemed to grow so well in the heat and the rain. Sister Carseldine recalled the problem her tent-mate created in attempting to grow roses.

> My tent mate was keen to grow roses. Some Nuns from Yule Island who were patients in my ward somehow managed to get her some rose bushes. One day there was a frightful odour round our tent. (We had on a previous occasion found a rat drowned in a tin of ablution water). Next day the smell was worse, but Scotty our batmen couldn't find the cause. When I said we'd have to report it he made a confession. Blood from discontinued transfusions was being used to help the roses along! The soil had to be removed and the roses never did flourish!

The tragic loss of so many members of the AANS taken into captivity by the Japanese had a profound effect on the Australian public. It reinforced political and military policy that nurses should not again be sent to areas where death, injury or capture was possible. When the Japanese advance in New Guinea threatened to overrun Port Moresby the nurses were withdrawn to the mainland, even though their presence was urgently required. As the war situation stabilized they were returned to Port Moresby but while the troops slogged it out with the Japanese for the next two years, nurses remained at the base hospital. It is true that there they performed under trying conditions, frequently under air attack. The treatment of casualties, however, in the initial Japanese attack over the Owen Stanleys and in the counter-attack back over the Kokoda Trail was the responsibility of other medical personnel. The RMO's, the Fd Ambs, the CCS's had a remarkable record of service during those torrid campaigns. Occasionally surgical teams from the AGH's were sent forward, consisting of a medical officer and some male orderlies. They, too, were able to treat casualties close to the front line. Even when the CCS's went forward the nurses were invariably left behind. Not until it was clear that the Japanese were defeated in New Guinea were the nurses allowed to join medical units outside the Moresby area. In retrospect, it is clear that their skills could have been used more effectively in the New Guinea campaigns, especially in surgical teams and in nursing malaria cases nearer to the scene of action. Such a policy might well have saved more lives. As it was, this valuable asset languished in the Port Moresby area for at least two years. This policy created further problems. Male orderlies, even though some of them did first-class work, could not be compared with trained nurses, skilled and experienced with many years in their profession. Moreover when the nurses were eventually returned to their units, they found that some orderlies were most unhappy that they were being replaced (and downgraded in status and salary).

This over-protection of nurses resulted in the development of other practices which were undesirable by any standards. Examples have been given of nurses being armed, contrary to the Geneva Convention. True, they were desperate days, but once they were armed, either to protect themselves or their patients they had to forego any protection under the Red Cross. It was a dangerous practice, which had it been known to the Japanese, could have had disastrous results. When the 2/12 AGH was at Warwick, for example, preparing for overseas movement to an undisclosed place in the SWPA they were given lessons not only in unarmed combat but in pistol shooting as well. They were issued with pistols in holsters and taken to a quarry for target practice.

2/1 CCS Milne Bay(74)

The historic action at Milne Bay in which the Japanese suffered their first major land reverse at the hands of the Australian troops was also noteworthy in that no members of the AANS were directly involved in those medical units allocated to the campaign. As early as June 1942 Milne Force had begun the task of fortifying the area to deny possible use of the base to the Japanese. The 11 Fd Amb and later 2/5 Fd Amb found that malaria was almost out of control, that personal hygiene was appalling and that the shocking weather aided the spread of disease. The arrival of the 110 CCS, 'a whittled-down 600 bed hospital', assisted by an American station hospital, was a blessing at this stage, but there was disappointment that the nurses had been left behind. There was still a feeling that the nurses should not be placed in positions of danger. As it happened, battle casualties were relatively light but malaria casualties were extremely high. It also happened that evacuation out of Milne Bay area was extremely difficult for a time and the above medical units were virtually acting as general hospitals, holding large numbers of patients. The presence of skilled nurses, particularly in the demanding task of nursing malaria patients would undoubtedly have provided more efficient medical treatment.

Some AANS were involved, however, on the *Manunda* in its precarious attempt to evacuate patients. It had left Brisbane on 2 September, 1942, arriving in Milne Bay four days later. During the night a Japanese cruiser and destroyer entered the harbour, flashed their searchlights on the *Manunda*, but miraculously left the ship untouched as it shelled the wharf area and sank a nearby vessel. The following night the same events occurred and next day a bomb landed close to the *Manunda*, which then left for Port Moresby.

2/2 CCS Port Moresby[(75)]

The 2/2 CCS with a record of service in the Middle East, Tobruk and Java behind them arrived in Port Moresby on 2 October 1942, with LtCol J.H. Stubbe in command. They established themselves at Koitaki in the plantation buildings vacated by the 46 Camp Hospital and within days this medical unit was frantically busy with casualties from the Kokoda Trail. The nurses, still under the command of Queensland Sister Vi Paterson did not arrive until November, but they were a welcome sight as the nursing demands were heavy indeed. They lived in tents in the rubber plantation with few amenities and amidst many dangers. Not the least of these were the vermin, rats, snakes, scorpions and even the occasional wandering cow. Queenslander Sister M.A. Marshall had an unnerving experience when a rat ran up the leg of her boiler suit and scampered round and round her body before exiting through the neck of her suit! As the Japanese were pushed back over the Owen Stanleys to the coastal towns, various surgical teams from 2/2 CCS, usually a surgeon and two or three male orderlies — but never nurses — provided urgent medical treatment close to the scene of action.

An accurate account of the conditions at Koitaki is provided by Sister Vi Paterson.

> The homestead was on a hilltop surrounded by rubber trees and jungle, with the misty blue Owen Stanley ranges to our north. A hessian fence on one cleared hillside enclosed the Sisters' lines — 2 tents for sleeping, one for mess and one for ablutions. We soon had all "mod cons", such as a bath in the tent, a wood copper outside and two dressing tables constructed by the natives. We also had a hessian enclosed cold shower with duckboard and oozing mud floor . . . Most afternoons and evenings there were tropical downpours and we slithered in mud as we worked in gum boots. Falls were an occupational hazard we accepted and expected. We worked twelve hour shifts 8.30–8.30 with time off for meals, to change into boiler suits and to draw our mosquito nets before dusk.

While at Koitaki the 2/2 CCS found one of their patients to be Sapper H.E. Beros, RAE who later achieved fame as the author of the poem the Fuzzy Wuzzy Angels of Kokoda Track. Included in his book of poems was one entitled Aussie Angels, which he dedicated to Sister Mary Wallace of 2/2 CCS.[(76)] It concluded with the lines

> For months no sight of women for these gaunt and hungry men,
> Now these bonnie nurses, eyes alight with mirth;
> We forget the war awhile, and things like Tommy Gun and Bren,
> And we thank God we have women on this earth.

47 Aust Camp Hospital[(77)]

A medical unit which does not appear in the official histories, one staffed mainly by Queenslanders, and which figured largely in New Guinea medical arrangements during 1943 and 1944 was the 47 ACH. It was originally established as a tent hospital in the grounds of the Goondiwindi District Hospital under Matron

Tent wards 2/2 CCS. Koitaki PNG. 1943.

47 ACH Koitaki PNG. 1944. AWM 59752

McMahon H.E.M. and later moved to the homestead of Walter Gunn at 'Kildonan'. Then it was divided with half going to Glen Innes and the remainder reformed under Queenslander Matron Cahill, K.

In July 1943 this unit was reformed at Chermside, with Major M.O. Kent-Hughes as OC; although female personnel were still held at Yeronga. It embarked on the *James W. Grimes* for Port Moresby in August, where Matron H. McMahon and 19 Sisters took over from the 2/2 CCS at Koitaki. The OC was very critical of the marquees used for accommodation as many of them had seen long service in the ME. The Sisters worked long hours in primitive conditions, with a heavy bed state of 258, very high for a Camp Hospital. Most of the patients were medical, with malaria, scrub typhus and dysentery cases requiring constant and heavy nursing. The 47 ACH had a remarkable record of service for a small unit until it returned to the mainland in July 1944. Many Queensland nurses saw service with this unit in New Guinea and subsequently in Cairns and Atherton.

5 CCS (105 CCS) Port Moresby(78)

Another medical unit in which nurses were to play a prominent part was the 5 CCS, formerly the 105 CCS, which arrived in Moresby in June 1942, the first medical unit in New Guinea to be staffed with experienced Middle East staff. It set up at 'Seventeen Mile', with access to an operating theatre at Murray Barracks. At this time it had to function as a hospital, in the absence of an AGH. Its major contribution in these early months was the initiative taken by the Commanding Officer, LtCol G. Swinburne in fighting the malaria epidemic. At this stage it was estimated that 50 per cent of troops were infected, with the rate even higher in some units. However, the nurses were not sent up from the mainland until 29 October when 4 nurses under Sister M.L. Dunman left Sydney on the *Manunda* and in November, 4 more joined in Brisbane to complete the establishment. This was the period when casualties from Kokoda began to arrive in large numbers. The CO recorded that many lives were saved because of the work of AANS members at this time. Nevertheless it must be noted that when surgical teams were sent out into the field with surgeons and male orderlies, no nurses accompanied them.

2/5 AGH Port Moresby(79)

By the end of 1942 the medical units in Port Moresby area were barely coping with the number of casualties. The 2/9 AGH had over 2,000 patients, the 2/2 CCS was constantly at a peak, the 46 Camp Hospital could take no more, while the Field Ambulances were frequently jammed full with patients awaiting evacuation. Most of the problems arose from malaria and the uncertainty of transport back to Australia. Further, there was the order that such cases should be returned to their unit as soon as possible. Some idea of the immensity of the problem may be gauged from Brigadier Disher's estimate that 100 per cent of the men at Milne Bay and at Buna had been infected. When the Sananada campaign came to an end early in January 1943, the Australian battle casualties numbered 6,154 but those due to malaria totalled 21,600. Even this was really under-estimated as it took no account of those who subsequently developed the disease on their return to Australia.

It was in this situation that it was decided to bring the 2/5 AGH to Moresby in January, 1943. Under Matron J.L. Cook it set up near Bootless Inlet. It too was an experienced unit, having seen service in the Middle East, Gaza, Greece, Eritrea and more recently at Armidale (NSW).

2/1 AGH Port Moresby(80)

Another very experienced general hospital in the 2/1 AGH was brought forward to Moresby on 8 September, 1943, presumably in anticipation of a forward move, as the allied advance into the islands of the South-West Pacific continued. Matron Hurley and the Sisters had to work with 2/5 AGH until their hospital was ready. They arrived the day after a Liberator bomber in taking off had crashed into members of the 2/33 Btn waiting at the airstrip. It was a tragic introduction to New Guinea for them.

Huon Peninsula Campaign

The mopping up operation against the Japanese in the Huon Peninsula and later up the northern coast of New Guinea to the Dutch New Guinea border took longer than expected, more because of the difficult terrain than because of any fierce resistance by the Japanese. The campaign had begun in September, 1943 with the capture of Lae and later Finschhafen, but not until April 1944 did the Australian forces enter

Madang. Alexishafen was the next to fall (April 44) enabling the 2/15 Fd Amb to set up a small detachment there. As the forces moved successfully up the coast, so medical units moved up behind them. Fortunately casualties were light and the work of the 11 Malaria Control Unit kept malaria cases to a minimum.

In so far as the medical arrangements were concerned in this Huon Peninsula campaign there was indecision and inconsistency on the part of army authorities whether nurses should be sent forward, particularly with casualty clearing stations. The nursing situation was exacerbated by confusion as to the respective roles of casualty clearing stations and general hospitals especially when evacuation problems caused a large build up in patients at CCS's. A small section of the 2/3 CCS (without nurses) took part in the landing at Lae. When it moved on, the 106 CCS (with nurses) moved in, its bed state rising to 1,100 at one stage during the last few months of 1943. In January 1944 advance units of the 2/7 AGH (but not the Sisters) moved up from Buna to Lae where the hospital was set up on the banks of the Busu River. The following month the nurses, under Matron E.F. Johns arrived. Meanwhile Finschhafen had been taken and the 2/3 CCS (minus the nurses) set up at Sinbang. Early in January 1944 the unit moved to Heldsbach and the nurses, under Sister J.M. Langham rejoined the unit. When the rest of the unit returned to Australia in March, the Sisters remained, being attached to 106 CCS.

At the end of May the 111 CCS arrived at Madang, but by then front line troops were at Hansa Bay, half-way to Wewak. It was on this stretch that the men contracted scrub typhus at an alarming rate. These were evacuated by sea to an ADS at Alexishafen and some to the 111 CCS at Madang. The ADMS made an urgent call for nurses to be sent to Alexishafen where the old mission station was taken over and converted into wards. LtCol K.B. Noad and three members of the AANS, Sisters T.L. Umpherstone, K.E. Lawrence and Queenslander F.E. Stevenson, were flown up to take over this critical assignment. At this stage they were further forward than any other Australian nurses in New Guinea. Reporting this on the ABC an officer stated that the presence of the nurses had probably made all the difference to the patients' chances of recovery.

2/11 AGH Buna, Lae, Madang, Aitape.[81]

After their sojourn in Queensland where they were part of the line of evacuation nursing casualties from New Guinea, Sisters from the 2/11 AGH were thrilled to learn that they were being posted to Buna, where members of their unit had already established the hospital in July–August 1943. They were taken up by DC 3 troop carrier, a new, even if at times unnerving experience for them. Sister M. Goldsmith described the event:

> The trip over the Owen Stanleys was wonderful and the views from the plane magnificent. On arrival at the airfield we were astounded at the number and size of the airstrips which stretched for miles, and the sight of the bombers and other planes lined up was most impressive. Because of the likely danger of air raids the wards were scattered over a large area and the Sisters were fortunate in having a "jeep" service to transport them. For the first few weeeks we were extremely busy receiving convoys from the battle areas, the majority of them taking two or three days to reach us, travelling by barge and by air. These included many cases of scrub typhus, most of them very sick and requiring constant nursing before evacuation to Moresby.
>
> One of our interesting occupations in our spare time was horse-riding, using the hacks left behind by the Japanese. Their condition improved on a diet of army biscuits, rice, dehydrated carrots and kunai. In time we even organized "picnic race meetings", which were very popular with the troops and patients.

A similar pattern emerged at Madang where the 111 CCS arrived at the end of March 1944 but the nurses under Sister E.M. Moriarty were not sent forward for another month. In August the CCS moved on to Alexishafen but three nurses had already been sent forward. The 2/11 AGH, after staging at Lae, took over from the 111 CCS in Madang in August and soon afterwards its nurses arrived, among them Sister Oakes, of Brisbane.

In January 1945 the 2/11 AGH which had seen long service in New Guinea, from Buna (1943) and Madang (1944) set up in Aitape. The site selected was not the best, for following torrential rain the nearby river flooded the camp site. While the male members filled sandbags, nurses set to work with shovels to

AANS New Guinea 1944. Visit by the Hon 'Eddie' Ward, Minister for External Territories. L. to R: Sisters Elms, McCallum, Bedford, Cottell, Bohm, Retschlag with 'Eddie' Ward.

dig drains to divert the water. Sisters in this unit regarded themselves as 'old campaigners' after five years in the army, ready to turn their hands to meet any emergency.

104 CCS Cape Wom, Aitape[82]

In the campaign to clear the Japanese from the coastal areas of New Guinea, the American forces by-passed Wewak and in April 1944 made simultaneous landings at Aitape and Hollandia across the border in Dutch New Guinea. In October the Australian 6 Division took over the responsibility for the area in which it was estimated 20,000 Japanese troops were cornered. This Division was supported by medical units, including Fd Ambs and the 104 CCS which arrived on the *Katoomba* on 12 November 1944. This was noteworthy in that for the first time nurses travelled with the CCS. Among the 8 nurses at the CCS were Queenslanders Murphy, M.M. (Brisbane), O'Neill, M.N. (Brisbane), Murch, K.J. (Cloncurry), Bowman S.B., (Boonah) and Leahy C., (Wallumbilla). The CCS was staioned at Cape Wom, while across the waters of the Bismarck Sea could be seen the Island of Muschu, still in Japanese hands. Battle casualties were heavy and there were problems at the CCS owing to the difficulties of evacuation by air. The six Sisters came face to face with the difficulties of being in the front line. Their quarters were 'two man' tents, equipped with folding canvas stretchers, set up on hard coral, close to the water's edge. Sister O'Neill found herself nursing scrub typhus patients, Sister Leahy the surgical. At this stage of the war, it was the responsibility of the nurses to collect and give blood under very primitive conditions.

Dr Walker paid a warm tribute to the dedicated work of the nurses in this campaign, particularly those of the 111 CCS in nursing scrub typhus patients.

Throughout the island campaigns the service of the nurses was particularly valuable where there were large numbers of cases of scrub typhus. Patients with this disease required constant attention, and before the arrival of the nurses in New Guinea a high percentage had died. Very often a nurse after being on duty all day would continue to work throughout the night, as she knew that the night sister could not give the scrub typhus cases the necessary attention and also perform her normal tasks in the ward. This devotion to duty was repaid by a steep decline in the mortality rate.[83]

With the war in New Guinea and the islands definitely swinging in favour of the allies, the consequent re-deposition of troops called for a reorganization of Australian General Hospitals in New Guinea. In March, 1944 LtGen Morshead, Commander, New Guinea Force issued a directive on this matter. The introduction of AAMWS released male personnel, while the reduction in casualty figures called for fewer hospital beds. The hospitals affected and the result in terms of AANS are shown in this table.

Table 4

Hospital Changes New Guinea 1944

	Beds	**AANS**	**Beds**	**AANS**
2/1 AGH	1,800	141	1,200	122
2/5 AGH	1,800	151	1,200	122
2/7 AGH	600	74	1,200	120
2/8 AGH	200	26	600	75
2/9 AGH	1,800	140	1,200	122
2/11 AGH	1,200	74	600	75
128 AGH	600	—	600	75

In so far as the AANS were concerned this meant a deal of cross-posting to other units, within New Guinea.[84]

Bougainville (Torokina)[85]

The allied victories in New Guinea during 1943 and early 1944 enabled bases to be established as a springboard for re-taking other islands to the north. The American forces, looking to the Philippines, had occupied the outer islands of the Solomons in March 1944 and soon afterwards established a base at Torokina on Bougainville. As Australian forces took over the responsibility of reducing the Japanese forces on the island in October 1944, they were faced with several campaigns, as each side of the island was divided by high mountains down the spine. The Torokina base had to be protected, while Japanese forces in both the north and south of the island had to be overcome. While Field Ambulances and ancillary medical units bore the initial impact three larger units were soon established — the 109 CCS which arrived at Torokina in October, 1944, the 106 CCS and 2/1 AGH which opened on 10 January, 1945. Nurses with the hospital were soon very busy and by March the establishment was increased to 1,200 beds. While battle casualties were relatively light, tropical diseases soon took their toll. Some confusion arose between 106 CCS, 109 CCS and 2/1 AGH as to the most appropriate line of evacuation. Some improvisation also became necessary in evacuating patients to Torokina, because of the geographical nature of the country. Jeep ambulances, native carriers and river transport were necessary. Where possible light aircraft were also pressed into action, particularly when heavy rain made roads impassible.

Matron Hurley and the Sisters soon adapted themselves to the environment of Torokina. Australian initiative showed out when 100 fowls were flown up by the Red Cross and the nurses set up their own fowl yard. Fresh eggs for the patients were indeed a luxury but the fowls had to be guarded carefully against surprise raids. Another innovation was to have an ice-cream machine brought up to provide ice cream for 100 patients a day. Dr Walker recalled another episode, which had amusing results:

> The nurses tried to solve their laundering problems by teaching the native girls to wash and iron, an ambitious project as none of their pupils had previously seen an electric iron. After a day's tuition the girls would remove their Mother Hubbards and set off into the jungle in their grass skirts, over which they would sometimes tastefully pull on a pair of worn-out step ins, salvaged from the garbage.

The security of the nurses was always a problem in any camp with thousands of allied troops who had been under battle stress for many months during which time they had had no contact with white women. At Torokina, for example, the nurses' quarters were protected by a 15 foot (5 m) barbed wire fence and a 24 hour armed guard. The Sister on night duty always had an armed escort. As in several camps in New Guinea, the American negro soldiers were a constant threat. Queensland Sister E.M. Allen described her impressions:[86]

> It was a tented hospital surrounded by dense jungle and flanked by two volcanoes. We were taken by LST from the ship to the beach and this time accommodated in huts with two sisters to each cubicle. During the next few weeks, all beds were occupied with casualties coming in from the forward CCS. With only 25 Sisters, it was a very busy time until the next ship arrived with the remainder of the unit personnel. As we were confined to a small strip of allied territory and it was feared that Jap infiltrators might penetrate the hospital lines, the Sister in charge of each ward was issued with a rifle, which seemed to us rather amusing at the time. (I stood mine behind the fridge). Fortunately we never had to use it.

When the Australian Sisters arrived in Bougainville at the end of 1944 one of the first signs they saw was a notice from the previous American occupants. 'On May 12, 1944 a WHITE WOMAN ate here'. This was apparently a big event in the lives of the all male American army, especially as the woman happened to be film actress, Carole Landis! The AANS were the first women of the services ever to come to Bougainville and they soon added their distinctive touch to soften the jungle around them. One of the traditions which grew up in war-time was that no matter where the hospital was establsihed, there must be flowers and gardens, not only to brighten the lives of staff and patients, but as a therapeutic aid to all in the camp. All this had a profound effect on the morale of men just back from the Jaba swamps or the

After the concert party at 2/8 AGH Jacquinot Bay, 1945. L. to R: Matron Wheeler, Matron E. McQuade White, Gracie Fields, Col. Sage.

The Duke of Gloucester visits 2/8 AGH at Jacquinot Bay, New Britain 1945.
L. to R: LtGen Robertson, Matron Wheeler, Duke of Gloucester, Sisters Weetman, Lunisden (partly hidden), McLennon, Jensen, Fowler, Campbell, Peacock, Gillies.

heart-breaking hills of the Numa Numa Trail. One report summed up the effect of the nurses' presence on patient recovery, thus:

> It is in the wards that you notice the greatest change. Flowers and a quiet feminine voice bring back the sane world to men just out of the line. In one of the surgical wards at the present time there lies a man shockingly wounded by Japanese fire. His morale is something to wonder at. It is not too much to attribute this high courage largely to the gentle, sympathetic treatment of the Sisters who are busy at the bedside all day long, chatting to him about the home he will return to soon.(87)

New Britain. Jacquinot Bay(88)

Australian army nurses attached to the 105 CCS and 2/8 AGH were involved in the campaign which freed New Britain in 1944–1945 and eventually led to the capture of Rabaul from the Japanese. Earlier, American forces had landed on the island and towards the end of 1944 Australian forces had taken over from them. As there were an estimated 40,000 Japanese on the island protecting the valuable and strategic base of Rabaul, it was anticipated that there would be a stiff campaign ahead for the Australian troops. The Field Ambulances bore the initial brunt of casualties who were evacuated as quickly as possible by air to Lae for admission to 2/7 AGH. There was need for more medical assistance so 105 CCS was sent to Jacquinot Bay where it opened on 9 January 1945. Members of the AANS arrived in February, to be attached to 105 CCS.

Sisters Jean Gracey (Cooyar), Gertrude Muller (Harrisville) and Dorothea Harrison of Beaudesert were among the eight army nurses to land at Jacquinot Bay, New Britain, the first white women to return there since the beginning of the Japanese occupation in January, 1942. At the end of March the 2/8 AGH was moved from Buna to Jacquinot Bay.(89)

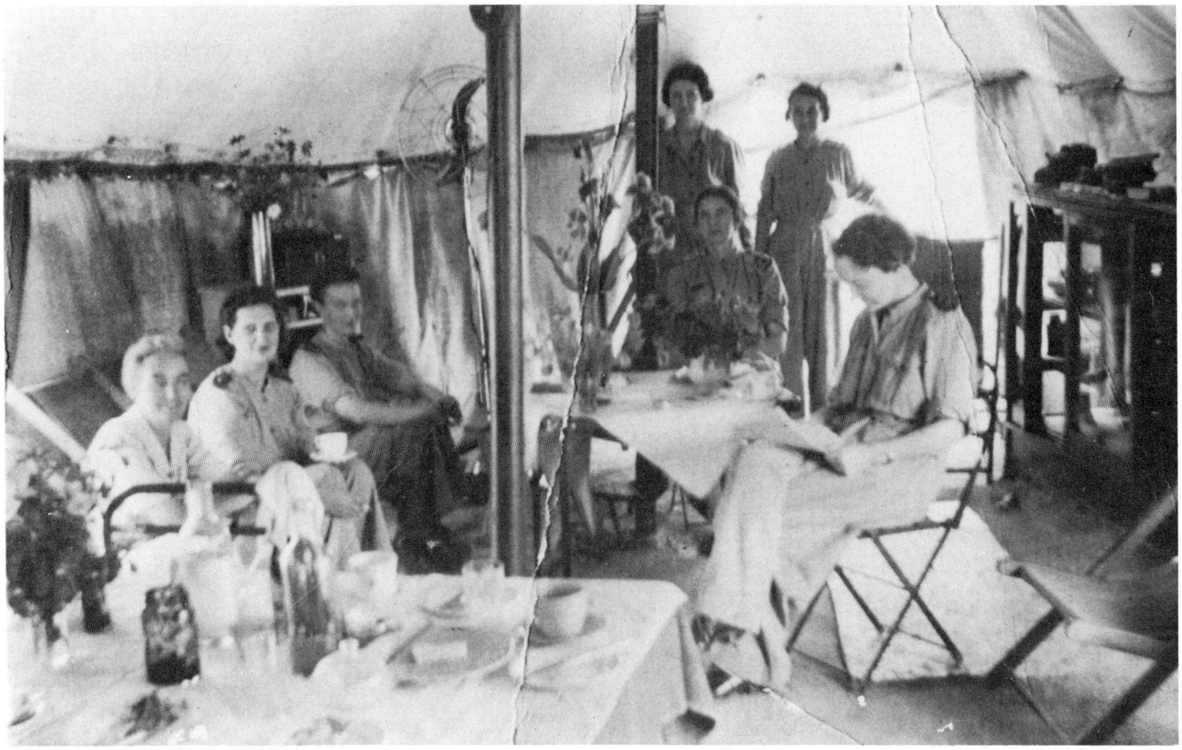

105 CCS Port Moresby 1945. L. to R: Sisters Gillanders, Muller, Harrison, Barnes, Damman, Richards and Leigh.

105 CCS Jacquinot Bay, 1943. L. to R: Sisters Dwyer, Harrison, Wilson, Gracey, Barnes, O'Loughlin, Muller and Smith.

Morotai, Borneo[(90)]

The final campaigns in which the AANS were involved were in Borneo and Labuan, but the major base known well to so many was Morotai, in the Moluccas, recaptured by the Americans in September, 1944. The holding units for these campaigns were the 2/4 AGH, 2/5 AGH, 2/6 AGH, 2/9 AGH, 2/12 AGH and the 2/1 CCS, 2/2 CCS, 2/3 CCS, and 110 CCS. In addition the 66 Camp Hospital in charge of Sister E.M. Mounsey and 5 nurses had arrived in Morotai in January 1945, followed by the Principal Matron Adv LHQ Matron E.J. Bowe. Nurses of the 2/5 AGH under Matron Cook and of the 110 CCS under Sister B.M. Millard provided staging facilities for those medical units about to embark for the Borneo campaigns.

One of the best known members of 2/5 AGH nursing staff was Sister D.A. Burnett, who had served with the hospital in Greece. For her outstanding service in the operating theatre in the New Guinea campaigns, and especially at Morotai, she was mentioned in despatches. (MID 1945).

Morotai was a cultural shock to the nurses for many of whom this was their first experience of an overseas campaign. It was the wet season when they arrived and conditions were extremely primitive. There were no roads, only slush and mud, in stark contrast to the tented wards among the coconut palms. The operating theatre was formed merely by four sheets dipped in disinfectant, with a table set up on a concrete slab, lit by an old shadow lamp. Water for sterilizing was obtained from primus stoves. Their tents were constantly in danger of flooding from high tides, so the legs of the beds were set in jam tins. It was quite an exercise to get out of the gumboots, undress and crawl under the mosquito nets. The nurses never let up on their military efficiency in so far as their uniforms were concerned. These were washed in salt water and pressed, either under the mattress or by whatever books or heavy objects were available.

The 2/6 AGH was another medical unit which 'staged' at Morotai on its way to other island campaigns. After waiting many months in Australia the unit eventually found itself on Morotai, housed by the 2/5 AGH. While the island itself was securely in allied hands, there were still nuisance bombing raids. Sister T.C. Troedson recalled those days:

> The Japs were still bombing the beaches and the alert was sounded by an antiquated gun which when fired at night meant everyone take cover. We were forced to take our helmets with us everywhere, not that there was anywhere to go. Because the Japs had not been cleared out, some of our Doctors got panicky and insisted that we learn to use pistols to protect ourselves. In spite of the Geneva Convention and the brassard, I thought this was a jolly good idea, as most of the Japs had apparently never heard of the Geneva Convention. Anyway I now know how to use a pistol.

Once again, the policy of not sending Sisters forward with the initial forces was followed. In the Tarakan landing on 1 May, 1945, 110 CCS provided medical support, but the nurses remained at Morotai with 2/5 AGH. On 10 June landings were made at Brunei Bay and Labuan Island, with 2/1 CCS ashore on the day of the landings. (The nurses arrived two weeks later). The 2/4 AGH advance party went ashore on 16 June and the hospital, with its nurses, was fully operative on 16 July. The 2/6 AGH also arrived in Labuan at this same time. Meanwhile landings had been made at Balikpapan on 1 July, with the 2/2 CCS and 2/3 CCS and a week later 2/12 AGH arrived. Nurses for all units came in by air or sea at the end of the month. At this stage, Australian and Allied troops realized that the Japs were in retreat. They envisaged many months, even years of island hopping, of bitter fighting, even an invasion of the Japanese mainland. There loomed ahead a long period of work for the medical units, with the probability of many casualties. It was therefore something of a shock that, following the dropping of some strange new bomb that inflicted massive damage and casualties, the Japanese unconditionally surrendered on 15 August 1945. Shock gave way to relief that at last the war was over and that the expected casualties would not occur.[(91)]

Life at Balikpapan was precarious for the first few weeks. Sister Kay was a member of 2/12 AGH who went to Balikpapan in the *Manunda*, which had to manoeuvre its way through the heavily mined waters of the Macassar Strait. Sister Kay related the experience of those times:

> As it was barely a week since the troops landed at Balikpapan there was still enemy action in the area. As night fell the Hospital Ship with lights ablaze was five miles off shore and during the evening sirens warned the approach of enemy aircraft. We were gathered below decks, with tin hats, emergency rations and water bottles to await the results of the air-raid. Fortunately the bombs fell between us and the shipping in the harbour. Next morning we disembarked

Hospital Ship, 'Manunda' at Lae (N.G.) 1944.

and were ferried ashore in the ship's lifeboats. As nursing Sisters found in other parts of the islands, their security from attack whether by the Japanese or by allied servicemen required constant surveillance. At Balikpapan the presence of Japanese in the surrounding jungle was a major threat to all in those first few weeks. Sisters of 2/12 AGH were housed in an area known to them as Chastity Hill, behind a ten foot high fence, constantly patrolled day and night. While Australian troops were firmly in control, the sound of 25 pounders and the crack of rifle fire were never absent, disturbing the sleep of nurses, by day and night. Sister Kay added,

> We were unable to leave our camp site unless we were accompanied by two other Sisters and three armed escorts, as this area was heavily booby-trapped and snipers were about in the hills and dug-outs.

RAAF Nursing Service(92)

Another feature of this campaign was the presence of medical staff and medical units of the RAAF and the RAN. Although the RAAF had no hospital in the area, RAAF personnel were nursed at their base and in their tents, although the more seriously ill were transferred to the army medical units, Fd Ambs or CCSs. Until this episode little had been heard of the RAAF Nursing Service.

The RAAF Nursing Service traced its origins back to a similar organization which had a long history with the Royal Air Force in Great Britain. Towards the end of World War I, with the development of the Royal Air Force, it was found necessary to establish a nursing service especially for air force casualties. By 1919 these Sisters had taken over from the army personnel who had been the original nursing staff. With the coming of peace and the development of the RAF between the wars, the RAF Nursing Service was established under Royal Warrant as a permanent branch of the RAF and in 1923 this service was honoured with the title, Princess Mary's Royal Air Force Nursing Service (PMRAFNS). For many years members enjoyed the status of officers but not until 1943 was commissioned rank granted.

At the outbreak of war all nursing in the RAAF was carried out by males, 'nursing orderlies' who had been recruited with some limited medical background and experience. When more serious medical or surgical treatment was required the patient was transferred to an army or civilian hospital and on occasions civilian nurses were brought into RAAF sick-quarters for emergency nursing situations, such as the epidemics of measles, mumps and influenza which swept through the major bases at Laverton, Point Cook and Richmond. With the rapid expansion of the RAAF in 1940 it became clear that a RAAF Nursing Service should be established, modelled on that of the AANS and of Princess Mary's Royal Air Force Nursing Service.

In July 1940 the Minister for Air, the Hon A. Fadden, announced the establishment of a RAAF Nursing Service, modelled on the lines of Princess Mary's RAFNS. IT would be an integral part of the RAAF but in the beginning would be staffed by Sisters seconded from the AANS and by civilian nurses. Miss M.I. Lang who had served at Salonica in the First World War was appointed Matron-in-Chief. RAAF Sisters were appointed to all major RAAF hospitals and sick quarters of RAAF establishments throughout Australia — in Queensland from Bowen to Amberley. Under RAAF administration Sister M.K. Doherty as Principal Matron, was responsible for the staffing in NSW and Queensland. Conditions of appointment

Visit by Governor-General's wife Lady Gowrie, Bundaberg 1943 — RAAF Base. Lady Gowrie speaking to F/O Eve Lear, W/0 Maud Woolcott and Sister W. Bruce.

Five of the first nursing sisters (RAAF) in Queensland watching pilots receiving their 'Wings' before being posted to combat active service.
Taken RAAF Base Amberley 1941.
L. to R: Sisters J. Buchanan (S.A.), W. Bruce (Qld.), A. Cleary (Qld.), J. Bacon (Qld.), A. Poland (Qld.).

were similar to those of the AANS, with commissioned rank, but not using RAAF titles. Their subsequent history included:

- Convoy duty with RAAF to the UK, Canada and South Africa.
- In Darwin during bombing raid.
- First posted to NG in Nov 42 and set up near Ioribaiwa.
- Part of MAETU.
- RAAF Medical Trg Unit.

With the development of air force stations in Queensland Matron E.E. Ward was appointed Matron No 2 Training Group, Brisbane on 21 June 1944.

RAAF Sisters served in all the major threatres of war, at home and abroad. At first they were posted to all the major air force bases throughout Australia, from Laverton to Fairbairn and Darwin. In Queensland, they soon became a familiar sight at Archerfield, Sandgate and Amberley and as the war front moved north to Townsville and the islands they too moved into forward bases. When the Empire Air Training Scheme began in 1940 members of the RAAF Nursing Service were part of the medical team allotted escort duties to Canada and Great Britain. Sisters W. Bruce and C. Rule were two of a number of Queensland Sisters involved in this work, while the Queensland *Daily Mail* of 7 Nov 1942 referred to Sister 'Katherine' Ida Rule, of Longreach as one of three RAAF Sisters who had returned from Canada. The local press also referred to Sisters E. Weber (Rockhampton), A.T. Cleary (Cambooya) and E.A. Hamilton (Mackay) as being among the first RAAF Nursing Service to arrive in New Guinea.[93]

One of the largest RAAF bases in Brisbane was at Sandgate which in the post-war years became the Eventide Home for the Aged. Sister Bruce related how one morning she saw an American fighter plane dive into Moreton Bay, not far from the base. Some of the patients rushed into the water to help rescue the crew, but in vain. That afternoon a friend of the dead pilot flew over the crash site and unfortunately he also crashed into the sea. RAAF Sisters while stationed in Australian bases were soon involved at first hand in the casualties of the war, receiving the Darwin casualties through Archerfield and the Centaur casualties through Greenslopes. One of the biggest scares was at Fairbairn soon after the Japanese break out from the POW Camp at Cowra. All staff were alerted, security tightened and machine guns mounted in strategic positions, as it was believed some of the Japanese were heading for Canberra to seize an aircraft. Fortunately for all concerned, none reached Fairbairn and after a few days it was business as usual at the hospital.

As far as is known, Queensland Sisters who served with the RAAF Nursing Service during the Second World War and with the RAF Nursing Service included: Armstrong, E.M., Bruce, W.J., Chandler, B.O., Cleary, A.T., Hamilton, E.A., Neilson, E., Richardson, I.B., Weber, E.M., Wroe, M. Bray, J. (PMRAFNS) Marlow, B., Bolger, H., MacAulay, E., Purtill, S., Rule, C., and Rule, L.(94)

The Medical Air Evacuation Transport Units(95)

This proved to be one of the most valuable contributions made by the RAAF medical services. It was modelled on the lines of the American system which had been operating in the SWPA for some time. The unit was formed at Sandgate on 28 February, 1944 in conjunction with US No 804 MAETU. A Sister and a medical orderly were attached to each transport plane responsible for the treatment and evacuation of sick and wounded of the three services, of Australian and allied forces. It was separate and distinct from RAAF ambulance units. Flt Lt (later SQn Ldr) Kiel, F.W. was in charge of the squadron, with Sister Kendrick, N.J. as the senior Sister. The original group of Sisters in this unit included Sister Wroe, M.H. (Qld), other members being Sisters Reed, N.J., McLeod, J., Mackenzie, L.G., Harbourd, V.E. and Scott, M.A. Subsequent members included Sister Chandler, B.O. (Qld) who was to play a significant role in rescuing POW nurses. A rigorous training programme was required, including ocean and jungle survival skills in case of forced landings or ditching at sea, the application of oxygen masks and problems of treating patients at high altitude. It proved to be dangerous work over the rough New Guinea terrain, a gruelling job indeed for nursing Sisters. Five lost their lives in this work, but none of them Queenslanders. They flew into areas where fighting was still going on to bring out the wounded. Later their task was to fly out the POW's. Sister Cleary, H.A. with Sister Braid, M.A. and Sister Wroe M.H. (Qld) as well as Sister Chandler, B. (Qld), were members of rescue teams who risked their lives to bring the prisoners back to safety. Sister Cleary described the risks they took both in the air and on land.(96)

> We were the first Australians to arrive in Singapore about 10 days after the surrender of the Japanese. The British were dismayed to find Sisters arriving and feared for our safety as there were still 40,000 Japanese roaming the island. They wanted us to return to our base at Morotai but we stayed. Accommodation was very primitive. We slept on the floor of a hotel on a blanket, with no light or water or toilet facilities. On the night of our arrival we met many Australian prisoners. They were hungry for news . . .
>
> Next morning we went out to Changi camp . . . it was impossible to realize all they had been through. It was our aim to get these prisoners back to Australia as soon as possible.
>
> We were flying every day, without normal navigational or radio aids . . . The planes were loaded beyond capacity. The Dakota had a loading capacity of 27 and we often had as many as 46 on board, but it seemed so hard to leave anyone behind, even for one day. Most of the evacuees were suffering from malnutrition and dysentery so the general conditions within the aircraft can be imagined. Practical nursing as we know it was impossible.

A fine tribute to the work of these Sisters was paid by an unknown writer in *RAAF Saga* (1944) when he referred to them as Nightingales of New Guinea.(97)

An unusual contribution to air force nursing was made by Sister Jean Bray, of Brisbane, who was in Britain at the outbreak of war in 1939. After trying unsuccessfully to join the QAIMNS or the AANS she was immediately involved in the evacuation of children from London. Soon afterwards she was accepted into Princess Mary's Royal Air Force Nursing Service. After serving in Britain she was sent to North Africa. Writing of her experiences there, Sister Bray commented;

> After Tunis we were stationed at Carthage, arriving there in the heat of a "Sirocco", the only shelter being the wings

of our plane. Soon a one ton truck arrived to take us to a convent which had been bombed out by the Germans, but we moved on to a large convent where we set up our hospital, No 1 RAF Hospital, CMF. It took some time to clean up the mess but sanitation of a kind was soon in order.

During her stay in Carthage, Sister Bray met many distinguished people, including Princess Mary and Lady Mountbatten. At one stage Winston Churchill was ill with pneumonia in Carthage and Sister Bray met Mrs Churchill, as she then was.

Sister Bray's unit was moved from Carthage to Italy in December, 1943 when it was established near Naples, in a building which was formerly a Fascist Convalescent Home. From this site the Sisters and patients had a remarkable view of Vesuvius particularly when it was capped with snow. In March the following year there were ominous signs that an eruption was imminent, something which had not happened since 1892. The Acting Matron takes up the story:

Ambulances had been assembled in readiness for evacuation if this sould be considered necessary and stretchers were prepared in the corridors for the helpless patients. Suddenly a most terrific explosion shook the whole neighbourhood and a loud continuous roar like an angry sea went on incessantly. The whole mountain appeared to be on fire and great red hot boulders were being ejected thousands of feet into the air with each explosion. As there was no improvement we had definite instructions to begin evacuating the stretcher cases within one hour. One by one ambulances drew up to the hospital door and were filled with our patients, each clutching his few belongings — one patient was still having his intravenous saline and a Sister walked beside him with the flask of saline held high as he was lifted up into the ambulance. By that afternoon all 360 patients had been distributed to different Army hospitals outside the danger zone.

Sister Bray and other members of Princess Mary's RAF Nursing Service were evacuated to a transit camp to await events, watching with some anxiety the magnificent sight of Vesuvius in eruption. After a month it was considered safe for all to return, but they found the district had changed remarkably. Grey volcanic ash covered all the fresh green leaves and fruit blossom, while buckets of ash had to be removed from the roofs of buildings. RAF Sisters had to cope with the vagaries of nature, apart from the effects of enemy warfare.

The Women's Royal Australian Naval Nursing Service[98]

The Milne Bay campaign was a reminder to other services that the Royal Australian Navy had its own medical service with facilities for the treatment of sick and wounded naval personnel. It also had its own nursing service, but this differed from that of the AANS and the RAAFNS in that traditionally the male sick berth staff had carried out the nursing role. This male sick berth rating had had much more training than the army nursing orderly, as he frequently had to undertake nursing duties on board ship, where it was often impracticable for female nurses to be appointed. Destroyers, corvettes and frigates, for example, carried sick berth attendants, male. However, this argument was not tenable in the case of shore based medical institutions for the navy. As the war developed and enrolments in the navy increased, additional hospital accommodation was needed. In 1940 the sick bay at Flinders Naval Depot (HMAS Cerebus) was enlarged into a naval hospital. So it was in Sydney (HMAS Penguin), Melbourne, Adelaide, Fremantle and Canberra. In Queensland, provision for sick-quarters was made in Brisbane, Townsville and Cairns. In Darwin HMAS Melville provided facilities initially but later the navy moved in to wards of the former hospital at Kahlin.

All these developments pointed to the need for a special nursing service for the navy, but only after considerable debate did this come about. In October 1942 the Women's Royal Australian Naval Nursing Service was established, with Miss Ina Laidlaw as the first Matron, an experienced nurse with the AANS in the 1914–1918 war. Conditions of service were based on those applicable to nurses in the army and air force, although there was a difference with respect to titles and rank. Nurses were given equivalent rank, for example a Matron had rank equivalent to that of Lieutenant-Commander. Their outdoor uniform was a traditional blue. Sisters in the WRAN Nursing Service soon found they were needed in all shore based medical establishments and in training sick berth attendants. They served with distinction in Australia and in the South West Pacific Area. In Milne Bay six Sisters were posted to the RAN hospital taken over from the RAAF. No member of the WRANNS served on sea-going ships, although on one occasion one Sister was posted to the *Manunda*. This may seem strange as logically this would seem to be an appropriate role for them. However the staff of Hospital Ships were not naval personnel but army.

Queenslanders who served with the Women's Royal Australian Naval Nursing Service included Sister Mary McKenzie who had trained in Roma and Sister Joan Conquest who later became Matron of Anzac Hostel, Brighton (Victoria). Sisters Margaret Saunders and Githia King were also listed as having enlisted in Queensland.[99]

For various reasons which need not be canvassed here, the Women's Royal Australian Naval Nursing Service was disbanded on 30 June, 1948.

The Surrender of the Japanese and the Recovery of POW's

The end of the war came with great suddenness. The Matron-in-Chief of the AANS, Colonel Sage, happened to be visiting 2/1 AGH at Torokina (Bougainville) at the time. Sister Hooper recalled the phone message for Colonel Sage after which the Matron-in-Chief broke the news to all that hostilities had ceased. There was stunned silence for a moment but then the reality of the news hit them all. Patients and staff rejoiced that the long ordeal of six years of war was over. The first thoughts of everyone were how long before they could get home, but there was much to be done before several hundred thousand troops could be returned to Australia. Life at the hospitals had to go on.

During these war years important contributions to morale were made by the Army Education Service and by the Amenities Service. Concert parties were always popular, so too were classical artists such as Isadore Goodman (pianist). Troops sat for hours, clad in tin helmet and gas capes, watching films on a distant screen, regardless of the pouring rain!

Sister Audrey Gray related the visit of Gracie Fields to 2/1 AGH at Torokina:

> The morning after the cessation of hostilities, ambulatory patients were transported out to the airfield to welcome the arrival of Gracie Fields. I was fortunate to be one of the Sisters accompanying them. After the plane landed and the door was opened, Gracie Fields stood in the entrance. She bid us good morning, then requested we sing "The Lord's Prayer" with her. The patients were thrilled, as she was dressed in full AIF uniform.
>
> Later in the morning, after Miss Fields had visited the wards and other parts of the hospital, morning tea was served in the Sisters' Mess. I must have been staring into space as Miss Fields leant over her table patting her hair and said "it's dyed love, it's dyed!"

The end of hostilities produced the same feeling of excitement in all medical establishments in Morotai, Borneo, Bougainville, New Guinea and in Australia. It was a time too, for many VIP's to visit the forward hospitals — Lady Blamey, Lady Mountbatten were there to meet the wounded and especially the recovered POW's. It was a time of historic ceremonies as Japanese army groups formally surrendered to allied commanders — in Tokyo Bay, Morotai, Rabaul, Torokina, Borneo, Labuan, Wewak and throughout the SWPA the procedure was the same: troops and senior officers lined up, the signing of the surrender document by the Japanese and the appropriate allied commander, the handing over of the Japanese general's sword. At all these ceremonies senior members of the AANS were invited to be present.

Sister Catherine Leahy, Sister Murphy and Sister Bowman were with 104 CCS at Cape Wom in the latter stages of the war. For about six weeks they were ferried daily to the ADS at the edge of the Wewak airstrip to tend seriously wounded men there until the Japs could be driven back far enough to make the airstrip safe for evacuation. With the surrender they were able to move to Wewak. Sisters Leahy and Murphy were two members of the AANS invited to witness the surrender ceremony of the Japanese at Wewak. On 13 September General Adachi signed the surrender before General Roberston and 3,000 troops, handed over his sword and the remaining Japanese laid down their arms. Sister Leahy described the ceremony in a letter:

> We were in the front row half way down the Strip. Major General Robertson with escort resplendent marched past and the official party were seated at a table. Then came Adachi looking five feet tall in baggy uniform with a ridiculous sword so long it was dragging on the ground most of the time. The Australians escorting him were all over six feet tall so splendid looking the contrast was farcical. As Adachi bowed low and laid his sword on the table I felt a twinge of pity . . .[100]

It so happened that later Adachi was charged with war crimes and sentenced to life imprisonment. Unable to face the shame of defeat and the loss of over 100,000 men in the New Guinea campaigns, he committed suicide while in detention at Rabaul.

Then came perhaps the greatest shock of all for the Nurses — the recovery and treatment of allied prisoners of war. The unexpected end of the war brought forward plans which had been made for such an eventuality, but the medical authorities had only an inkling of the enormity of the task ahead. True, in the island re-conquests of the previous months some small groups of allied POW's had been recovered, Sikhs, Chinese and Indians. The Indians were recovered in Borneo and in the words of the ADMS 'their condition was pitiful'. Malaria, dysentery, malnutrition, beriberi, tropical ulcers and a variety of diseases brought about by neglect, starvation, ill-treatment and poor hygiene had reduced these men to virtual skeletons. All medical units in the islands were placed on alert with the cessation of activities. The Australian army was made responsible for recovering prisoners from Rabaul, the Moluccas, Borneo, Timor and the Celebes. It was arranged that these should be brought to a central point at Morotai where they could be given appropriate medical treatment.

Many Australians were known to be POW's in Malaya and Singapore and the task of recovering these was the responsibility of Mountbatten's South-East Asia Command. At first it was planned to evacuate these, when rescued, through Labuan where the 2/4 AGH, the 2/6 AGH and the 2/1 CCS we were established. However, in view of the probable urgency of medical treatment the 2/14 AGH was rushed to Singapore.[101]

In August 1945 the 2/14 AGH at Townsville was re-organized in preparation for a mission in the islands to care for the prisoners of war likely to be released with the imminent defeat of the Japanese. As it happened, the unit embarked on 27 August on the H.T. *Duntroon* and disembarked at Singapore on the 13 September, setting up hospital in St Patrick's College, of bitter memory to those who experienced the

An historic photo Morotai 1945. Matron Bowe, General Blamey, Matron Hanrahan, LtGen Berryman.

last days prior to the fall of Singapore in February, 1942. Some of the Sisters had gone from the islands on the *Manunda* to Singapore where they arrived five or six days before the *Duntroon*. As the *Manunda* was the first ship to sail across from the islands, the mine fields had to be 'swept' before 'she' could proceed. While waiting for the *Duntroon*, the sisters spent their time visiting Kranji and Changi. At Kranji one of the sisters, Sr M. Setschell, who had been with the 2/13 AGH, was overcome when she was recognized by a soldier whom she nursed in 1941 and who thanked her for coming back. In a short time over 1,200 released POW's and internees were admitted, including those members of the AANS rescued from Sumatra. The CO of 2/14 AGH, Queenslander Col W.E.E. Langford, was fortunate in that despite the hurried reorganization he was able to gather an experienced staff of medical and nursing personnel. Matron M. Brown noted with pleasure the efficient way members of the AANS went about their task, especially tending their colleagues after their ordeal in Japanese prisoner of war camps. Among those she singled out for special commendation was Queenslander, Sister E. Doig who already had a long record of outstanding service with the AANS. By November the task allotted to 2/14 AGH had been completed and the unit sailed for home and demobilization.[102]

In New Britain, Rabaul became a centre for the collection and treatment of POW's. Nurses were flown in to assist the 4 Fd Amb in urgent nursing treatment, especially for several hundred Indian, as well as many female POW's, including nuns. Some of these patients were sent on to the 2/8 AGH at Jacquinot Bay. The work at Rabaul was so demanding that 105 CCS was also sent there. One of their first tasks was to set the Japanese to work cleaning up the filthy mess in the camps, in order to prevent epidemics.

Manila

At the end of August 1945, 24 Sisters were detached from army hospitals in Morotai and Borneo and attached to an American Medical Corps Unit in Manila in the Philippines. They worked with American personnel treating the rescued allied POW's. Sisters Ewing and Reid were involved in this work.

Another area where Australian POW's were recovered was at Kuching in Borneo. Medical and surgical teams, including nurses, from 2/4 AGH and 2/6 AGH went in small craft from Labuan up the river estuary to locate the prison camps. There they found the same story — hundreds of allied troops most suffering from the effects of malnutrition, plus malaria, dysentery and other illnesses. Many were too ill to move, all required skilful nursing if they were to survive. Nursing sisters, with their good nursing, good humour and sympathetic understanding played an important role in speeding the recovery of the sick and wounded POW's.

The Rescue of AANS POW's

Behind the dramatic rescue of the surviving members of the AANS captured in the Singapore tragedy was another story in which Queenslanders played a remarkable role. This time it concerned the RAAF Nursing Service and in particular Sister Beryl Chandler who had trained with some of the nurses in Brisbane. No 1 Medical Air Evacuation Transport Unit (MAETU) of the RAAF had been set up in 1944 and had established a fine record of service in New Guinea and the islands. At the end of the war three such units were operating in the recovery and evacuation of rescued POW's. Sister Chandler had made several such trips into Borneo during August and September, 1945 when she was suddenly despatched to Singapore. On the 14th September she found herself rostered, by a remarkable coincidence, to go out with a team 'to find the missing Australian Army Nursing Sisters', who were last heard of in Palembang, (Sumatra).[103]

Haydon Lennard, the ABC war correspondent in the area, had been pressing Brigadier J.E. Lloyd, to initiate a search for the nurses. In this first flight, Matron Sage, Matron-in-Chief of the AANS, was in Singapore and expressed a wish to go, but this was refused as it was considered to be too dangerous.

Sister Chandler was asked if in view of the probable medical need ahead, she would like a Medical Officer to accompany her. In her reply, she said:

> Now in all the time I had been engaged in Flying Duties I had never flown with a Medical Officer and had carried casualties from all parts of the SW Pacific Area and Borneo to the various Australian General Hospitals in New Guinea and Australia. As far as I know none of the original fifteen Flying Sisters of the RAAF ever needed Medical Officers aboard their aircraft. We had been trained for the work so the necessity for a doctor had never arisen.

As it happened the Medical Officer appointed was Major Harry Windsor, (AIF) well-known Brisbane cardiac surgeon. When the aircraft arrived at Palembang on 15 September it was immediately surrounded by 30 fully armed Japanese. It was some time before they were satisfied about the credentials of the team and the purpose of their visit. However, all enquiries as to the nurses' whereabouts drew a blank, but Haydon Lennard went off to follow up some new leads. In the meantime Sister Chandler had visited a POW camp and arranged for a plane load of patients to be evacuated to Singapore. Miss Sage was there to greet them but was extremely disappointed to find no nurses among them. While at Singapore Sister Chandler received a vague message from Haydon Lennard stating that he had located some Australian women POW's and that he was bringing them to Lahat. Preparations were made immediately for departure, even though the state of the airfield at Lahat was unknown. On this trip Matron Sage and Sister Floyd accompanied them. The dramatic scene on their arrival has already been described, but much of the credit for their rescue and recovery must go to Haydon Lennard of the ABC and the members of the RAAF Nursing Service.

Sister Betty Jeffrey described the dramatic meeting on that remote airstrip in Sumatra when those who were left finally were re-united with their liberators.[104] They had been driven in trucks to a railway station then by train to Lahat, a journey of some hundred miles from the prison camp. Then they sat down in the shadow of some bushes to wait for the plane. Then it arrived.

> The engines stopped, there was complete silence, nobody spoke, we just stared in disbelief. Then a door opened and out stepped two women who smiled and waved to us. This was completely unexpected, who were they? Both were dressed in the same good old army grey we knew so well and were then wearing — not skirts as before but SLACKS. This was a new thing but it made us feel shabbier than ever in our patched up old grey dresses. As they came closer we knew it must be Matron Sage, our Matron-in-Chief, followed by Sister Floyd, who was one of our 2/10 AGH nurses who had miraculously got home from Singapore in March 1942 with another group of nurses . . . It was a dramatic moment, like a tableau, nobody moving or speaking for a minute. Suddenly we all started talking and laughing together, with Matron trying to talk with each one of us at once. She kept saying "I am the mother of you all" — and we liked it. "Where are all the others?" she asked. Poor Sister James, we let her do the talking. "This is all", she said simply.
>
> "There are twenty-four of us".

Sixty-five had left Singapore in 1942. The rescued included Queensland Sisters Florence (Flo) Trotter, Joyce Tweddell, C.S.M. (Chris) Oxley, C.E.M. (Del) Delforce, Jessie Blanch, E. Short, Val Smith, Sylvie Muir and Violet McElnea.

Matron Sage, in her account of the meeting with the rescued POW Sisters at Lahat, was obviously too overcome, even to write about it later.

> As we stepped out of the plane we were thrilled to see 24 Sisters in their grey cotton uniform each wearing the rising sun badge. The meeting was one too difficult for me to describe and the occasion, the fulfilment of a desire cherished over the years since the fall of Singapore.[105]

Major Harry M. Windsor, (2/14 AGH) who flew with Matron Sage to Lahat airstrip referred to the 'almighty reunion' that Matron Sage had with the Sisters.

With the rescue of the surviving 24 members of the AANS held in captivity by the Japanese the full story of their ordeal became known. They endured the most primitive conditions, lack of adequate food and water, indescribable sanitary conditions and without the common decencies and courtesies normally afforded to prisoners of war, especially those of officer rank. They survived the brutalities of Japanese guards, the physical punishments, the mental torments. Personal details of their treatment were given to Army Intelligence and later to the War Crimes Commission. More vivid accounts were written by Sister Betty Jeffrey. (*White Coolies*) and Jessie Simons (*While History Passed*). It was inevitable that under such conditions illness would take its toll and without adequate medical facilities some would not survive. As with all POW's in Japanese hands, physical debilitation brought about by malnutrition accentuated by malaria and dysentery reduced their human frames to virtual skeletons. The amazing part of this episode is that so many survived. Had the war not ended so suddenly with the dropping of the bombs on Hiroshima and Nagasaki it is doubtful if many would have survived a few more months.

The story of the AANS in captivity is one of the darkest of the war years, yet one of the most inspiring in the history of the nursing profession. Even in those days of inhuman treatment they retained their dignity and their courage; in keeping with the highest ideals of their profession they went out of their way to do what they could to help the sick among the women and children imprisoned with them. Then too they had to face up to the grim task of nursing their dying colleagues and in the end of burying them in the prison cemetery. Such was their courage that eight of them died only in the last few months before the capitulation of the Japanese, including Queensland Sisters Pauline Hempsted (d. 19 Mar 1945) and Pearl Mittelheuser (d. 18 Aug 1945).

One of the most extraordinary and moving stories to come out of the AANS captivity was the creation of the 'Women's Vocal Orchestra'.[106] Some 600 women and children were housed under appalling conditions in a camp near Palembang, in south-east Sumatra and for a time their spirits were lifted by the singing of hymns and songs. Then a wonderful thing happened. Among the prisoners was Margaret Dryburgh, a Presbyterian missionary, who wrote down with amazing accuracy and detail pages of Beethoven, Debussy, Chopin and Dvorak. With the help of another musician, Norah Chambers, a graduate from the Royal Academy of Music in London, they arranged the music for four-part choral singing, using salvaged scraps of paper and stumps of precious pencils. After many careful rehearsals the choir gave its first concert on December 27, 1943. The glorious sounds of Dvorak's New World Symphony ('Going Home), Tchaikowsky, Debussy and Beethoven soared through the compound. Many wept openly. They had not expected such beauty among the bedbugs, the cockroaches, the rats, the smell of the latrines, the malaria, the dysentery, the hunger of the camp. This vocal orchestra gave concerts all through the years of imprisonment until early 1945, by which time 15 of the original choir of 30 had died, including Margaret Dryburgh. But these women had unlocked the remarkable power of music to uplift the human spirit in adversity, which proved to be a critical factor in their ultimate survival.

Although the survivors scattered at the end of the war, the music manuscripts came to light in 1980 and plans were developed to make a documentary and tape of this 'Women's Vocal Orchestra'. Among the survivors were three members of the AANS, Betty Jeffrey (Melbourne), Mickey Syer (Sydney) and 'Flo' Trotter of Brisbane. In 1983 they journeyed to California to meet other survivors and to hear the Peninsula Women's Chorus sing the choruses they sang so long ago. Of all these songs, none captured the spirit of the prisoners more than The Captives' Hymn, the words and music written by Margaret Dryburgh in 1942 and sung at each Sunday church service.

The Captives' Hymn

Father, in captivity we would lift our prayer to Thee;
Keep us ever in Thy Love, grant that daily we may prove
Those who place their trust in Thee more than conquerors may be.

Give us patience to endure, keep our hearts serene and pure;
Grant us courage, charity, greater faith, humility,
Readiness to own thy will, be we free or captives still.

For our country we would pray, in this hour be Thou her stay,
Pride and selfishness forgive, teach her by the law to live,
By Thy grace may all men see that true greatness comes from Thee.

For our loved ones we would pray, be their guardian night and day.
From all dangers keep them free, banish all anxiety,
May they trust us to Thy care, know that Thou our pains doth share.

May the day of freedom dawn, peace and justice be reborn,
Grant that nations loving Thee o'er the world may brothers be,
Cleansed by suffering, know rebirth, see Thy Kingdom come on earth.

So the Second World War, 1939–1945, came to an end. Demobilization proceeded apace for all troops, including the AANS. All patients were transferred as quickly as possible to Repatriation Hospitals in each State, in the case of Queensland this was at Greenslopes. Six years of war-time nursing had taken its toll on all nurses, for some it was not possible to return to the profession. Others were enriched by the experience and returned to civilian nursing. In the years ahead they were to be appointed to senior positions throughout Australia. Some took advantage of the Commonwealth Re-Construction Scheme, completing special post-graduate nursing courses or training in allied medical fields. They were part of an enduring tradition of service in all theatres of war in all circumstances and all conditions. Official figures list 3,477 members of the AANS, including 522 from Queensland. The return of most of these to civilian nursing meant that civilian nursing would never be the same again. As for army nursing the future was to make further demands on the service now accepted as an integral and essential part of the defence forces.

Victory march of AANS through Brisbane City Streets 1945.

Table 5

AANS (Qld) World War II

Members of the Australian Army Nursing Service, enlisted with Q, QX, or QFX numbers in World War II. This list does not include those Queenslanders who may have enlisted in other States, in other services or in overseas forces. See Index for additional names.

QFX 40927	Abbott, A.L.	QFX 42349	Althaus, E.E.
QFX 700118	Abbott, J.S.	QFX 40929	Anderson, M.A.
QFX 23505	Adams, E.M.	QFX 50136	Anderson, N.J.
QFX 51704	Adams, J.T.	QFX 46881	Anderson, O.M.
QFX 19075	Adams, M.	QFX 23682	Anthony, H.J.
Q 142003	Ahern, M.G.	QFX 58206	Antonelli, D.
QFX 40927	Alcorn, J.I.	QFX 54898	Arbuckle, E.C.
QFX 25360	Alcorn, M.E.	QFX 55951	Archer, M.V.
QFX 33283	Alfredson, B.M.	QFX 41032	Asmus, D.M.
QFX 19161	Allen, E.M.	QFX 41041	Austin, N.L.
QF 123190	Allison, F.M.		

QFX 19149	Backhouse, B.D.	QFX 40960	Bohm, A.D.
QFX 40988	Baker, A.M.	QFX 44617	Bonner, A.R.
QFX 19158	Baldock, A.M.	QFX 48779	Booth, E.W.
QFX 46902	Ball, P.M.	QFX 40929	Bowman, S.B.M.
QF 119729	Banks, G.E.	QFX 57767	Brady, M.K.
QFX 45279	Barnard, L.B.	QFX 40938	Brake, M.E.
QFX 46882	Barns, I.M.	QFX 48924	Brandt, R.
QFX 22821	Barrett, S.M.	QFX 42665	Bray, W.
QFX 700078	Bartlett, C.L.	QFX 54949	Breen, R.M.
Q 141150	Batt, E.O.	QFX 41039	Briant, M.E.
QFX 51709	Beard, V.M.	QFX 44629	Brook, I.
QF 143295	Beardmore, F.E.	QFX 42686	Broomhead, M.P.
QFX 6395	Beattie, T.J.	QFX 63337	Brosnan, F.J.
QFX 48908	Bebbington, M.R.	QFX 55881	Brosnan, G.J.
QFX 46905	Bedford, V.M.	QFX 19063	Brown, G.E.
QFX 40958	Beer, M.D.	QFX 40932	Brown, M.E.
Q 142200	Beetham, E.G.	QFX 22469	Brown, O.M.
QFX 44630	Bell, M.H.H.	QFX 42666	Brown, R.D.
QFX 40989	Bellert, M.B.	QFX 42683	Browne, A.E.
QX 6484	Bennett, M.A.	QFX 42397	Browne, O.A.
QFX 40985	Bennett, P.E.M.	QFX 48921	Brownsdon, M.K.
QFX 48904	Bere, E.M.	QFX 54905	Bruce, G.K.
QFX 55161	Beverley, B.J.R.	QFX 46883	Bryce, A.N.
QF 843176	Binns, E.	QFX 46879	Budge, D.M.
QFX 6125	Blain, E.M.	QFX 43171	Burbidge, B.E.
QFX 19074	Blanch, J.J.	QFX 6431	Burnett, D.A.
QFX 59847	Blanch, M.R.	QFX 48914	Burns, E.C.
QFX 44161	Bleney, A.W.	Q 70277	Butler, L.M.
QFX 51705	Blundell, B.O.M.	QFX 23684	Byers, J.
QFX 41035	Boatfield, M.G.		

QFX 50137	Cahill, K.	QFX 25366	Clifford, I.L.
Q 70220	Caldwell, M.K.	QF 119914	Cocking, F.M.
Q 267509	Callaghan, E.D.	QF 142025	Coghlan, E.M.
QFX 419072	Calnan, E.	QFX 6105	Coleman, D.M.
QFX 43169	Cameron, A.B.H.	QFX 54956	Collins, M.A.
QFX 44631	Cameron, J.A.	QFX 44638	Collyer, A.L.C.
QFX 23683	Campbell, A.T.	QFX 46895	Compton, H.V.
QFX 35390	Campbell, R.M.	QFX 54962	Condon, M.
QFX 46547	Cannel, B.J.	QFX 40970	Conheady, I.C.
QFX 48778	Carey, A.M.	QFX 44618	Connole, M.M.
QFX 57443	Carey, M.M.	QFX 25354	Cooke, E.
QX 6375	Carmichael, E.H.D.	QFX 40653	Coombs, M.J.
QFX 6488	Carmody, M.C.	QFX 50155	Cooper, A.A.
QFX 46896	Carseldine, D.C.	QFX 44640	Cooper,L.M.
QFX 25356	Carseldine, E.M.	QFX 44619	Cooper, R.I.
QFX 54985	Carson, A.	QFX 40650	Corrigan, K.
QFX 54955	Carter, E.	QFX 41043	Cottell, G.M.
QFX 6374	Casey, E.L.	QFX 6415	Couche, J.W.
QFX 40654	Casey, J.B.	QFX 40933	Craig, B.
QFX 33287	Cashin, J.M.	QFX 50139	Cranley, M.L.
QFX 43177	Chadwick, E.A.	QFX 55876	Creagh, A.O'D
QFX 33372	Chadwick, I.C.	QFX 54951	Crow, U.C.
QX 6485	Chambers, E.E.	QFX 40934	Cuddihy, E.M.
QFX 40639	Chandler, M.H.	QFX 40967	Cumming, B.D.
QFX 44636	Chew, M.C.B.	QFX 41046	Cumming, M.J.
QFX 48922	Clements, M.E.	QF 140272	Cundy, A.V.
QFX 25365	Clements, P.	QFX 54960	Curtin, M.
QFX 46548	Clements, R.	QF 124385	Cutmore, E.A.

QFX 40935	Dalby, E.M.	QFX 43178	Dewar, F.F.
QFX 40936	Daley, M.E.G.	QFX 40966	Dickson, E.D.
QFX 50138	Daly, E.A.	QFX 6106	Dickson, J.C.
QF 119946	Daly, E.A.	QFX 50146	Dobbyns, I.E.G.
QFX 55159	Dark, P.Y.	QFX 54899	Doherty, M.M.R.
QFX 42327	Davidson, E.M.	QFX 6112	Doig, E.N.
QFX 44620	Davidson, H.R.	QFX 40937	Dolgner, P.N.L.
QFX 6126	Davis, R.E.	QFX 46884	Donohue, E.
QFX 44621	Dean, A.W.	QFX 55877	Dorward, J.
QFX 6115	Dean, E.M.	QFX 62759	Dowie, J.I.
QFX 42396	Deane, A.C.	QFX 40986	Doyle, C.E.
QFX 25358	Deen, A.S.	QFX 48923	Doyle, M.F.
QFX 19071	Delforce, C.M.	QF 142780	Drynan, K.C.
Q 140154	Delpratt, E.W.	QFX 55056	Duffield, A.M.
QFX 23507	Denman, J.L.	QFX 19159	Duffield, E.M.
QFX 48907	Denton, A.A.	QFX 54943	Duncan, E.E.
QFX 6487	Desbos, M.A.		

QFX 46878	Eagles, F.A.	Q 267779	Evans, N.A.
QFX 54957	Eales, D.K.	QFX 42395	Everingham, K.
QFX 40984	Edwards, E.J.	QFX 40642	Ewing, A.M.
QFX 59031	Egan, G.M.	QF 140270	Ewing, L.
QFX 43173	Elliott, K.	QFX 6357	Exton, A.M.
QFX 41031	Elms, J.		

QFX 48776	Fairhall, C.G.
QX 19148	Faulkner, I.M.
QFX 40939	Fechney, J.A.
QX 40940	Fenton, E.M.
QFX 6428	Finlay, M.E.F.
QFX 55874	Fitch, D.M.
QFX 55872	Fleming, V.
QFX 48920	Fletcher, D.E.
QFX 40941	Flower, D.H.
QFX 56297	Fogarty, M.N.
QFX 23785	Foley, B.M.
QFX 48782	Foster, P.M.
QFX 50133	Fowler, I.M.
QFX 54958	Francis, D.E.
QFX 59032	Francis, M.O.
QFX 54900	Fraser, J.
QFX 41038	Fraser, J.E.
QFX 40973	Fraser, M.
QFX 48909	Freeman, E.L.
QFX 6127	Freeman, M.
QFX 50135	Freeman, T.
Q 70272	Fullerton, J.I.

QFX 60233	Gallen, J.I.
QFX 40942	Gant, B.A.M.
QFX 25359	Garbutt, N.C.
QX 19147	Geraghty, M.
QFX 53206	Geraghty, M.P.
QFX 45120	Gilbert, E.M.
QFX 37039	Gillanders, M.D.
QF 268218	Gillies, C.B.M.
QFX 6128	Glasgow, M.M.
QFX 50151	Glen, I.M.
QFX 51706	Golden, M.A.
QX 6377	Golden, R.J.
QFX 48903	Goodger, G.L.
QFX 19157	Goodman, H.M.
QFX 41047	Goos, J.W.
QFX 6416	Gordon, M.M.L.
QFX 42685	Gorman, M.G.K.
QFX 22498	Gorrie, J.M.
QFX 40962	Gould, K.A.
QFX 44641	Gracey, C.J.
QFX 33462	Graham, G.H.
QFX 55880	Gray, A.E.
QFX 40651	Gray, B.P.
QFX 40943	Griffith,M.A.
QFX 19076	Grigg, I.E.
QX 6486	Grudington, A.
QFX 25363	Guilfoyle, B.M.

QFX 58204	Hailes, L.C.
QFX 25362	Haines, E.L.
QFX 58205	Hair, G.G.
QFX 63805	Hale, M.T.
QFX 6245	Hall, E.M.
QFX 50485	Hambleton, D.A.
QFX 6491	Hamlyn-Harris, M.B.
QFX 44610	Hancock, M.B.
QFX 48915	Hanlon, I.A.
QFX 6107	Hanrahan, E.F.
QFX 46893	Hansen, K.I.
QFX 6108	Harland, V.M.
QFX 40944	Harris, E.J.
QFX 37036	Harrison, D.M.
QFX 6101	Harvey, F.E.
QF 123045	Hawkins, M.F.
QX 37038	Haworth, M.C.
QFX 6356	Hayes, B.E.
QFX 48919	Hayes, P.M.
QFX 6422	Hood, J.M.
QFX 6258	Hooke, J.S.
QFX 50145	Heathwood, I.
QFX 42350	Heers, H.O.
QX 22499	Hitchings, M.E.F.
QF 143028	Hobill, A.R.
QFX 6102	Hely-Wilson, U.
QFX 22714	Hempsted, P.B.
QFX 43179	Henning, M.
QFX 43180	Henry, C.M.
QFX 48772	Henry, M.V.
QFX 48916	Heybroek, S.
QFX 6109	Hoey, I.
QFX 48901	Holmes, M.
QFX 6118	Hooke, H.D.
QFX 23669	Hooper, K.E.M.
QFX 48905	Howlett, A.L.
QFX 48780	Hulley, M.V.
QFX 55875	Hutton, I.G.

QFX 43181	Imhoff, J.
QFX 54103	Imhoff, R.
QFX 46899	Irwin, M.J.

QFX 22497	Jackson, D.A.	QFX 44611	Johnston, N.M.
Q 140158	Jackson, M.L.	QFX 124390	Jolly, W.R.R.
QFX 42682	Jenkin, R.	QFX 40968	Jones, G.L.
QFX 6421	Johnson, M.A.	QFX 54906	Jones, M.I.
QFX 61751	Kamp, M.O.W.	QX 6111	Keys, L.E.
QFX 43172	Kavanagh, F.	QFX 54952	King, E.I.
QFX 43182	Kavney, L.M.	QF 144649	King, G.E.
QFX 41040	Kay, M.	QFX 50143	Kingsford, M.N.
QFX 51701	Keliher, J.M.	QFX 22818	Kipsett, A.
QFX 40649	Kelly, M.B.	QFX 33282	Kirwan, C.
Q 70279	Kemp, E.M.	QFX 40652	Knowles, M.B.
QFX 41415	Kenny, E.T.	QFX 55160	Knox, D.R.
QFX 44622	Kenyon, E.M.	QFX 54950	Kochevatkin, M.
QFX 25357	Kerr, G.M.	QFX 35390	Kurren, R.M.L.
Q 70293	Lambert, L.E.	Q 266250	Lipstone, R.A.
QX 19156	Lang, V.C.	QFX 55158	Lloyd, F.M.
QFX 41045	Lanksey, E.A.	QFX 50150	Lucas, M.J.
QFX 44633	Larsen, E.M.	QFX 44609	Lucas, M.K.
QFX 44637	Laughren, E.F.	QFX 43183	Luce, M.M.
QFX 50156	Leahy, C.	QFX 48912	Luck, D.M.
QFX 50134	Leahy, M.M.C.	QFX 44639	Luddy, M.B.
QFX 22820	Lear, M.A.	QFX 40983	Luke, M.
QFX 46880	Leitch, M.	QFX 46885	Lynn, M.M.H.
QFX 40971	Lennon, J.S.	QFX 59643	Lyon, A.C.
QFX 42584	Leslie, N.B.M.	QFX 41037	Lyon, E.W.
QFX 6424	Levaring, J.A.		
QFX 6426	McAlpine, H.B.	Q 124815	McDonnell, M.M.
QFX 26082	McCabe, H.M.	QF 271253	McDougall, M.
QFX 54104	McCallum, A.E.	QFX 22822	McElnea, V.I.
QFX 50154	McCarthy, I.M.	QFX 40947	McGeary, B.A.
QFX 64116	McCauley, K.	QFX 19056	McGhie, E.R.
QFX 700083	McConnell, A.B.	QFX 46903	McGuigan, B.
QFX 40645	McCormack, E.A.	QFX 54944	McKim, I.D.
QFX 44160	McCready, E.N.	QFX 40948	McKinnon, E.J.
QFX 55059	McDonagh, M.A.	QFX 46887	McNalley, N.
QFX 22815	McDonald, G.M.	QX 25367	McNee, M.
QFX 35251	McDonald, M.M.	QF 119783	McPherson, L.G.

QX 23508	Macartney, E.
QFX 6278	Machon, I.F.
QF 141966	Mackenzie, M.M.
QFX 43185	Maldas, M.M.
QFX 6103	Marks, I.A.
QFX 6427	Marshall, M.A.
QFX 55058	Marshall, M.G.
QFX 48775	Martin, J.T.
QF 268999	Martin, M.E.
QFX 40945	Masters, M.A.
QFX 40969	Matheson, A.J.C.
QFX 46894	Matthews, A.M.
QFX 44612	Matthews, L.I.
QFX 55882	Maudsley, J.E.K.
QFX 40648	Messer, I.R.
QFX 54901	Meston, J.T.
QFX 46901	Miller, C.D.
Q 70300	Miller, E.L.
QFX 6104	Miller, P.B.
QFX 23558	Milne, C.
QFX 19068	Mittelheuser, P.B.
QFX 54897	Mollard, M.F.
QFX 40946	Moore, M.A.
QFX 44613	Moran, F.P.
QFX 48774	Moran, R.M.
QFX 56382	Morrison, N.L.
QF 140475	Morton, V.M.
QFX 54963	Moss, D.A.
QFX 43170	Moss, M.
QFX 22816	Muir, S.J.M.
QFX 55055	Muirhead, S.D.
QFX 37037	Muller, G.C.
QFX 50157	Mundell, M.M.
QFX 54961	Murdock, M.A.
QFX 54953	Murphy, E.M.
QFX 44623	Murphy, G.M.
QFX 44634	Murphy, M.M.
QFX 51708	Murphy, Y.M.

QFX 54959	Neill, C.
QFX 6489	Netterfield, M.M.
QX 19160	Neville, J.P.L.
QX 6420	Newitt, E.I.
QFX 6279	Newman, J.I.
QFX 50152	Newton, M.

QFX 22817	O'Connor, A.T.
QF 70294	O'Connor, C.
QF 141854	O'Connor, F.E.R.
QFX 6280	O'Keefe, E.M.
QFX 42398	O'Leary, B.
QF 142782	O'Leary, H.M.
QFX 44635	Oliphant, H.C.
QFX 22717	O'Loughlin, D.C.
QFX 22501	O'Loughlin, M.E.
QFX 46888	O'Mara, D.C.
QFX 40646	O'Neill, J.B.E.
QFX 46886	Origlasso, M.E.
QFX 19073	Oxley, C.S.M.

QFX 44162	Page, F.M.
QFX 6299	Palmer, A.E.
QFX 44624	Park, J.
QFX 48777	Parker, B.
QFX 25361	Parker, V.V.
QFX 6119	Paterson, V.M.
QFX 6162	Patrick, E.E.
QFX 48913	Patterson, J.
QFX 48911	Pearce, I.E.H.
QFX 6110	Petersen, F.M.
QFX 59772	Philp, B.
QFX 46897	Philp, M.S.
QFX 48902	Pierpoint, D.M.
QFX 40655	Pollock, G.I.
Q 142121	Postgate, A.E.
QFX 22715	Powell, J.E.B.
QFX 40643	Powell, J.H.H
QFX 41044	Price, L.C.
QFX 25364	Price, P.A.M.
QFX 40647	Prichard, B.L.
QFX 56381	Prideaux, N.A.
QFX 48918	Pryde, A.N.
QFX 22716	Pugh, P.
QFX 55057	Purkess, M.J.
QFX 6300	Pym, P.

QFX 62883	Quinn, M.F.

QFX 19069	Ralston, I.D.
QFX 42399	Ramsay, M.
QFX 40949	Ramsey, M.J.F.
QFX 44614	Rattray, C.M.
QFX 40950	Redman, E.M.
QFX 23681	Reeve, E.F.M.
QFX 50153	Reid, M.C.A.
QFX 6116	Reid, M.V.
QFX 40963	Reid, R.A.
QFX 40951	Reid, T.
QFX 40952	Retschlag, D.M.
QFX 51703	Rice, R.C.
QX 19145	Richardson, E.M.
QFX 44628	Richardson, H.M.
QFX 40982	Ricketts, E.M.R.
QFX 6120	Riddell, G.E.
QFX 54954	Ridge, J.E.
Q 124719	Robertson, V.L.
QFX 46889	Roden, E.M.
QFX 6121	Roe, D.G.
QFX 48917	Ross, H.C.
QFX 46892	Ross, I.
QFX 40959	Rossiter, J.M.
QFX 23506	Rowland, J.I.
QFX 59644	Russell, J.M.
QFX 63286	Ryan, J.I.

QFX 48773	Schultz, I.B.
QFX 41414	Scott, A.E.
QFX 41050	Scott, L.C.
QFX 46906	Scragg, B.
QFX 55879	Seamark, R.
Q 70257	Seymour, G.M.
QFX 48910	Searle, M.A.
QFX 40961	Seegar, B.A.T.
QFX 22814	Selwood, M.C.
QFX 55285	Shambrook, M.A.
QFX 55001	Sheahan, C.
QF 140423	Sheahan, G.
QFX 48906	Sheridan, C.E.
QFX 40972	Sheridan, K.W.
QFX 22911	Short, E.M.
QFX 6282	Short, W.O.
QFX 45280	Sier, E.J.
QFX 40953	Simpson, V.M.
QFX 6272	Sinclair, L.C.
QFX 44615	Sinnott, S.E.
QFX 46890	Skerman, E.L.
QFX 46900	Sketchley, N.V.
QX 6281	Skyring, T.A.
QFX 50144	Smith, A.
QFX 41048	Smith, D.E.
QFX 50142	Smith, L.W.
QFX 48925	Smith, P.G.B.
QFX 48781	Smith, P.M.
QFX 22819	Smith, V.E.
QFX 6122	Soorley, J.
QFX 41042	Spearritt, A.
QFX 42687	Speer, M.M.
QFX 6423	Steen, M.L.E.
QFX 55878	Stephens, D.K.
QFX 50141	Stevens, E.D.H.
QFX 44625	Stevenson, F.E.
QFX 61754	Stevenson, H.L.
QFX 6114	Stewart, H.
QFX 55873	Stewart, M.
QFX 40640	Stirling, A.
QF 70310	Stobert, F.
QF 70286	Stone, S.
QFX 63508	Stormouth, M.J.
QFX 41034	Stuhmcke, M.P.
QFX 54902	Sullivan, R.M.
QFX 19155	Suttie, C.M.
QFX 40965	Swallow, G.
QFX 64029	Swan, B.I.

QFX 40964	Tannan, A.
	Taylor, M.J.
QFX 46891	Templeton, L.G.
QFX 6117	Thompson, I.
QF 22500	Thiedeke, C.M.
QFX 41033	Thompson, K.B.
QFX 40925	Thorburn, A.
QFX 6429	Thorpe, M.F.
QFX 54903	Tierney, K.M.
QF 142106	Tindale, D.K.
QFX 40954	Tinney, H.A.
QX 35347	Tinney, M.I.
QFX 6277	Tomlins, D.M.
QFX 55060	Tranter, E.M.
QFX 44632	Troedson, T.C.
QFX 19077	Trotter, F.E.
QFX 59428	Turnbull, T.M.
QFX 43175	Turner, E.
QFX 19070	Tweddell, J.

QFX 56298	Veivers, G.M.
QFX 19146	Vigar, E.J.

QFX 6161	Wallace, M.	QFX 6492	White, P.A.M.
QFX 22495	Warfield, N.M.A.	QFX 61041	Whitehead, G.C.
QFX 44616	Warren, M.E.B.	QF 70288	Whiting, E.
QFX 40926	Weaving, J.M.	QFX 43168	Willett, M.K.
QFX 41030	Webb, A.S.	QFX 40955	Williams, M.V.
QFX 43174	Webster, M.G.	QX 6430	Wilson, H.
QFX 25355	Wells, M.L.	QFX 40641	Winter, E.V.
QF 142962	Wham, C.A.	QFX 6490	Woodward, N.E.
QFX 41049	Whelan, J.A.	QFX 43184	Woolley, A.
QFX 40956	White, M.E.	QFX 46898	Wright, D.M.
QFX 63338	White, M.E.	QFX 40957	Wright, P.M.
Q 70307	White, M.M.	QFX 44626	Wynter, A.O.
QX 6298	Zielke, J.B.	QFX 54904	Zoeller, B.J.
QFX 63382	Zillman, P.G.		

References

1. For an account of social conditions in Australia at this time see
   HASLUCK, P. *The Government and the People 1939–1941*. Canberra. Australian War Memorial.
   HASLUCK, P. *The Government and the People* 1942–1945. Canberra. Australian War Memorial.
2. For details of the submarine attack on Sydney Harbour:
   CARRUTHERS, STEVEN, L., *Australia Under Siege*. Solus Books, Sydney, 1982.
3. AANS Files 417/20/28 and 21/720/332 AWM.
4. Commissioned Rank AANS, File 21/720/144. AA (Melb). Readjustment of ranks of Sister and Staff Staff Nurse AANS 1942. File 21/721/427 (AA Melb).
5. Status of AANS. File 509/1/1 and File 339/1/200. AA (Melb).
6. Marriage of AANS. File 21/720/195 AA (Melb). AANS, Married Discharge Policy 1943–44, File 21/3/223. AA (Melb).
7. Reports on Military Hospitals. File 403/1/14, (AWM).
8. Amendments to mil Forces (Women's Services) Regs 1943–44. File 21/3/208, AA (Melb).
   BONNER, ANNA RUTH, trd Ispwich Base Hosp 1922, Lady Chelmsford, Bundaberg & Infant Welfare, Bris. Staff Mt Perry Hosp, Stanthorpe, pte hosp Stanthorpe. AANS Reserve 1935. AANS 1939, 8 Fd Amb Nth Qld; 1942 NG, Murray Barracks; evac to Aust; post Enoggera, Holland Park & Cairns. 1946 retd as LtCol.
   SHEAHAN, GRACE IRENE. Gen nursing Mater 1970 and Brisbane General; Rosemount Repat 1922–23; Maryborough Hospital 1924–1939, including Matron after 1925; Matron Dubbo 1941–42; Matron 112 AGH (Greenslopes) 1942–45;
9. WALKER, A.S. *Australia in the War of 1939–1945*. Series Five Medical. Vol 11. Middle East and Far East. Chap 4.
10. Queensland Sisters appointed to C.H. Chermside at this time included Sisters White, M; Scott, L; Matheson, A., Ramsay, M., Whitehead, G., Dean, A., Everingham, K., Browne, D., and O'Leary, B.
11. C.H. REID'S RIVER (Cairns & Atherton). Queensland Sisters with this Hospital were Sisters Althaus, E., Cocking, F., Condon, M., Guilfoyle, B., Heers, H. and Murdoch, M.
12. 8 CH Warwick. Sr Adams, E.M.
13. 47 ACH Goondiwindi. Unit History War Diary. File 11/3/34 AWM. Among Queensland Sisters associated with this unit were the following:—
    COMPTON, HANNA VIOLET, b 9.2.08. Trd Mackay Dist Hosp, gen & midwifery, chn's Welfare Bris; AANS 1942–45, service Aust & NG, 47 ACH, 112 AGH. Theatre Sister Townsville Gen Hosp, Matron Mt Isa Gen Hosp, Matron Nambour Dist Hosp.
    DICKSON, ELLA DOUGLAS. b 19.9.15. Trd Bris Gen Hosp & Women's Hosp 1937–42; AANS 1942–46, 2/2 AGH. CMF tutor Sister 1951–53. Staff Repat Hosp, Greenslopes 1946–76.
    Bedford, V.M., Beer, M.D., Bohm, A.D., Cahill, K., Casey, J., Chadwick, E., Cooke, E., Cottell, G.M., Cumming, B.D., Elms, J., ARRC., Gould, K.A., Hambleton, D.A., Hansen, K.I., Jones, G.L., McCallum, A., Matheson, A., Reid, R., Retschlag, D.M., Rossiter, J.M., Seegar, B.A.T., Scott, L.C., Scragg, B., Smith, D.E., Smith, T.A., Swallow, G., Tanwan, A.,
14. 56 ACH Cowan Cowan (1st Fortress). Among Queensland Sisters who served with this unit were the following:— Banks, G.E., Boatfield, M.G., Davidson, E., Graham, G.J., Kamp, M., Lambert, L.E., McCallum, A., O'Connor, C., Roden, E.M., Pollock, G., Ross, I., Stone, S., Whiting, E.
15. 22 ACH, Canungra, 74 ACH, Mt Isa and ACH, Toorbul Point.
16. CASEY, JOFFRETTE BRIDGES, b 23.6.1915; trd Maryborough Gen Hosp 1935–39, Women's Hosp Bris, 1939–40. 1940 — Sister in C, TSS Zealandia evac civilians from Hong Kong & Manila. AANS 1941–45, Aust & SWPA.

17. 7 ACH, Redbank, Queensland Sisters appointed to CH Redbank at this time included Sisters O'Neill, J., Briant, M., Cottell, G., Gant, B., Goos, J., Lyon, E., Miller, E., Templeton, L., Robertson, V., Webb, A., Price, L., Kerr, G., Chadwick, I., Stirling, A., Chandler, M., Dewar, F., O'Connor, F., Ewing, A., Campbell, R., Powell, J., Winter, E., Murphy, Y., McCormack, E., Murphy, E.

    KING, GLADYS ELIZABETH, b 16.6.10. Trd Nambour Diamantina Hosp, 1937-40; Staff Maryborough Gen Hosp, 1941-42; AANS 1942-46, 4 ACH, 7 ACH.

18. Among those Queensland Sisters appointed to CH Enoggera at this time were Sisters Whiting, E., Carseldine, D., Bell, M., Bere, E., Booth, E., Brook, I., Butler, L., Cocking, F., Howlett, A., McCallum, A., McKim, I., Parker, B., Philp, M., Pryde, A., McCabe, H., Grigg, I., Batt, E., Wright, D., Sheridan, C., Coghlin, E., Freeman, T., Stewart, M., Denton, A. and Bebbington, M.

    6 CCS Ipswich and ACH Ipswich, among Queensland Sisters known to have served in these units were the following:

    ALFREDSON, BERTHA META, b 15.2.1917. Trd Bris Gen Hosp, Maryborough Mat Hosp and Chns Welfare; AANS 1941-46. Service Aust, NG.

    BURBIDGE, BERYL EMMA, OBE. b 4.3.02 Gympie, educ Gympie. Trd Bris Gen Hosp 1923-27; obsts Q Eliz Hosp, Hobart; AANS 1942-46, 6 CCS & MRU; 1946-52 Staff Bris Gen Hosp; Dep Matron 1952-58; Matron PA Bris 1956-58; Gen Matron Nth Bris Hosp Bd. OBE retd 1968.

    BROWNE, AUDREY ELLEN, b 6.4.17; trd Charters Towers Dist Hosp, Women's Hosp Brisbane; Staff of Stanthorpe, Babinda, Mareeba, Bowen; Dist Hosps; AANS 1942-44, Aust, NG & SWPA, 2/6 CCS, 11 ACH, 2/9 AGH, 2/14 AGH.

    RICKETTS, EDITH MAUD, b 1912, educ Clermont & Yeppoon; trd Clermont Hosp 1928-33, Staff Clermont Hosp; AANS 1942 — service Atherton Tablelands, 2/4 AGH Redbank. After war retd to Clermont and on staff hospital. d 10.3.65.

    Sisters Ahern, M., Baker, A., Bellert, M., Bennett, P., Bonner, A., Bridge, D., Cashin, J., Cranley, M., Crow, V., Daly, E., Doyle, C., Eagles, F., Eales, D., Edwards, E., Haworth, M., Kavanagh, F., King, E., Lambert, L., Luke, M., McDonald, M., Mackenzie, M., Smith, P., Tinney, M., Turnbull, T., Turner, E., Webster, M.

19. CAHILL, K. see fn (8).
20. GILBERT, EDNA. Staff Ipswich Gen Hosp; AANS 1942-1945; correspondence and personal reminiscences.
21. 1 ACH (Horn Island) Queensland Sisters who served at Horn Island included:

    Sisters Althaus, E., Clements, P., Condon, M., Davidson, E., Gray, B.P., Heers, H., Prichard, B.L., Thorburn, A. and Williams, E.,

    MESTON, Joyce Toressa. b 26.5.1916; trd Gympie Base Host, midwifery Women's Hosp Bris; AANS 1942-45, Toowoomba, Horne Is., Holland Park, Concord. d. 2 Nov 1983.

22. Sister Gilbert reminiscences, supra.
23. 77 ACH. Unit War Diary AWM. Queensland Sisters with this unit included:

    FLETCHER, DOROTHY. b 21.11.14; trd Ipswich Gen Hosp, midwifery Women's Hosp. AANS 1942-46, 2/6 AGH.

    REID, REBIE ALISON. b 28.8.11. Trd Tamachy Hosp, Rocktn 1929-32; midwifery Sth Syd Women's Hosp 1933-32; mothercraft, Tresillian Hosp 1935-36; Act Matron Rocktn Hosp; Staff Nurse, Paddington Host; Melb Univ (Mildura Annexe), sick bay 3 yrs; AANS 1941-47. ACH Tenterfield, 2/12 AGH Balikpapan. After war on special UN mission to Manila to help repatriated POW's.

    BARNARD, L.B., Beardmore, F.E., Daly, E.A., Lipstone, R.A., Moss, M., Page, F.M., Prideaux, N.A., Searle, M.A., Sier, E.J., Stevens, E.D.H.

24. SHEAHAN, G. MATRON. Report of 112 (Brisbane) Military Hospital Greenslopes (Typescript) 8 Feb 1942 - 13 Nov 1945. History of 112 Australian General Hospital Base Hospital, now known as the Repatriation General Hospital, Greenslopes. *Anzac Day*, 1971.

    Many Queensland Sisters served with this unit at varying times during the war years. The following are known to have served with 112 AGH.

    BROOMHEAD, PHYLLIS. Trd Bris Gen & Women's Hosp 1941-42. AANS 1942-1946. Served Qld & NSW. Eighth Camp Hosp, 2/2 AGH, Concord & Greenslopes. After war, staff Maternal & Child Welfare, Royal Coll Nurse (Lond), Dep Social Security 23 yrs.

    COLLINS, MARY ALMA. Trd Rockhtn Gen Hosp and maternity course at Lady Goodwin Hosp Rockhtn; AANS 1942-1946; 1946 Dep Matron Rockhtn Hosp; 1962 Matron Rockhtn Hosp; d 10 Dec 1971.

    CUMMING, MARY JOSEPHINE; trd Bris Gen Hosp 1928-32; Lady Bowen & Women's Hosp, Bris; Infant Welfare, Sydney. Staff Cunnamulla & Roma Hosps. AANS 1941-45, Aust, NG & SWPA; 112 AGH (Greenslopes), 116 AGH (Charters Towers), 2/11 AGH Buna, Lae, Madang, Aitape. Post-war — geriatric nursing.

    PHILP, BETTY. b 11.6.15. Trd RBGH then Rockhtn Hosp for midwifery — private nursing. AANS 1940-1945, 112 AGH, Hosp Trn, Rockhpn, 2/6 AGH. After war Child Health Nursing & Matron Toowoomba Home.

    SEAMARK, RAE. b 20.5.12; trd Atherton Dist Mem Hosp 1930-34, Innisfail Maternity Hosp 1936, Child Welfare, Bris 1937. AANS 1941-46; 112 AGH, 102 CCS, 110 CCS, Aust & SWPA, Torres St, Tarakan, Borneo. Dip Nurs Admin 1948, Civil Defence Course Mt Macedon, Private & public nursing.

    ALCORN, M.E., Allison, F.M., Cameron, J.A., Creagh, A.O'D., Hawkins, M.F. (Nicholls), Stobert, F., Thorburn, A., Weaving, J.,

25. ibid.
    AMPS, F.M.
    MAHER, E.
    NEILSON, E.F.
26. Personal correspondence from Sr Gray.
    GRAY, AUDREY ETHEL. Trd Mt Isa Dist Hosp, 1935–40; Women's Hosp Bris 1942; AANS 1942–47; service Aust & SWPA.
27. History of 112 AGH, *op cit*.
28. Matron Sheahan's Report, op cit.
29. President, Union Jack Club to the Hon E.M. Hanlon, Premier of Queensland, 11 July 1946.
30. Reports on Military Hospitals, 403/1/14 AWM. Hospital Provisions Mainland 403/1/17 AWM.
31. Unit War Diary 2/1 AGH, 11/2/1 AWM.
32. 2/1 AGH Sisters. FOLEY, B.M. Trd Rockhtn Base Hosp; midwifery cert Rockhtn; AANS 1941–45, service 2/1 AGH Gaza, WA, NG & SWPA.
    KAVNEY, LILLIAN MARGARET. b 1918, Trd Toowoomba Gen Hosp 1941, Mothers Hosp Tba 1942, Maternal & Child Welfare; AANS 1942–47, Aust & NG, 112 AGH, 116 AGH, 2/1 AGH. After disch Repat Hosp Greenslopes until 1951. Further 9 yrs RAANC; Dip Nurs Educ (Coll Nurs Aust); 1961-65 Occup health nurse; 1967 Nurse Educator Greenslopes.
    Baldock, A.M., Byers, J., Duffield, E., Geraghty, M.P., Goodman, H.M., Hooper, K., McConnell, A.B., Suttie, C.M.
33. Unit War Diary, 11/2/2 AWM.
34. 2/4 AGH Unit War Diary 11/2/4 AWM. Australian Women's Hosp File, 21/720/279 AA (Melb).
35. Queensland Sisters with 2/4 AGH included:—
    Carson, A.E., Doherty, M.M.R., Glen, I.M., Ramsay, M., Redman, E.M., Smith, A.A., Tinney, H.
36. Files 399/17/-, 403/7/11, (AWM).
37. Report on Atherton Tableland area as base hospital scheme for NG Force. File 422/7/8 AA (Melb).
38. 2/6 AGH Unit War Diary, File 11/2/6 AWM. Queensland Sisters included:—
    CORRIGAN, KATHLEEN, b Toowoomba 6.6.09; trd Bris Gen Hosp 1929-33, Lady Bowen Maternity Hosp, Child Welfare Bris, Psychiatric Nursing Goodna; Matron at Augathella, Charleville & Julia Creek; AANS 2/2 AGH Rocky Creek, d 7.8.72.
    McALPINE, HELEN BAYNE, b 1.2.01 (Scotland); trd Bris Gen Hosp; AANS 1940–45; 2/5 and 2/6 AGH, ME, Greece, Crete, Aust; Post war, Staff Bris Gen Hosp & Matron Garden Settlement Retirement Village, Chermside, d 13.6.82.
    TROEDSON, THEA CELESTE, b 3.10.13, trd RBH Bris, 1932–36; 1947–48 Staff Sister Comm Bk Bris.
    WHITEHEAD, GRACE CLARIS, b 18.3.1906, d 16.10.1977. Trd Diamantina (Bris), Midwifery, Lady Bowen (Bris), Maternal & Child Welfare (Bris): AANS 13.2.42 - 19.3.46. Served on 4th & 5th Amb Train from July 43. July 45 left on Manunda for Kuching for repatn of POW's Post war career in Mat & Child Welfare in many Qld provincial centres.
    Backhouse, B.D., Carseldine, D.C., Chew, M.C.B., Cooper, A.A., Couche, J.W., Curtin, M., Davidson, H.R., Fraser, J.E.,
39. 2/14 AGH, 116 AGH Unit War Diaries. 2/14 AGH — File 11/2/14 AWM. 116 AGH — File 11/2/31 AWM. Hospital Site Townsville 1943 — File 211/7/123 AA Melb.
40. Queensland Sisters included:
    HENRY, C.M. Trd Bris Gen Hosp; AANS 1942. Service 112 AGH Greenslopes.
    MARTIN, J.T.
41. KAY, MURIEL, b 25.6.16; trd Stanthorpe Gen Hosp 1936–40. Sister Priv Hosp; AANS 1942–45, Aust & SWPA, 116 AGH, 2/12 AGH, Balikpapan 1945. 1963–68 Lady Davidson Hosp.
42. Queensland Sisters at 116 AGH, Charters Towers:
    DUNCAN, EILEEN ETHEL, AANS 1941–46, 116 AGH Charters Towers, 2/7 AGH Cairns.
    FRASER, JEAN. AANS Nov 1942–April 1945. Served 116 AGH, 114 AGH, 1 Ortho Hosp Aust. 110 CCS Tarakan & Aust.
    GRACEY, C.J. From Cooyar District. AANS Jacquinot Bay, New Britain, SWPA.
    MESSITER, F. Trd Bris Gen Hosp. 10 yrs Bris Gen 1930-39; 4 yrs Bundaberg Gen Hosp; AANS 1940-46, 116 AGH, 117 AGH, 119 AGH (Prom Capt) 1947-51, RAANC.
    RICHARDSON, HELENA MARY. b 7.10.16; trd Rockhtn Hosp 1936-40, midwifery 1940-41; AANS 1942-46, 116 AGH (Charters Towers), 1944-46, 2/7 AGH Lae, NG, 1947-48 Repat Hosps Greenslopes & Heidelberg. 1948-56 Mat & Child Welfare Nursing, 1957-58 Nursing UK. 1958-80 School Health Serv Rockhtn.
    SPEARRITT, AMY, MID. Trd Ipswich Gen Hosp, 1933-37; Lady Chelmsford Hosp Bundaberg, 1938; Child Welfare, Qld Baby Clinics 1939. AANS 1941-46, Aust & NG; 116 AGH, 2/1 CCS, 47 ACH, 2/7 AGH, MID.
    SCOTT, LILIAN CLARE, b 20.11.13; trd Maroochy Dist Hosp, Nambour 1932-36, midwifery Crown St Syd; AANS 1941-47, served 116 AGH, 47 ACH (Koitaki, NG), Cairns, Atherton. After war service Repat Hosps, retd 1978 Asst Dir Nursing, Sydney.
    SCHULTZ, IVY BARTZ. b 14.12.1912; trd Brisbane 1931-35; AANS 1942-45. Awarded MBE, FNM and FACN.
    Sisters Allsop, M.F., Asmus, D.M., Austin, N.L. Bell, M.H.H., Boatfield, M.G., Brandt, R., Briant, M.E., Brosnan, B.A., Brownsdon, M.K., Burns, E.C., Cameron, I.J.A., Chew, M.C., Clements, M.E., Collyer, A.Z.C., Cooper, L.M., Dight, C.J., Dowie, J.I., Doyle, M.F., Elms, J., Everingham, K., Finlay, M.E.F., Hanlon, I.A., Hanrahan, E.F., Hayes, P.M., Heybrock, S., Kay, M., Kenny, E.T., Lanksey, E., Larsen, E.M., Laughren, E.F., Luck, D.M., Luddy, M.B., Lyon, E.W., Morton, V.M.,

Murphy, M.M., McCabe, H.M., McGuigan, B., Oliphant, H.C., Paterson, V.M., Patterson, J., Pearce, I., H., Pryde, A.N., Pym, P., Ross, H.C., Smith, D.E., Smith, P.G.B., Thompson, K.B., Thorpe, M.F., Troedson, T.C., Wallace, M.A., Webb, A.S., Wilson, H.

43. 2/11 AGH Unit War Diary, File 11/2/11, AWM.
2/12 AGH Unit War Diary, File 11/2/12 AWM.
1 Ortho Hosp Unit War Diary, AWM.
Among Queensland members of the AANS known to have served with the 2/11 AGH at this time were Sisters Campbell, R.M.C., Chandler, M.H., Creagh, A.O'D., Cumming, M.J., Drake, M.E., McDonagh, M.A., McNally, N., Neill, C., Newton, M., Sier, E.J., Smith, P.G.B., Stevenson, F.E., Wells, M.L.
Queenslanders with 2/12 AGH included Sisters Barnes, B., Bell, M., Brown, O., Craig, B., Cuddihy, E., Dalby, E., Flower, D., Kay, M., Lanksey, E., Larsen, E.M., O'Loughlin, M., O'Loughlin, D., Searle, M., Templeton, L., Warfield, N.
44. Unit War Diary 117 AGH, File 11/2/32, AWM. Some of the Queensland Sisters known to have served with the 117 AGH and 1 ACH at Toowoomba included:
CRAIG, BLANCHE, b 29.7.17; trd Toowoomba Gen Hosp, gen & obsts; AANS 1942-46, Aust, SWPA, 117 AGH (Toowoomba), 7 ACH (Cairns), 2/12 AGH (Warwick), Balikpapan, Borneo. Post-war — Dip Indust & Soc Welfare, R.Vic Coll Nurs; 1949-59 Bris Clinic.
CARSELDINE, EFFIE MARY, b 25.7.13; trd Thursday Is 1931-36, midwifery Townsville 1936-37, nursing at Palm Is, Emerald & private nursing Qld country towns; AANS 1941-46, Aust & NG, SWPA,
45. MOONEY, J.F. *Downlands. The First Fifty Years. 1931-81* Downs Printing Co. T'ba, March 1981, pp 92-99.
46. MCCALLUM, A.E. (Nan), b 11.1.17; trd Bris Gen Hosp 1935-39, Midwifery Cert Maryborough Base Hosp 1940-41, Child Welfare cert Bris 1941, Staff Bris Gen Hosp 1939-40; AANS 1941-46, Camp Hosps Enoggera & Cowan Cowan, No 1 Amb Trn, 47 ACH NG, 112 AGH (Greenslopes).
47. BREEN, REGINA MAY, b 6.7.17; trd Rockhtn Gen Hosp 1937-39, Diamantina Hosp 1939-40, Women's Hosp 1941-42; AANS 1942-43, No 1 Amb Trn, 116 AGH T'ba.
PARKER, BESSIE, b 25.1.12; trd Nambour Hosp 1935-39. Maternal & Child Welf. Vic. Staff Gen Hosp Cunnamulla, Stanthorpe; AANS 1941-46, NT, NG, NQ.
POLLOCK, G.I. b 20.1.11; trd Maryborough Gen Hosp & Bris Women's; Staff Maryb & Cairns Hosp, School Health Servs, Cairns & Innis; AANS 1941-46, PNG, Port Moresby, Murray Barracks, Finschhafen, 2/2 & 106 CCS, Cowan Cowan Camp Hosp, No 1 Amb Trn, 112 AGH.
Sisters Brown, R.E., Graham, G., Thorburn, A., Weaving, J.
48. 1. 2/1 Hospital Ship *Manunda*, File 11/7/1, AWM.
2/2 Hospital Ship *Wanganella*, File 11/17/2, AWM.
Hospital Ship 1 PL *Oranje*, Files 11/17/3 & 253/4/1, AWM.
2. Hospital Ships 1943, File 53/402/2 AA (Melb).
Hospital Ships Reorganization File 59(402) 5AA (Melb).
3. Australian Nurses on loan to Hospital Ship *Oranje* 1941-42. File 21/720/130 AA (Melb).
WYLLIE, JOYCE, d. Ernest and Doris Wyllie, of Bundaberg; trd Sydney Hosp and served on staff: AANS 2/3 Aust Hosp Ship, *Centaur*. Went down with *Centaur*, 14 May 1943. Name appears on the Bundaberg War Memorial and on the Sydney Memorial, erected near Sydney to commemorate the names of those servicemen and women who have no known graves.
49. *Centaur* Report No 1 1943, File 59/402/6, AA(Melb).
Sinking of *Centaur*, 1944, File 59/402/36, AA(Melb).
50. LUCAS, MARY, b 6.6.1910; trd St Martin's Hosp Bris, 1934-38; obsts trg Bris Maternity Hosp, 1939; staff of Innisfail and St Martin's Hosps; AANS 1941-46, SWPA.
51. President's Address. *The Medical Journal of Australia*, Oct 11, 1952, pp 522-524.
52. FOWLER, ISMAY, trd Bris Gen Hosp, 1937-41; Midwifery Women's Hosp Melb 1942; Staff Women's Hosp Melb; AANS 1943-45, RAANS and RAANC 1946-50; Service 1 Sea Amb Transpt 2/14 AGH, Townsville, 116 AGH, Cairns, 2/8 AGH Jacquinot Bay; 112 AGH, Greenslopes & 113 AGH Heidelberg.
53. Correspondence from Sister Seamark. Another Sister who served at Jacky Jacky was Sister Alice Woolley.
WOOLLEY, ALICE, b 2.2.1912; trd Charters Towers Hosp, Women's Hosp Bris, Infant welfare, Sydney; AANS 1941-47, 110 ACH, NG, Jacky Jacky (Cape York), Tarakan and SWPA.
54. Gas experiments. Routine Orders 2/2 AGH, 30 Oct, 1944. Unit War Diary 2/2 AGH, 11/2/2, AWM.
55. PARKER, B. Sister — see fn (47).
56. Malaria Experiments. HQ AMF AG 78075 of 21 Sept 45. Also Files 417/20/28 AA(Melb), 211/6/1114 AA (Melb).
57. Reference to civilian work performed by AANS is also mentioned by Dr Walker (Medical Services of the RAN and RAAF), pp 456,460.
58. MURPHY, F. *Desert, Bamboo and Barbed Wire*. Ollif Pub Co., Syd, 1983.
59. McQUADE WHITE, E. Sister, b 28.1.01. Altho' an Australian spent nearly 20 yrs of her early life in South Africa. On her return to Aust trd at Bris Hosp and on staff there for a number of years. Matron pte hosp Ayr. Joined NT Med Service.

Enlisted Sept 1939. AANS 1939-46. Matron of first Army Hosp in NT, 119 AGH (1941), later Matron NT, 1943 PM NSW; author of *Reminiscences of an Australian Army Nurse*. After discharge nursing in Eng and in Bahamas.
Hospital Accommodation, Darwin 1939-40. File 33/402/8 AWM and 1940-41, 39/401/89. Establishment of 600 bed hospital, Darwin, 1941. File 19/401/101 AWM.
Medical Services, Darwin to Katherine 1942, file 33/401/203 AWM.
60. McQUADE WHITE, E. *Reminiscences of An Australian Army Nurse*. George & Lamb, Bris. nd.
61. Ibid. p 43.
62. Ibid. p 46.
63. Ibid. p 49.
64. AANS Monthly Reports 1943-44, File 21/3/132, AA(Melb).
HURFORD, CATHERINE MAUD, b Bris, dau Capt & Mrs Hurford. Trd Bris Gen Hosp; Matron Slade Sch. Warwick; AANS 1940-45, PA to Princ Matron NT; staff 111 AGH (Tas).
After war, Matron Strathaven P. Hosp Hobart until retirement 1963. d Hobart 2/10/82.
65. PURKESS, MARY JOAN, b 12.11.18 (England); trd Bris Gen Hosp, midwifery at T'ba Mothers; AANS 1941-46, 112 AGH, 74 ACH, Aust.
and Sisters Hanlon, Isobel; Luddy, Mary; Lyon, Elizabeth; O'Leary, Bridget; Smith, Phyllis.
66. Wigmore, Lionel, *The Japanese Thrust* pp 653-674
67. AANS Rabaul—Japan. File 509/2/4 AWM.
68. Correspondence from Sister Pollock.
69. 46 Aust Camp Hosp.
WALKER, A.S. *The Island Campaigns*, op cit, pp 5, 40, 42, 44, 87, 269, 276, 465, 487. See also 46 ACH Unit War Diary (AWM).
70. MULLER, G. Sister Gertrude Catherine. Trd Gen Hosp Gympie. Midwifery Women's Hosp. Maternal & Child Welfare, Bris & Maryborough. Staf of Mat Welfare 17 yrs. Served Monto & Home Hill Qld. AANS 1941-46. CMF 41-42. Redbank, Camp Hosp; Murray Barracks, Port Moresby. Two tours duty NG. 5 CCS. Later to New Britain.
ALFREDSON, B. Sister, See fn (18).
71. 2/9 AGH, Port Moresby. Unit War Diary, File 11/2/9, AWM.
ANDERSON, OLGA MARCELLA, b 14.1.17. Trd Bris Gen Hosp, 1937-41. Midwifery cert King George V. Camperd NSW, 1948. Repat Dep 1947-77. AANS 1942-47, Aust & NG.
HORAN, EILEEN. Trd Stanthorpe Dist Hosp & Midwifery Women's Hosp, Bris; AANS, service Aust & NG.
HUTTON, IRIS GRACE, b 22.11.15. Trd Rockhtn Gen Hosp 1935-39 and Women's Hosp Bris, 1940; AANS 1942-47; Aust. NG and SWPA. RAANC 1951-70, service in Malaya.
NOBLE, MARCIA ELLEN, from Bundamba, Qld. AANS 1940-46, NG & Aust.
72. MARSHALL, N.M. Matron.
73. CARSELDINE, M.E. see fn (44).
74. Unit War Diary 2/1 CCS. File 11.6.1, AWM.
75. Unit War Diary 2/2 CCS, File 11.6.2, AWM.
76. Aussie Angels, by Sapper, H.E. BEROS in *The Fuzzry Wuzzy Angels and other Verses*. F.H. Johnston Pub Co, Syd. n.d.
77. 47 ACH. Qld Sisters included Hansen, K.I., Compton, H.V., McCallum, A.E., Cottell, G.M., Scott, L.C., Smith, D.E., Elms, J., Retschlag, D.M., Casey, J.B., Chadwick, E.A., Seegar, B.A.T., Beer, M.D., Swallow, G., Bohm, A.D. Tanwan, A., Bedford, V.M., Jones, G.L., Smith, T.A., Hambleton, D.A., 47 ACH Koitaki see Files 405/7/3, 11/3/34, AWM.
JONES, GWENDOLINE LILIAN, b 24.9.16; trd Bris Gen Hosp, 1937-41; AANS, 47 ACH, Aust & NG; Post war service Concord Repat, Hollywood Repat and Kilcoy Hosp 1948-1963.
SMITH, THELMA. Trd Bris Gen Hosp & T'ba Mothers; AANS & RAANC, 1942-64; Served with Camp Hospitals at Goondiwindi, Glen Innes, T'ba, Canungra, Koitaki (NG); 2/6 AGH Buna, 2/7 AGH, Lae, 101 Lt Fd Amb, Port Moresby, 112 AGH (Greenslopes). BCOF (Japan) 1952-53; RAANC 1955-60; Dip Nurs Educ, Army Health School, Healesville 1964.
78. Unit War Diary, 5 CCS (105 CCS), File 11/6/8, AWM.
79. Unit War Diary 2/5 AGH, File 11.2.5 AWM.
TOMLINS, DAPHNE MAY, Sister, b. 30.4.06; trd Bris Mater Hosp 1932-34; midwifery Crown St 1935; Child Welfare, Bris 1936. AANS 1940-46, mostly with 2/5 AGH Darwin, Middle East, Greece & Crete, Eritrea, Armidale, New Gunea, Moresby, Lae and Madang.
80. Unit War Diary 2/1 AGH. File 11/2/1, AWM.
81. Unit War Diary, 2/11 AGH. Files 509/1/1 and 417/20/28 AWM. For 2/11 AGH at Buna, File 481/12/39.
CHRISTIE, DOROTHY MAY. Trd NSW Hosps 1937-41; AANS 1943-45, service Aust NG, SWPA; 103 AGH, 113 AGH, 2/7 AGH Lae, 102 AGH. Name appears on Bundaberg War Memorial.
EGAN, GERTRUDE MARY, b 28.3.06. Trd Mackay Dist Hosp 1931. Challinor Centre, Ipswich, 1933, Obsts 1935, Child Welfare 1936, Theatre Sister Glenrowan PH; AANS 1942-46, NG and Aust, Madang, Lae. After war Matron Challinor Centre and Wolston Park Hospital.

ENGLAND, MOREEN CECILIA, b 1910 Brisbane, trd PA, Sydney 1937–41; AANS 1942–46, Aust and NG, Port Moresby & Lae. 1946–47, Maternity trg, 1948–9 obsts, 1949–76 pte nursing.

OAKES, ELWYN, b 7.2.20, Childers (Qld); trd St Martin's Hosp Bris 1938–42; AANS 1942–1947; service Aust & SWPA.

82. Unit War Diary 104 CCS, file 11/6/7 AWM. One of the Sisters who renderd distinguished service in this Buna campaign was STEVENSON, FLORENCE EVELYN, Sister b 8.7.13. Trd Rockhtn Gen Hosp 1933–37, Women's Hosp 1939, Staff Blackall Hosp 1940–41; AANS 1942–46, service 112 AGH, 2/11 AGH Buna, Madang, Aitape, Alexishafen (special scrub typhus nursing). Post-war mat & child welfare & school health services.

83. WALKER, A.S. *Medical Services of the RAN and RAAF.* pp 468–9.

84. AANS, March 1944, file 422/7/8, AWM.

85. The Australian Nurses in Bougainville. File 509/11/4, AWM.

BLANCH, MYRA, FCNA, AFAIM. b 21.10.1910; trd Nambour Gen Hosp, 1935–39, and obstetrics at Epworth Hosp Melb; 1940–42 Aust Inland Mission Hosp Innamincka (SA); 1942–45 AANS, Aust & SWPA, Moreton Is, 2/2 CCS, 2/1 AGH Torokina, 112 AGH Greenslopes. 1945–46 R Flying Doctor service. Post-war yrs Matron various hosps NSW. 1966–79, Matron St Andrew's Hosp, T'ba.

86. ALLEN, E.M. MID, BEM. Trd Bris Gen Hosp; AANS, service 2/1 AGH.

87. File 509/11/4, AWM, supra.

88. 105 CCS Jacquinot Bay. Unit War Diary. 11.6.8, AWM.

COOKE, EVELYN, b 7.3.04; trd T'ba Gen Hosp. 1921–24, and maternity trg 1924–25; Matron country hosps Qld; AANS 1941–47, ACH's Exhibition, Goondiwindi, Glen Innes; 1942 NG. 105 CCS, 112 AGH (Greenslopes);

HARRISON, DOROTHEA MARY, b 6.9.14; trd Beaudesert Hosp 1938, midwifery Bundaberg Hosp, Staff nurse at Kingaroy & Gladstone Hosps; AANS 1941–46, Aust & NG, 105 CCS. Post-war service at Charleville, Turrawan PH, St Andrew's War Memorial Hosp; Dip Nurs Edu 1966; Nurse educator St Martin's & Holy Spirit Hosps, Bris.

89. 2/8 AGH. Unit War Diary. File 11/2/8, AWM.

90. The Borneo Campaigns;

WALKER, A.S. *The Island Campaigns.* op cit, Chap 16.

ANDERSEN, MARY ADELE, b 2.5.18; Trd Mater, Gen and Chns Hosps before 1944. AANS 1942–46, 2/3 CCS. 7 Div, Balikpapan, Borneo. After war — Crown St Women's & RNS, Syd.

STEVENSON, SADIE, b 2.8.14, T'ba. Trd RPA Sydney; private nursing Brisbane. AANS 1942–46, Aust & NG, 2/6 AGH Labuan.

91. One of the longest serving members of the AANS in the Labuan campaign was Sister Margaret Steen, 2/6 AGH.

STEEN, MARGARET, b 13.6.11; trd Mackay Base Hosp 1932–35; midwifery Women's Hosp; ophthalmological nursing dip, London 1949; AANS 1940, 2/6 AGH Gaza, Greece, Crete, Kantara, Atherton 1943–45; Labuan & Borneo 1945; disch 1946.

92. RAAF Nursing Service. File 509/1/2, AWM.

93. WEBER, E.; trd Rockhampton 1925–29, private nursing, Matron Muttaburra Hosp, RAAF Nursing Service 1941–47, including service in N.G.

CLEARY, A.T.

HAMILTON, E.A.

RULE, C.I. b 1915, trd Longreach Base Hosp 1932, Brisbane Women's 1940; RAAF Nursing Service 1941–46; Service NSW, Qld and on escort duty to Canada.

94. RAAF Nursing Service.

ARMSTRONG, EDNA MAY, trd Maryborough Gen Hosp 1933–37, and on staff there after trg. Sister at Gregory PH 1938, Midwifery trg Women's Hosp 1939, Newmarket PH 1939–41, Atherton Dist Hosp 1941–42. RAAFNS 1942–1946. Served Richmond, Amberley, Greenslopes, Sale, Southport, Con Dep. 1948–50 Dunedin PH, R Melb Hosp, Mat & Child Wel trg 1951; Tutor Sister Bris Gen Hosp 1954. Sister PA Bris, 1958–76.

BRUCE, WINIFRED JEAN, b 4.9.14. Trd Bris Gen Hosp 1937; Staff BGH; 1941–44 RAAF Nursing Service. Served Richmond, Amberley, Archerfield, Sandgate, Greenslopes. Escort duty to Canada, Fairbairn (Canberra) and Bundaberg.

DOHERTY, MURIEL KNOX, RRC. b 19.7.96; trd RPA Sydney. Studied nursing educ Brit & Europe 1929–33; AANS 1939. Matron RAAF NS 1940; PM 1941; RAAFNS NSW & Qld; Comm reorg NS in NSW; 1945 Member UNRRA.

MARLOW, BERYL DAWSON. Trd Bris Gen 1935–39 and on staff 1939–42. RAAF Nursing Service 1942–46. Bradfield Park, Port Moresby, Milne Bay, Townsville, Concord, Greenslopes. Later Greenslopes and Rosemount 1946–48. Perry Park Rehab Centre Bris 1948–51.

NEILSON, E., b 8.11.08; Gen and Child Welf Cert Bris Gen Hosp, Midwifery Cert Lady Goodwin, Rockhtn; Comm Gov Hosp NT 1936–40; 1941–46 RAAF Nursing Service, Senior Sister (F/L). After war pte nursing and 15 yrs R Melb Dent Hosp 1973.

RODWELL, JOAN MAVIS, b 17.4.1918; Trd Mater Hosp Bris. 1936–40 attached to No 2 MAETU. Post-war, pte nursing in Melb and in South Africa.

RICHARDSON, ISABEL BESSIE, b 29.11.13; 1933–36 Triple Cert, Bris Gen Hosp; Pediatrics Johns Hopkins Hosp USA, Med Pract GP Practice. RAAFNS 1941–46. Pilot Officer.

BRAY, JEAN, b 26.10.07; trd Bris Gen Hosp 1928-32, Child Welfare Fortitude Valley 1934, Maternity, Lady Bowen 1934-35. 1941-46 Princess Mary's Royal Air Force Nursing Service (UK); Service England, Scotland, Italy and North Africa. Pte nursing and at Rosemount Repat Hosp.

TAYLOR, EILEEN, Mt Larcom, NC Line. WAAF. Trd Sandgate, Bradfield Park, Sydney — then to Greenslopes, 1945-48. 1948 Bundaberg Gen Hosp and completed trg after war.

LANG, MARGARET, OBE. AANS WWI. Served Salonica. 1940 Matron in Chief RAAF NS. d 14 Feb 1983, aged 89 yrs. *Grey & Scarlet* 1983, p 35.

PURTILL, STEPHANIE. RSM, OBE. Trd St Vincent's Hosp, Toowoomba; Private nursing Qld & Vic; RAAF NS 1941-1945, Amberley, Townsville & escort duty to Canada; Rehab Vic Hosps; Sisters of Mercy, All Hallows' 1947, professed as Sr Pauline Mary 1950; 1956-1982 Sister in Charge Xray and radiology departs.

95. The Medical Air Evacuation Transport Unit (MAETU).
WROE, M.H.
CHANDLER, B.O.
See also, 'Experiences of a Flight Nurse', Joan Rodwell. *The Misericordian*, 1946, pp 42-53.

96. Correspondence from Sister Cleary.

97. *RAAF Saga*. The RAAF at War, RAAF Directorate of Public Relations, Sydney, nd, p 102.
See also *Victory Roll*. The RAAF at War, p 42 for references to Sisters Beryl Chandler, Marie Wroe and other RAAF Sisters serving with the MAETU.

98. The Women's Royal Australian Naval Nursing Service. File 509/2/3/, AWM.

99. McKENZIE, MARY, Trd Roma Dist Hosp; Member Navy Nursing Service.
CONQUEST, JOAN, Trd in Qld; Member Navy Nursing Service; Later Matron Anzac Hostel, Brighton, Vic.
SAUNDERS, MARGERET.
KING, GITHIA.

100. Correspondence from Sister Leahy.
LEAHY, CATHERINE, b 3.6.15, trd T'ba (Gen & Midwifery) AANS 1943-47, 112 AGH Greenslopes, 2/7 AGH Lae, 104 CCS Cape Wom. Post-war Repat Hosp Heidelberg, Matron Wallumbilla Hosp (Qld), Yarrawonga (Vic) 1949-54.

101. 2/14 AGH Unit War Diary, File 11.2.14, AWM.
See also File 420/2/2. AWM.

102. Ibid.

103. Correspondence from Sister Chandler.
LENNARD, WALLACE HAYDON, b 4.7.09. Newspaper journalist. Joined ABC News 22.5.39; War correspondent 1942-46. Arrived Port Moresby Feb 1942 and reported first raid and 109 subsequent raids. Went with 7 Div to Kokoda, Buna, Gona. Dec 1943 flying to Cape Gloucester, plane crashed, others killed, Lennard badly burnt. 1944-45 in Darwin, NG, Solomons, Burma and SEAC. Returned ABC. Ret 1947.
EWING, AUDREY MARY. Trd Maryborougn Base Hosp. 1935-39. Bris Women's Hosp. 39-46 Lady Chelmsford. AANS: CH Redbank, 5 CH Rocky Ck, Cairns, (Edge Hill). 2/12 AGH Warwick, Balikpapan. 2 Reception Group. Manila (for reception of POW's — very secret mission!). After war staff Bundaberg Hosp.

104. JEFFREY BETTY. *White Coolies* op cit, chap 25.

105. File 336.1.1289 AA(Melb). Ill-treatment of POW's and internees including Australian Nurses massacred on Banka Island.

106. Peninsula Women's Chorus presents *Song of Survival*. A Women's Vocal Orchestra. Dr Patricia Hennings Director. Programme 1983.

Chapter Six

New Generations of Service:

Japan to Vietnam and Beyond, 1946–1984

The use of helicopters combined with the 'triage' system showed gratifying results in saving lives and in no mean measure reduced battle fatalities. The Sisters' contribution was outstanding and they continued to serve under the most trying conditions until the withdrawal of the Australian Force late in 1971.

Report on Vietnam.

1946 BCOF Japan

For many, the Second World War did not finish with the surrender of the Japanese in September 1945. When it was decided that there would be a British Commonwealth Occupation Force (BCOF), provision was made for medical services, including a hospital for which some members of the Australian Army Nursing Service would be required. Thus it was that the war establishment for the 130 Australian General Hospital was laid down and preparations made for its immediate equipment. Of necessity it had to be in the beginning a makeshift affair, depending on which medical and nursing personnel were available and what stores and equipment could be secured at short notice. So on 18 October, 1945, 130 AGH was formed at Lae, New Guinea on the site of 2/7 AGH, moving then to Morotai where equipment from the 2/5 AGH and 2/9 AGH was taken over. Sister Monica McMahon,(1) then Matron of 110 Military Hospital, Perth was appointed first Matron of 130 AGH, taking with her volunteers from among the AANS. Together with other medical and nursing personnel, the unit embarked on the *Manunda* and disembarked at Kure, Japan on 25 March 1946.

The former Japanese Naval Academy at Eta Jima was the building allocated for the hospital. As happened many times during the Second World War, a building not built for a hospital required a great deal of improvisation before it was suitable and even then there were many difficulties. Members of the AANS were quartered in austere dormitories, 22 in each, with stretchers and steel lockers. Japanese washrooms were converted into shower rooms, but it was almost two years before sleeping and messing arrangements could be said to be civilized.

The sudden change from tropical warmth to the sharp Japanese winter produced its own problems. The nurses found the Australian issue uniforms quite inadequate. It was some time before 'coats, windproof women's' became available, as well as 'boots, leather, brown, fleecy'. The incidence of recurrent malaria among the troops, precipitated by the sudden switch of seasons, soon overwhelmed the 300 bed hospital,

Facio-Maxillary Staff, Greenslopes 1946–47.
L. to R: Sisters Kavney, Casey, newman, Wright and Physio Miss Phillips. At the rear members of the AAMWS.

causing the DGMS, MajGen S.R. Burston to approve that the size of the hospital be doubled, with 400 general and 200 special beds. A change in army policy of permitting families to join their husbands in Japan gave a new and added responsibility to the nursing service. As well as caring for the civilian dependents and their children the nurses found themselves looking after babies — over 100 were born in the hospital.

During October–November, 1948 the hospital transferred to a new site overlooking Anzac Park and Kure Harbour. In 1949 it became the British Commonwealth General Hospital with a small detachment of Sisters at a CDS in Tokyo. As the hospital became a British Commonwealth Hospital rather than an Australian General Hospital, it accepted patients from all allied forces. Medical and nursing personnel also had the opportunity of working with staff from other forces, including the 92 Indian General Hospital and Sisters from the QAIMNS.

Queensland's nursing Sisters were well represented in this occupation force hospital. Sister Edna Doig, with long service in the Middle East behind her was Deputy Matron. Sisters Elizabeth Lyon, Cameron Bartlett, Bridget Guilfoyle, Jean O'Neill, Thelma Smith, Margaret Steen, M. Geraghty, A. Duffield and P. Zillman were among those to serve in Japan during the years up to 1952.On the social side, Queensland Sister Foley became the first of BCOF members to be married in Japan.[2]

Queensland Sisters also had close contact with their colleagues in the air force nursing services. They met up with the Royal Air Force Sisters at Iwakuni, members of PMRAFNS and with the RAAFNS. Queensland Sisters who distinguished themselves at this time included Sisters Lucy Rule (Rockhampton), Ethel Morgan and Eunice Feil (Toowoomba), Lou Marshall (Yelarbon) and Mab Wilson (Brisbane).[3]

Some well-known Queensland Nurses (Greenslopes 1946).
L. to R: D. Roe, E. Hanrahan, Miss Caldwell, B. Stein, Matron Abbott, Miss Sorensen.

At first there was some concern about the safety and security of soldiers (including Sisters) in the occupation force. No one was quite sure what the reaction of the Japanese would be to having their conquerors in their midst. Strict precautions were taken to ensure the safety and welfare of all personnel, especially those of the women's services. The CO of 130 AGH, Col C.W. Nye, promulaged a detailed list of Standing Orders for their protection. For example,

- Between the hours of 0900 and 1700 any female member may leave the camp area, provided she is accompanied by a male escort.
- After 1700 hours members will not leave or remain away from camp in parties of less than two, accompanied by at least two male escorts.
- Parties of not less than four — 2 female members and 2 male escorts will travel in any one vehicle.

The nurses found this highly amusing, but after a year of so experience proved these precautions were unnecessary and they were gradually relaxed.(4)

All members of the occupation forces, including Sisters, had to run the risk of disease, with primitive Japanese sanitation, poor hygiene standards and the occurrence of epidemics among the Japanese population. It spoke well for the high medical and hygiene standards of the allied forces that they were free of smallpox, typhus, relapsing fevers and cholera even though they were in areas infested by flies, rats, mosquitoes and vermin. In 1948 an encephalitis outbreak in the civilian population resulted in 3,000 deaths, but there were no cases among BCOF personnel who had been inoculated against the disease. The high standards demanded by BCOF in turn helped in raising health standards among the Japanese. Japanese doctors were invited to watch operations at 103 AGH, while Japanese nurses were given a course of instruction in nursing and hygiene by the AANS and Matron McMahon.

New Conditions of Service

In these early post-war years subtle changes were made to the status of all service personnel including nurses. They had enlisted as volunteers to serve in a war that had now ended. Many were demobilized, others volunteered to stay on and serve with BCOF. In July, 1947 a 'post-war interim army' was formed and all members of the AANS still serving were transferred from AIF to IA. A further change came in 1949 when they were again transferred to the Australian Regular Army (ARA).

In this changeover period from an army on active service fighting a war to the unusual status of an occupation force in the homeland of the defeated enemy, the Department of the Army took the opportunity for reviewing the conditions of service of all personnel, including nurses. Rates of pay, hours of duty, recreation leave, etc were considered in the light of awards made to nurses in civilian hospitals in Australia. It found that rates of pay for Captains and Lieutenants in the AANS at that time compared favourably with those awarded to Sisters and Staff Nurses but that for Majors and above, in senior positions of responsibility, army rates fell below those of Matrons in civilian hospitals. Perhaps the biggest difference lay in the hours worked and in the attitude to work. In army hospitals the minimum working fortnight for nurses was 108 hours, in civilian life 88 hours. Moreover, in emergencies army nurses worked even longer hours, without additional financial gain, while civilian nurses received payment of time and a half rates for overtime.(6)

One of the difficult problems discussed at this time with nurses in either the permanent military forces or the citizen military forces, concerned the age at which they should be retired to the Reserve of Officers (R of O). The Navy and the RAAF had already settled for 55 years but the army authorities were of the view it should be 57 years for the Matron-in-Chief. The DGMS had other views. He pointed out that an army nurse who retired at 55 had little chance of a civilian appointment. His greatest concern was to have a sufficient number of nurses below the age of 40 who would be eligible for active service overseas. If nurses were encouraged to stay on to the age of 55, young inexperienced nurses, without army training, would have to be recruited in an emergency. The AG was not convinced by these arguments and recommended a common retiring age of 55 years for nurses and 57 for the Matron-in-Chief.(7)

1948 Royal Title

While these events were taking place in Japan, other decisions were being taken in Canberra and Melbourne which were to have important effects on the future status of the Australian Army Nursing Service. In November 1948, King George VI was pleased to grant the title Royal, in recognition of the services of the AANS in two World Wars. Thus the RAANS came into being. Further changes were in store. The post-war reorganization of the Australian Army Medical Services did not provide for the establishment of a General Hospital, consequently the appointment of female personnel to the Regular Army was not envisaged in the early years. However with the development of the Regular Army, the introduction of National Service, the war in Korea and the fear of a communist uprising in countries in South-East Asia it became necessary to maintain a General Hospital overseas and Camp Hospitals within Australia. To provide nursing staff for these medical units, approval was given in July 1949 for the appointment of female officers and for the enlistment of other ranks (female) in the Australian Regular Army.

1951 Nursing Corps

A major change to the status of army nurses came in February 1951, when the service was designated a corps and became the Royal Australian Army Nursing Corps, (RAANC). The new Corps absorbed the former Australian Army Medical Women's Service which had been established in 1943 to provide nursing orderlies and other roles in military hospitals. In the new organization commissions were granted to fully qualified nurses while other ranks were enlisted to take over nursing orderly and ward orderly duties within hospitals. Conditions for appointment as officers (Lieutenants) in the Royal Australian Army Nursing Corps were gazetted and in general these varied little from long standing practice in the acceptance of army nurses — trained nurses registered by a Nurses' Registration Board, of good character, with acceptable 'personal attributes', a British subject, single or widowed without dependents. Nurses had the option of applying for a long service commission, to make a career within the army, up to the retiring age of 55 years or for short service commissions of five years. On appointment nurses were given a six weeks indoctrination course at a school of army health. Conditions of appointment made it clear that if a nurse married the appointment would be terminated immediately.

While these changes established the Royal Australian Army Nursing Service as an integral part of the Australian Regular Army, there was still need to make arrangements for the training of a pool of reservists who could be called up in an emergency. Accordingly, a Royal Australian Army Nursing Corps Company, Citizen Military Forces, (CMF) was raised in each Command towards the end of 1952. These volunteers were required to attend 12 days compulsory training each year, 14 days of continuous camp training with an additional 7 days optional. Training was not only geared towards nursing but towards an understanding of army administration and the role of the medical services therein. In this, the objective was little different from that of 1910–1914 when nurses were classified as 'efficient'. By February, 1954 there were over 600 CMF nurses on the active list. Moreover, in view of the uncertainty of the times, a further reserve of qualified nurses willing to serve in any national emergency, was built up.

1951–1961 Royal Australian Army Nursing Corps Honours

Meanwhile other changes came to the RAANC and numerous honours were bestowed on the Corps and on individuals within the Corps.

- In 1953 at the time of Her Majesty's Coronation, Her Majesty the Queen was graciously pleased to accept appointment as Colonel-in-Chief of the Royal Australian Army Nursing Corps.
- In 1953 the RAANC obtained its own corps badge. It is in silver with a scarlet backing, surmounted by a crown, with a lamp in the centre. The lamp is the Greek lamp of learning, sometimes seen in the statuary of Hygeia, the Greek Goddess of Health. The motto *Pro Humanitate*, in the service of all mankind, exemplifies the devotion and service of nurses for sick and wounded of all nations.
- 1957 An alliance between the RAANC and Queen Alexandra's Royal Army Nursing Corps was approved by Her Majesty the Queen.[8] Nursing Sisters from both corps served side by side in two world wars and in the post World War II campaigns in South-East Asia. In 1971 this was taken a step further with exchange postings, members of the RAANC gaining experience by serving in BMH's in Britain and Germany while QA's served in Australia and New Guinea.
- In 1958 approval was granted for the adoption of a Corps Flag. The grey and scarlet flag, defaced by the corps badge, is now flown proudly in units where the RAANC serve.
- It became customary for the Matron-in-Chief, RAANC to be appointed Queen's Honorary Nursing Sister. Queenslander Colonel E.M. Doig, RRC, FNM, was appointed to this position in 1961.

1950–1953 Korean War

The Korean War erupted suddenly, although there had been tension in the area since the end of the Second World War when the Russian and American armies met at the 38th parallel. The division of the country into two, with the communists in the north and the capitalists in the south was contrary to United Nations policy. The UN wanted UN supervised elections for the whole of Korea, but when the communists refused UN access to the north, elections were held only in the south. Subsequently the Republic of South Korea, with Dr Syngman Rhee as President, came into being in August 1949. Border incidents continued until 25 June 1950 when the North Koreans invaded the South. When the UN resolution calling on the North Koreans to withdraw was ignored, a UN force was organized under General Douglas MacArthur. Soon Australian army, navy and air force contingents were involved, along with Great Britain, New Zealand and other UN forces. It was a bitter war which raged up and down through North and South Korea until the Armistice of July, 1953. The Australian army suffered 1,538 casualties in Korea, including 281 killed or missing.[9]

The first plan for the treatment of Korean casualties was to air lift them as quickly as possible to the British Commonwealth General Hospital at Kure with all its modern facilities. In September, 1951 the

First Nurses and Sisters of the A.R.A. to go to Japan 1952.
L. to R: Pte Silk, Lt Haynes, Lt Probyn-Smith, Pte Worthington.

British Commonwealth 'Z' Medical Unit was established in Korea and together with medical personnel from other parts of the Commonwealth, Australian Sisters were posted to this unit.

The extra work involved in nursing casualties from Korea placed a greater strain on the limited nursing facilities at the BCOF Hospital. Longer hours had to be worked and recreational leave cut. In her monthly report for November, 1950, Matron McMahon described the health of the RAANS as being only FAIR for the month. In a letter of explanation to the DGMS she explained.

> During the month of November a very high percentage of RAANS were suffering from very heavy colds, and due to increased numbers of patients, the staff were working much longer hours than they had been accustomed to and therefore members of the RAANS were not looking quite as well as usual.[10]

The arrival of additional nursing staff from Queen Alexandra's Royal Army Nursing Corps (QARANC) enabled hours to be cut and recreational leave restored.

A major contribution to the medical services in the Korean War was made by the RAAF Transport Unit in ferrying casualties from the Korean battlefields to the base hospitals in Japan, particularly at Kure. This was no routine task. First there were the rugged mountains of southern Japan, then the 100 miles of open sea, then the rugged bare skelton of Korea, the collection of wounded and the equally perilous return journey. Weather was often severe, with Korean airstrips often caked with ice and snow. Frequently airmen had to hose down the wings of their Dakotas to get rid of ice which had accumulated during the night. In six years some 14,900 medical casualties were evacuated on RAAF transports with nursing sisters caring for them. In a remarkable safety record not one patient was lost or suffered injury in these flights.

Queensland nurses were well represented in these RAAF evacuation flights. Those known to have been involved included Sisters L. Marshall (Yelarbon), M. Wilson (Brisbane), E.L. Feil (Toowoomba), L. Rule

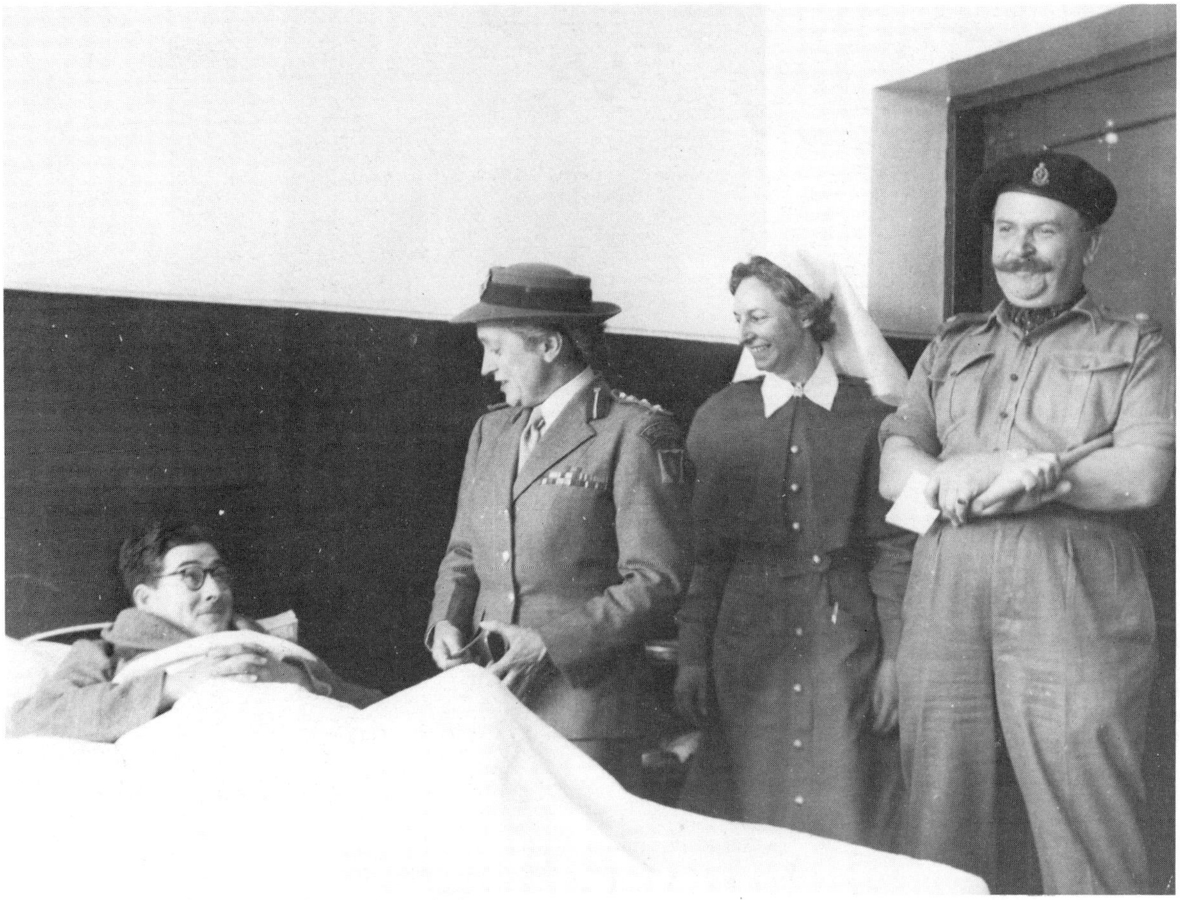

Matron Bowe and Sister Ford at the British Commonwealth Communications Zone Medical Unit, Seoul, Korea. (British Army Photo).

The ugly and cheerless view outside the same army hospital in Townsville.

Sister B. Probyn-Smith laying a wreath at Anzac Square, Brisbane, shortly after her return from Korea.

(Rockhampton), A.B. Edwards (New Farm)[11] and E. Morgan (Toowoomba). Sister Feil recalled the highlights of that experience:

> I remember evacuating British, Australian, Canadian, New Zealand and Turkish casualties. After we arrived back at our base at Iwakuni (Japan) the patients rested briefly at the RAAF Hospital before being placed on a hospital train for the journey to Kure. We were also called upon to accompany the sick and wounded back to Australia, usually on Qantas with an overnight stop at Manila in an American army hospital there. Sometimes we took British boys to Singapore where they were air lifted to Britain by RAF aircraft.

Sister Wilson brought them back by a different route — Iwakuni, Guam, Port Moresby (where they were received by a native choir), Sydney.

Prince Philip, Duke of Edinburgh with Matron E. Hanrahan at the Heidelberg Repatriation Hospital.

1948–1971 The Malayan Emergency

A period of instability emerged in South-East Asia immediately after the war, as communists sought to overthrow 'capitalist' and 'imperialist' regimes. By using guerilla tactics, ambushing and killing rubber planters and by assassinating key political leaders, they created a situation which threatened the stability of the whole South-East Asian region. Australian forces joined the Commonwealth forces in Malaya in a campaign to combat the guerillas. In July 1950 RAAF bombers operating from Tengah airfield in Singapore joined the attack. Australian soldiers remained in Malaya until 1960, losing 15 killed and 27 wounded, with many casualties due to illness.

In September 1955 six members of the Royal Australian Army Nursing Corps were sent to Malaya where they nursed in British Military Hospitals at Kamunting (Perak), at Kuala Lumpur and in the Cameron Highlands. Later the number was increased to eight, the Sisters nursing members of the Commonwealth Forces (and their dependents), locally enlisted members and even Iban trackers from Borneo.

Although the Malayan Emergency officially ended in 1960, political instability in the area remained, as communist guerillas opened up other 'little wars'. Commonwealth troops remained in Malaya as part of the South-East Asia Strategic Reserve, ready to meet any new emergency. This reserve force was based on Terendak (Malacca) where there was also a fully equipped British Military Hospital. RAANC Sisters continued to serve until 1971 when the whole complex was handed over to the Malayan Government and the force moved to Singapore. Here Australian nurses as members of 28 Medical Unit continued nursing duties until the British Military Hospital was closed in August, 1971.

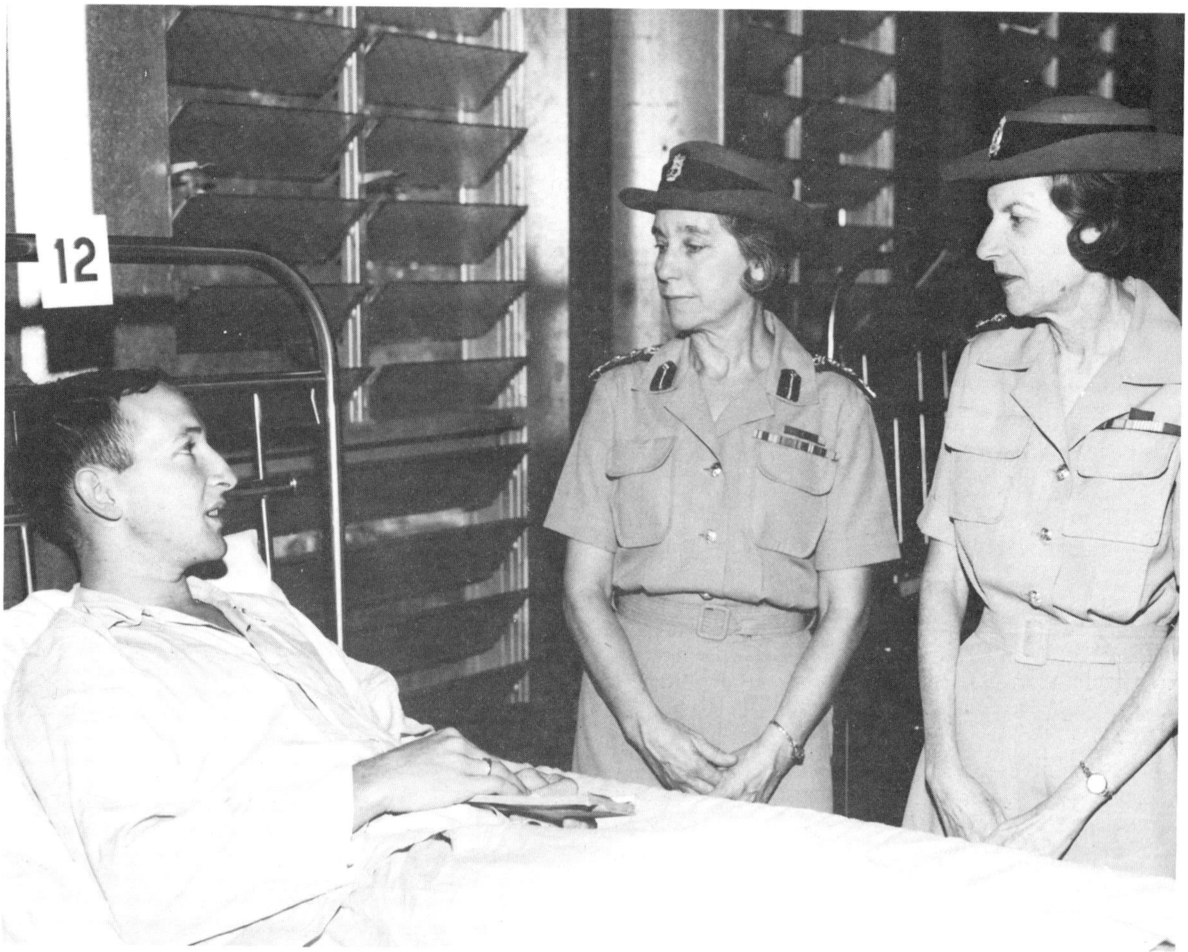

Col E. Doig, Matron-in-Chief and Director, Army Nursing Services with LtCol Guilfoyle, Asst Director Army Nursing Services, Northern Command visiting army hospital at Pallarenda, Townsville 1969.

New methods of bringing in the wounded, Malaysia 1958.

Casualties also came from the confrontation between British and Commonwealth troops and Indonesian guerillas in North Borneo during the period 1963–65. The Federation of Malaysia, which came into being in 1963 with UN approval, included Malaya, Singapore and the North Borneo territories of Sabah and Sarawak. President Sukarno, of Indonesia, as part of his crush Malaysia campaign, encouraged terrorist activities, especially in Borneo. Australian troops were committed to assisting the Commonwealth force in this campaign, losing 7 dead and 10 wounded. The abortive Communist coup in Djakarta, which eventually toppled Sukarno, brought an end to the fighting between Australia and Indonesia. Australian casualties were again evacuated to the British Military Hospital at Terendak, to be nursed by RAANC Sisters.

1966 New Guinea

After the end of the Second World War Australia still had responsibilities for administering Papua and New Guinea, rebuilding the country after the devastation of war. It was also charged with the responsibility of leading them first of all to self-government (1972) and then to independence (1974). Australia also assisted in the training of indigene servicemen and in the development of a Papua New Guinea force. In this peace-time situation two Australian Sisters were invited to return to Port Moresby in 1966. They found the scene vastly different from that when another generation of Australian Army Nursing Service Sisters had laboured there until 1945. However, they also found that the Officers' Mess at Taurama Barracks, where the 1st Battalion of the Pacific Island Regiment was stationed, had been constructed on the 'hard standing' of the 2/5 AGH operating theatre more than 20 years previously! By 1969 the number of Australian Sisters had increased to 4, with one Sister engaged on classroom and bedside teaching, training Papuan and New Guinea nurses and orderlies. They were assisted in their work by other Sisters from unexpected sources. There were wives of Australian servicemen who had previously been Army Nursing Sisters, there were also civilian nurses both Australian and indigene.

ANZUK Force 1971–73

As the British forces withdrew from Malaysia and Singapore in 1971, it was clear that a strong presence was still needed in the area to deter communist aggression and to deal with any emergency which might arise. So Australia, New Zealand and UK forces brought into being the ANZUK Force. RAANC Sisters who had been serving with 28 Medical Unit at the British Military Hospital were moved to the former RAF Hospital at Changi which became the ANZUK Military Hospital. A few Sisters were sent to the Royal Navy Sick Bay which became the ANZUK Woodlands Hospital. This experience was noteworthy for Australian Sisters for the nursing staff of Australian, New Zealand and British nurses was completely integrated. Secondly, the hospital included maternity and children's wards, a new experience for many army nurses. In fact more than 1,200 babies were born in the first two years of the hospital's existence. Changing political philosophies with a change of Australian government in 1972 brought about the withdrawal of Australian commitments in South-East Asia and this episode came to an end.

In these years of what might be seen as minor skirmishes in South-East Asia during the 1950's and 1960's, prior to the war in Vietnam, several Queensland members of the RAANC saw service in Malaya and Singapore, notably Sisters Bridget Guilfoyle, Jean O'Neill and Edna Doig. The latter continued a distinguished administrative career in army nursing, becoming Assistant Director, RAANC, Northern and Eastern Commands 1956–1961. When Jessie Bowe retired in 1961 Edna Doig became Matron-in-Chief, RAANC.

Vietnam 1962–1972

For ten years Australia was involved in this bitter and controversial war, an undeclared war between the communist forces of North Vietnam and the American–supported South. The rights and wrongs of that campaign and the various battles which eventually led to the withdrawal of American and Australian troops and of the ultimate defeat of South Vietnam need not be canvassed here. Indeed the official Australian history of this episode had not been published by 1984. It will suffice to indicate Australia's involvement, the casualties incurred, the medical facilities provided and the role played by members of the Royal Australian Army Nursing Corps, particularly of Queenslanders.

Australia's commitment began in August 1962 when 30 military advisers went to Vietnam to train members of the South Vietnamese army fighting the communist insurgents from the north. By 1965 the Australian Army Training Team Vietnam had grown to 100. In that year the Australian Government despatched the first fully integrated military unit, the First Battalion, Royal Australian Regiment, together with logistical support. It was based at Bien Hoa, with a total of 1,500 men. A steady increase in numbers continued, until by 1967 the Australian Force Vietnam represented the three armed services. The Task Force was situated in a rubber plantation at Nui Dat while the Logistical Support Group was at the sea port of Vung Tau. By the end of 1967, some 8,000 Australian troops were in Vietnam and casualties began to mount. Pressure from the Australian electorate forced the gradual withdrawal of Australian troops from Vietnam by 1972. In a decade of fighting over 50,000 Australians served in Vietnam, with 2,821 casualties, including 496 dead.

Medical arrangements for Australian sick and wounded relied heavily on Field Ambulances, as the kind of guerilla fighting and the nature of the terrain precluded the setting up of a base hospital for some time. While the usual kinds of illness common to jungle fighting had to be treated, an increasing number of surgical cases occurred, not from bombing or artillery but from land mines, booby traps, mortar fire, sniping and 'hit and run' encounters. While at first Australian casualties were treated by US medical units, 8 Aust Fd Amb was established at Vung Tau in 1967, including 4 members of the RAANC. The following year 1 Aust Fd Hosp was established at Vung Tau and over the next few years 43 members of the RAANC were to serve in Vietnam. One report summarized their work:

> Here a sophisticated and highly successful "triage" system was developed which involved reception, assessment, resuscitation and surgery of multiple casualties. The use of helicopters combined with the "triage" system showed gratifying results in saving lives and in no mean measure reduced battle fatalities.
>
> The Sisters contribution was outstanding and they continued to serve under the most trying conditions until the withdrawal of the Australian Force late in 1971.[12]

This Field Hospital was a radical departure from the standard general hospital or casualty clearing station of World War II. It was small in bed state, usually about 150, but compact and comprehensive in its facilities. It operated much closer to the scene of action, where the modern helicopter service was able to bring the casualty from the battlefield to the hospital within a couple of hours or less. No longer was there a delay of days on end as the wounded were transported by cumbersome means back to a hospital. Moreover, sophisticated transfusion systems and modern drugs were available immediately. Sisters at the Field Hospital, operating under the 'triage' system were able to bring their professional service into action as part of the medical team, at a very early stage in the patient's treatment. They had to be highly competent technically and yet adapt themselves to the needs of different groups of casualties. Unlike their colleagues in civilian life they had to be 'generalists' rather than 'specialists', yet their 'generalism' was highly 'specialized'. Another important change in this Vietnam situation was the rapidity of evacuation. With helicopters and aircraft readily available, casualties were airlifted as soon as possible back to a base hospital, either to Japan, to Butterworth (Malaya) or to the mainland. No longer was there the problem of overcrowding at the hospital waiting for a hospital ship or other means of evacuation.

All operations in Vietnam were eventually overtaken by the propaganda, casting doubts on the legitimacy of the war and questioning whether Australian troops should be involved. Knowledge that 'back home' there was mounting criticism of the war, with large 'moratorium marches', created a guilt feeling as much as the fighting units as in the RAANC. To their credit, it must be said that members of the RAANC put these doubts aside, as they had a task to do, day and night. Sister Shale recalled that

> the work was rich and rewarding in retrospect. One developed a close bond with patients, especially the critically ill. We worked 6 days a week, 8 hours a day (the girls before me worked 12 hours). We also had an exchange system working with the US nurses and I found living and working with Americans a new and interesting experience. By far the most important aspect of my tour of duty in Vietnam was the companionship I experienced — something one doesn't experience outside of a war situation.

Queensland members of the RAANC known to have served in Vietnam included the following:[13]

Clayton, G.K.
Ferguson, P.K.
Findlay, M.A.
Hall, E.A.
Long, T.J. ARRC (Matron)
McLean, D.L.
Minchow, J.E.
O'Neill, J. (Matron)
Pittendreigh, A.M., ARRC (Matron)
Potts, L.K. ARRC.
Raby, M.E.
Ruddle, F.M.
Shale, E.M.
Sorley, V.M.
West, P.H.
Wright, A.C.

Two additional Queensland Sisters of the RAANC to go to Vietnam were Sister Ann Wright of Redcliffe and Sister Patricia Ferguson of Coorparoo. Sister Wright trained at the Royal Brisbane Hospital Nursing School and Sister Ferguson from the Holy Spirit Hospital, Brisbane. They both served for 12 months at

Colonel Edna Doig. RRC, FNM, FCNA, Matron-in-Chief and Director, Royal Australian Army Nursing Corps 1961-1970.

the 130 bed 1 Australian Field Hospital at Vung Tau. At the time of their arrival in Vietnam 5 of the 12 members of the RAANC serving there were Queenslanders. Another event of note at this time was the wedding of Sister Barbara Green, from Mt Isa, to Lt Jim Heard of the NZ Army Training Team. They were married in the Anzac Chapel at Vung Tau, the first wedding of a service woman in Vietnam. Her bridesmaid was another Queenslander, Sr Mary Shale, ably assisted by Sr Gillian Clayton, of Ipswich.

Army Nursing — in Retrospect

The period from the Boer War to Vietnam and into the 1980's has been a remarkable period in Australian history — and indeed in world history. It has been dominated by wars, two world wars and many minor skirmishes, insurgencies, guerilla wars and confused campaigns. Almost every decade Australia has been involved, helping 'the mother country', defending freedom, assisting the United Nations or combining with long established allies such as the USA or New Zealand. It has fought in the Middle East, the Near East, the Far East, on land at sea and in the air.

Wherever the Australian troops have gone, the Australian Army Nursing Service has gone, as part of the Australian Army Medical Service. At first, in the Boer War, they went tentatively, even apprehensively, uncertain what their role would be. Of their dedication and determination to be there to assist the sick and wounded there can be no doubt. Those who were not sent officially found other ways of getting to the battlefields of South Africa. They went as amateurs in a military sense, virtually ignorant of the demands and requirements of army life. They nursed alongside professional nurses of the British army and acquitted themselves well, a reflection on the standard of training and the discipline of civilian hospital life in Australia. This experience was a critical factor in the establishment of the Australian Army Nursing Service in 1902.

From these uncertain beginnings the AANS established itself as an essential part of the army medical services in the First World War. By sheer dedication to their task and by their professional efficiency they overcame 'passion and prejudice' in high places. Their record in Egypt, France, England, India and other battlefields is now part of history. While several hundred Queensland nurses 'answered the call' and all performed at a high level, several were decorated for exceptional bravery under fire and for dedicated nursing services. Many returned to civilian nursing giving inspiration and leadership between the wars.

The Second World War saw an enormous response from nurses, some 3,500 serving in the AANS (600 from Queensland) while others served in other branches of the services and in allied forces. Important changes took place in status, rank, uniform and conditions of service although their role in 'front line' situations remained unclear. The fate of nurses captured by the Japanese in Singapore and Rabaul was a horrifying shock to civilian and army authorities, bringing about some caution and over-protection for nurses in New Guinea. Despite varied conditions of warfare and climate in Britain, the Middle, the South-West Pacific Area and in Australia, the morale of the AANS remained high, their tradition of faithful service enhanced by their record in many campaigns. It is customary these days for military historians to debunk traditional heroes, the Gallipoli veteran, the bronzed Anzac digger, the 'Sixth Divvy' veteran, especially 'The Thirty-Niners'. The RSL, Anzac Day, army generals such as Blamey have not escaped these attacks. The British, the Americans, the Dutch and other allies have also come in for their share of critical devaluation. Thus far members of the nursing service have escaped their treatment. No doubt in the future some writer seeking notoriety will attempt to describe them as they were known in pre-Florence Nightingale days. Such an assessment would be far from the truth and it would lack credibility.

The RAANC has entered a period of peace-time activity but its formation continues both in the Regular Army and in the CMF. In times of war military nursing is the pacesetter in the profession, with hundreds of nurses enlisted for specific purposes. New environments, some created, some the backdrop of warfare, form a new framework for the nursing profession. New situations arise, new medical and nursing problems have to be met. New techniques emerge to meet different kinds of illnesses and battle casualties. The nurses return from their service more mature, better experienced, more highly equipped technically. Their impact on civilian aspects of nursing is immediate and far-reaching. However, in times of peace military nursing, unless action is taken, may become divorced from developments in civilian nursing. Professional isolation

AANS. An historic photo which includes many famous Queensland nurses.
Seated in front, L. to R: Sister Hanrahan, Miss Grace Wilson, Miss Caldwell, Miss Sorensen, Miss Paten.
Standing at rear, L. to R: Miss Burbidge, Sister Kenyon, ______Sister Jacobs, Sister Marks, Sister Tuddy, Sister Dean, Sister Murphy.

can be a disaster for both the nursing profession and for the defence forces. This is overcome, in part, in Queensland and no doubt elsewhere, by regular contact in professional medical and nursing associations by members of both civilian and army groups. The CMF, properly developed, is also an essential part of this exchange of professional information and ideas. In Queensland (1984) the responsibility for all nursing matters in 1 Div falls to SO1 Lt Col Jan McCarthy.

The modern nurse is now a highly trained and skilled professional. When she enters army life she is working within the requirements of the profession of soldiering. As with other professionals in medicine, law, engineering there has to be a re-assessment of the professional role. A similar re-assessment takes place when a professional enters the public service or some other highly structured bureaucratic organization, a large hospital or a tertiary educational institution. The basic role of the nurse is to provide skilled professional care to medical and surgical cases in peace or war, but the army environment both enhances and inhibits the prospects of her carrying out her professional role. In army hospitals and in smaller medical units, she works within the framework of army medical staff and senior nursing staff. Her professional training stands her in good stead when she must take the responsibility for making important nursing decisions. Frequently she does not have the 'back-up' staff available in civilian hospitals. This may well indicate the nurse's role in future military campaigns.

The traditions of military nursing are maintained in Queensland and elsewhere into the 1980's, nearing a century of proud and dedicated service. At 1 Military Hospital, (Yeronga), 11 Fd Amb (Brisbane) and 4 Camp Hospital (Townsville), Queensland members of the RAANC (both male and female) are embarking on a new career, conscious of the illustratious traditions behind their Corps.[14] True, some traditions have been lost, with the changeover from scarlet and grey to green service uniforms. This of course was not achieved without great opposition. The introduction of registered male nurses into the Corps is a step back into history, as men were involved in military nursing long before it was fashionable or acceptable for women, as nurses, to be part of the army medical services. It may solve the problem of sending nurses with surgical teams into forward battle zones. This new breed of Queensland military nurses retains the essential qualities of earlier generations — dedication, compassion, discipline, but these are supplemented by the confidence born of higher training and a knowledge of modern medical and nursing technology. Moreover they have the advantage of learning something of the art and science of military nursing in the context of a military hospital. No such provision existed for Australian nurses at the outbreak of either World War. To this can be added the training available through CMF units. Queensland nurses will continue to make a valuable contribution to the Royal Australian Army Nursing Corps in the decades ahead, wherever the call of duty may lead them.

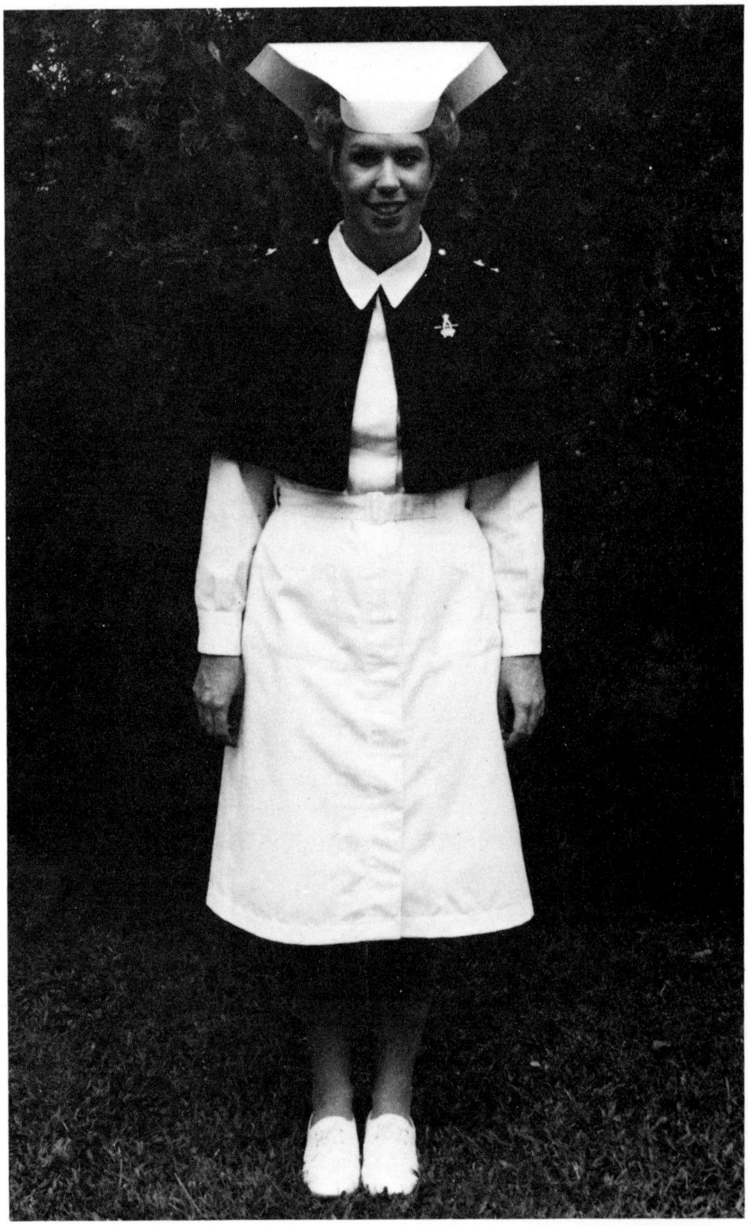

One of the most recent recruits to the Royal Australian Army Nursing Corps, Lt Erin Small.

References

1. McMAHON, MONICA, Matron 110 AMH (Perth); Matron 130 AGH (Japan).
2. DOIG, EDNA, see fn 9 Chap 3, p

   LYON, ELIZABETH WILSON ('Tige'), b Glasgow 1907, educ St Margaret's Bris & St Mary's Herberton; trd Cairns Dist Hosp, Child Welfare Bris; AANS 1942, Qld hosp, 1944 2/1 AGH Bougainville. 1946 to BCOF Japan for 5 yrs with family unit; 1952 ARA, Matron 3 Aust Camp Hosp Yeronga until retirement 1962. Service P.A. Bris.

   LYON, AGNES CRAWFORD, b Toowoomba 1912; trd Cairns Dist Hosp; AANS 1943, Horn Is, Thursday Is, various military hosps NSW and Qld, including 112 AGH Greenslopes; 1947–1970 Greenslopes Repat Hosp.

   GUILFOYLE, BRIDGET MARY, ARRC, FCNA. Trd Bris Gen Hosp 1934–39; Dip Nurs Admin; AANS & RAANC 1941–1971. Japan, Korea, Malaysia. Dep Asst Dir Army Nursing Services, Matron RMC Duntroon, 1 Mil Hosp, 2 Mil Hosp; Asst Dir Army Nurs Service Northern Command; Col RAANC (retd). Sister, Comm Health DepBris, 1972–81.

   O'NEILL, JEAN, AM. Trd Cairns Base Hosp 1952; RAANC 1953–1977. Service in Japan, Korea (1954–55), Malaysia (1963–65), Vietnam (1968–69). Dip Nurs Edn (1959), Dip Nurs Admin (1974). Member Order of Australia 1976.

   SMITH, THELMA ALICE. See fn 77 Chap 5, p 252.

   STEEN, MARGARET. See fn 91, Chap 5 p 253.

   EARWALKER, ELIZABETH. Trd Ipswich Gen Hosp, completed 1951; RAANS and RAANC 1953–55, service Japan.

   ZILLMAN, P. Trd Bris Gen Hosp 1937–1941; AANS 1945–1949; 2 ACH Chermside, 130 AGH Japan; post grad course Heid Repat 1949; CMF 1954–1960, retd LtCol; Repat Depart until 1976.
3. RULE, LUCY, ARRC. b 4 Sept 1916; trd Longreach Gen Hosp 1934–39; Women's Hosp Bris 1940; Private nursing 1940–42; RAAF Nurs Serv 1942–53, Service Japan & Korea, Sandgate, Amberley, Pt Cook, Laverton.

   FEIL, EUNICE, b 3 Dec 1923; trd Toowoomba Gen Hosp 1942–45; midwifery trg Women's Hosp 1945, maternal & child welfare 1947; nursing Biloela 1947–49. RAAF Nurs Serv 1949–53; Japan & Korea 1951–52; Post RAAF serv post-grad TB nurs in Vic; Chest Clinic Bris 12 yrs; Child Care Off Dep Children's Services, Bris.

   MARSHALL, LOU, FNM. b Buderim (Qld); RAAF Nurs Serv 23 yrs; FNM for evacuation work in Japan & Korea, with 11 months service on medical evacuation flights.

   WILSON, MAB. b 24 April, 1918, Qld; trd Bris Gen Hosp 1939–43; Staff nurse Eidsvold 1944; Women's Hosp 1945; pte nurs 1945–50; RAAF Nurs Serv 1950–54; served Japan & Korea 1951–52, medical evacuation service; Post RAF service, pte nurs until retirement 1979.

4. Brief History of Australian Women's Services, Japan, 1947. File 130/1/42 (AWM).

5. History of Australian Nursing Staff, 1948. MP 742 File 21/3/651 (AA Vic).

   AANS Personnel, BCOF Gen Hosp 1949. MP 742 supra. File 21/3/707 (AA Vic).

   Monthly Report Nov 50: RAANS in BCOF. MP 742 Depart Army Corres Files 1943–51, File 21/3/746 (AA Vic).

   Conditions of Service AANS, MP 742 Depart of Army. Correspondence Files 1943–51, Box 88. File 21/3/575. (AA Vic).
6. The Royal Australian Army Nursing Corps. (An outline of the foundation and development of the Australian Army Nursing Service). Aust Army HQ. nd.

7. Retiring age, AANS. 1948–49. MP 742 supra. File 21/3/644 (AA Vic).

   AANS personnel retired against their wishes. MP 742 supra File 21/3/606. (AA Vic).
8. Army Order 120/1957, Army Orders Oct 1957.

   Her Majesty the Queen has been graciously pleased to approve an alliance between the Queen Alexandra's Royal Army Nursing Corps and the Royal Australian Army Nursing Corps.
9. BARTLETT, NORMAN (ed) *With the Australians in Korea*. AWM Canberra, 1954.

   McCORMACK, GAVAN. *Cold War, Hot War*. An Australian Perspective on the Korean War. Hale & Iremonger, Syd, 1983.

   O'NEILL, ROBERT, *Australians in the Korean War*, 1950–1953. AWM Canberra, 1981.
10. Monthly Reports AANS, BCOF 1949–51. MP 742 Depart of Army, Corres Files, Box 89. File 21/3/746 (AA Vic).

    PROBYN-SMITH, BARBARA. b 22 June '29, Trd. Bris Gen '47-'51; Army Nursing 1951-63; service Japan, Korea, Perak, Malaysia; Matron Comm Rehab Centre Mt Martha.
11. EDWARDS, AILSA BETTY. Gr Capt RRC (1981), FNM (1983). b 5 Sept 1926, educ St Mary's Convent Sch (Qld); trd Mater Hosp Bris 1944–47; RAAF Nurs Serv 1952–83; Nurs officer in charge of aeromedical flights Vietnam Aust, 1965–67; Dep Matron 4 RAAF Hosp Butterworth Malaysia 1965–67; Snr instructor aeromedical trg medical operational support unit RAAF, Richmond, NSW 1969–73; Matron 4 RAAF, Hosp Butterworth, 1973–74; 6 RAAF Hosp Laverton, Vic, 1975–79; Dir RAAF Nurs Serv 1979–83. Queen's Hon Nursing Sister 1979.
12. *The Royal Australian Army Nursing Corps*. Supra. Part 2 p 9.

    COE, JOHN J. (ed) *Desperate Praise. The Australians in Vietnam*. Artlook Books, Perth, WA, 1982.

    MACKAY, IAN, *The Australians in Vietnam*. Rigbys Adelaide, 1968.

    BONDS, RAY (ed), *The Vietnam War*, Lansdowne Press, Sydney, n.d.

MACLEAN, MICHAEL, *Vietnam, The Thousand Day War*. Thomas Methuen, London, 1981.

THOMPSON, SIR ROBERT (ed). *War in Peace. An Analysis of Warfare Since 1945*. Orbis Pub.Co. London, 1981.

13. CLAYTON, GILLIAN. Trd Ipswich Gen Hosp 1961–65; Midwifery Ips Mat Hosp 1966–67; Psychiatric trg Challinor Centre 1968–69. RAANC (ARA) 1969–75; Vietnam 1969–70. CMF (FT) 1977–79; ARA (Supplement) (1979–80.

POTTS, LORRAINE, ARRC. b 23 Mar 45; trd Royal Bris 1963–67; RAANC, 1967–1971, Vietnam 1969–70.

SORLEY, V.L. Trd Toowoomba Gen Hosp 1960–64; RAANC 1970–72; Vietnam 1971; 2 Mil Hosp Ingleburn, 1 Fd Hosp Vung Tau, 1 Mil Hosp Yeronga.

WEST, PAMELA, b 28 Sept 43; trd Royal Bris Hosp 1960–64; Bush nurs Vic, Repat Hosps Hobart & Greenslopes. RAANC 1965, Vietnam 1969–70. Currently member RAANC, 6 Camp Hosp Bonegilla.

MINCHOW, JUNE, b Mackay; trd Mackay Hosp 1957–61; midwifery Q Vic Hosp Launceston 1961–62; Theatre Cert R Perth Hosp 1965; RAANC 1967. 1 Aust Fd Hosp Vietnam 1969–70; ARA 1967–77; 1979 Appd Sci Nurs Dip Admin; Mater & Greenslopes Hosps.

SHALE, ELIZABETH MARY, b 13.2.46 Bris, educ Maryborough; 1964–67 nurs trg Footscray Dist Hosp (Vic); 1967–68 Midwifery Mater Mothers, Bris; RAANC 1970–72, SVN 1971; Royal Melb Hosp 1973, RAANC 1976–78.

KAISER, LYN. Trd Mater Hosp Bris 1966–70; RAAF NS.

Queensland members of the RAANC in 1984 include the following:—

14. COULSON, ROBYN, b 1957, trd Princess Alexandra Hosp Bris 1974; postgrad trg Eng, midwifery Nambour Hosp. Serving member RAANC, 1 Mil Hosp Yeronga.

McBRIDE, CYNTHIA, b 1942, trd Adelaide, midwifery Bris 1975–76; staff various Qld hosps; CMF 1967–70, RAANC 1982, 2 Fd Hosp, 1 Mil Hosp Yeronga.

NEILL, SUSAN, b 1959, trd P.A. Hosp Bris 1977, midwifery 1977, staff Thursday Is & Bamaga; RAANC 1983, 1 Mil Hosp Yeronga.

RYAN, DILYS, b 1947, trd Ipswich Gen Hosp 1969–72; staff Qld hosps; RAANC 1980, 1 Mil Hosp Yeronga.

SMALL, ERIN, b 1958, trd Royal Bris Hosp; midwifery Royal Women's maternal & child health cert; staff Cunnamulla & Cloncurry Hosps; RAANC 1983, 1 Mil Hosp Yeronga.

Abbreviations

AAMWS	Australian Army Medical Women's Service.
AAMC	Australian Army Medical Corps.
AANS	Australian Army Nursing Service.
ABC	Australian Broadcasting Commission.
ACCS	Australian Casualty Clearing Station.
ACH	Australian Camp Hospital.
ADGMS	Assistant Director-General of Medical Services.
ADMS	Assistant Director Medical Services.
ADS	Advanced Dressing Station.
AG	Adjutant-General.
AAH	Australian Auxiliary Hospital.
AGH	Australian General Hospital.
AHS	Australian Hospital Ship.
AIF	Australian Imperial Force.
AMF	Australian Military Forces.
ARA	Australian Regular Army.
ARCS	Australian Red Cross Society.
ARRC	Associate Royal Red Cross.
ASH	Australian Special Hospital. Australian Stationary Hospital.
BCOF	British Commonwealth Occupation Force.
Bde	Brigade.
Bn	Battalion.
BGH	British General Hospital.
Capt	Captain.
CCS	Casualty Clearing Station.
CO	Commanding Officer.
Col	Colonel.
DADMS	Deputy Assistant Director Medical Services.
DDMS	Deputy Director Medical Services.
DGAMS	Director-General Australian Medical Services.
Div	Division.
EPIP Tent	European Privates Indian Pattern Tent.
Fd Amb	Field Ambulance.
GOC	General Officer Commanding.
GRO	General Routine Order.
GSW	Gunshot wound.
HQ	Headquarters.
Hosp	Hospital.
HS	Hospital Ship.
Inf	Infantry.
KIA	Killed in action.
LCT	Landing craft, tank.
LHQ	Land Headquarters.
L of C	Line of Communication.
LSI	Landing ship infantry.
LST	Landing ship tank.
LT	Lieutenant.
LtCol	Lieutenant-Colonel.
MAC	Motor ambulance convoy.
MAETU	Medical Air Evacuation Transport Unit.
Maj	Major.
MBDS	Main beach dressing station.
MD	Military district.
MDS	Main dressing station.
ME	Middle East.
MO	Medical Officer.
NAD	No appreciable disease. Nothing actually diagnosed.
NCO	Non-commissioned officer.
NSW	New South Wales.
NYD	Not yet diagnosed.
OC	Officer Commanding.
OC Med	Officer-in-charge, medical.
OC Surg	Officer-in-charge, surgical.
OR's	Other ranks.
PM	Principal Matron.
PMRAFNS	Princess Mary's Royal Air Force Nursing Service.
POW	Prisoner of war.
PUO	Pyrexia of unknown (or uncertain) origin.
QAIMNS	Queen Alexandra's Imperial Military Nursing Service.
Qld	Queensland.
QL of C	Queensland Line of Communication.
QM	Quartermaster.
RAF	Royal Air Force.
RAAF	Royal Australian Air Force.
RAAFNS	Royal Australian Air Force Nursing Service.
RAANC	Royal Australian Army Nursing Corps.
RAMC	Royal Australian Medical Corps.
RAN	Royal Australian Navy.
RAP	Regimental Aid Post.
RMO	Regimental Medical Officer.
RO	Routine Order.
RRC	Royal Red Cross.
SA or S.Aust	South Australia.
SBA	Sick berth attendant.
SIW	Self inflicted wound.
SMO	Senior Medical Officer.
SO	Staff Officer.
SWPA	South-West Pacific Area.
Tas	Tasmania.
URTI	Upper respiratory tract infection.
VAD	Voluntary Aid Detachment.
Vic	Victoria.
VD	Veneral disease
VDC	Volunteer Defence Corps.
WA or W Aust	Western Australia.
WE	War establishment.
WET	War equipment table.
WO	Warrant Officer.
WRANNS	The Women's Royal Australian Naval Nursing Service.

Index